METHODS ENGINEERING

john wiley & sons, inc., new york · london

EDWARD V. KRICK
Associate Professor
Department of Industrial Engineering
Lafayette College

METHODS ENGINEERING

design and measurement of work methods

Library of Congress Catalog Card Number: 62–8775

Printed in the United States of America

To Mother and Dad

Preface

This book differs from existing textbooks on the subject of motion and time study in the following important respects.

1. A variety of new and more enlightened theories and techniques are introduced. Although there is need for a fresh textbook approach to the subject, I recognize that in a field burdened by tradition there is a limit on the rate of change the market will bear. Therefore I have avoided departing from tradition beyond what I believe the field will accept readily.

2. In general, extreme detail has been avoided, especially with respect to the manner in which various techniques and procedures are to be executed. Spelling out the how-to-do-it particulars to the ultimate in detail is wasteful and in fact futile in a college-level course. The details of execution differ radically from company to company, and in fact are readily assimilated on the job. Furthermore, such details are ordinarily long forgotten by the student if and when he eventually has need for them. Perhaps worse, they often obscure the more important matters concerning when and where to employ a given technique, its relative worth, the flaws and pitfalls associated with it, and other aspects of a broader and more crucial nature. The field of industrial engineering is experiencing an era of extraordinary progress, a period of upheaval, of increasing rigor and objectivity, of improvement in perspective. Students in a course of this nature know or certainly sense this, and rightly resent having substantial amounts of relative trivia forced on them. This resistance is likely to be severe if the students are aware of some of the major problems confronting an enterprise and the newer methodology available to industrial engineers for attacking them.

3. Overall perspective is given much stress in this volume. If the student is prepared so that when the need arises he can make an intelligent decision as to the best policy, technique, or course of action, this is the important matter and mainly where the battle is won or lost. If he is not capable of doing this, even though he knows details of *how* to do things, the educational process has failed.

4. More emphasis is put on evaluation of principles and practices. Where practical, the evaluation material has been separated from the expository content, so that if it is desirable to avoid the former and reserve it until the latter portion of the course or for an advanced course, it is feasible to do so. A moderate amount of critical appraisal is incorporated for several reasons. The major one is to raise the student's level of discrimination above that which so often results from the ordinary motion and time-study course. Another is to stimulate interest. An effective means of enhancing student motivation is to include some of the controversial aspects of the subject. It is a mistake for a textbook to mask this material by presenting it as a cut-and-dried, flawless, scientific, noncontroversial subject, which of course it is not. I have created no such illusion, hoping instead to instill an objective, realistic, middle-of-the-road attitude on the part of the reader.

5. The philosophy and procedure advocated herein constitute more of an engineering or design approach than a cost reduction or work simplification point of view. In the past, too much emphasis has been given to principles, tools, and techniques and too little to the manner of applying these. Specification of the method of work calls for the same rigorous problem-solving process an engineer would apply to any other type of design problem. Therefore, considerable attention is devoted to the design process as a framework for application of principles and techniques.

6. A knowledge of elementary statistics is presumed. Familiarity with elementary sampling theory is an important prerequisite to any course treating a measurement process. Extensive application of statistics in work measurement methodology is long overdue.

7. In addition to presenting principle and technique and an evaluation of them, considerable attention is devoted to their successful application. A number of suggestions are made to assist in obtaining the best results from each technique presented and to minimize the difficulties arising from the use thereof. This counseling on implementation is as important as the how-to-do-it instructions.

8. A number of common claims made about the techniques and principles presented in this book are brought to light and discussed.

This is thought to be one of the most valuable attributes of this work. One purpose of this feature is to fortify the reader against a number of unrealistic, unreliable, biased assertions commonly encountered in work-measurement literature and from other sources. Perhaps a more important purpose, however, is to provide a very enlightening experience for the student and to thereby raise his "threshold of blind acceptance" by objectively examining the claims cited. Among the assertions discussed are some glaring examples of gross disregard of professional objectivity, accepted rules of scientific inference, and practicality from which the student can profit.

This book is intended to serve primarily as an introductory textbook. However, since material on evaluation of theory and technique has been included and in the main is separate from the expository material, this plus the section on administration of the methods engineering function, plus some of the more advanced subjects will provide a substantial amount of material for an advanced course in methods engineering. To provide a well-rounded introduction to the subject, I have attempted to achieve the following.

1. Familiarize the student with the fundamental principles, procedures, and techniques.
2. Provide an appraisal of these.
3. Provide a basis for intelligently choosing the most appropriate technique or course of action for a given set of circumstances.
4. Provide advice on effective implementation of the procedures and techniques presented.
5. Achieve a higher than usual level of discrimination and thus a superior ability to distinguish sound from unsound theories, procedures, and claims, as the reader is confronted by them in the literature and in practice.

Briefly, the material content and organization is as follows.

Introduction. Although the ordinary undergraduate seems to thirst for applied subject matter, he is usually unprepared for it in maturity. In recognition of this situation, a major portion of this section is directed toward bringing to the fore at the very outset some of the features of an applied subject that frequently cause difficulties in the mind of the ordinary undergraduate who has yet to be thoroughly immersed in the real life business world.

Part I: The design process. This, the first of six major sections, deals with the general approach to a design problem.

Part II: Introduction to methods engineering. A general orientation to industrial engineering and then to one of its subspecialties: methods engineering.

Part III: Methods design. Application of the design process to specification of the work method. Here, the traditional as well as new principles and techniques of this engineering specialty are introduced, with the general design process as a framework. One significant feature of this section is the blending of what have heretofore and illogically been held as two separate bodies of knowledge: a wealth of knowledge derived by psychologists, pertaining to integration of the human being into a system; and the principles derived and traditionally applied by methods engineers, exemplified by the classic principles of motion economy.

Part IV: Work measurement. Especially in this section dealing with work measurement (time study),[1] emphasis has been put on evaluation and successful implementation. A variety of new ideas are presented especially on the incessantly troublesome problem of rating. Noteworthy are means of coping with deviations in method during the observational period and a basically different rating sampling scheme. Some useful aids to the development of standard data are introduced, including regression analysis and use of the electronic computer.

Part V: Special methods engineering problems. There are no two characteristics of a production system that cause more problems for designer and manager than the unbalance in production capabilities of the system's components and the variation in their behavior. The effects of unbalance and variation on all phases of design and operation of a manufacturing system are far-reaching and severe enough to warrant a separate chapter devoted to the methods engineer's approach to each. Since indirect labor activities are extensive and different from direct labor jobs in a number of significant respects, a separate chapter is delegated to the methods engineer's approach to this type of problem.

Part VI: Administration of methods engineering function. This section, a major innovation, consists of a series of chapters dealing with various methods engineering administrative problems and means of coping with them. The administrative section offers perspective to what has preceded, introduces a variety of important matters previous

[1] The variation of terminology in this field is extreme. As a consequence, when alternative terms are known to be prevalent, such terms will be included in parenthesis immediately following the expression used, as in the case of work measurement and its alternate, time study.

textbooks have neglected, and provides an intelligent basis for management and appraisal of a methods engineering department.

Published descriptions of the Methods-Time Measurement system of predetermined motion times are either a very extensive and detailed description of the system or merely a presentation of the tables of time values. Neither of these extremes is satisfactory if a student is to be able to make an intelligent trial application of the system without having to read a textbook of detailed material or the instructor having to verbally present a myriad of details. Appendix A has been specially prepared to satisfy the need for a condensed, simplified description of the MTM system that will enable the student to satisfactorily apply the technique on a trial basis without having to invest a prohibitive amount of time and without having to learn an objectionable number of detailed rules and definitions in the process.

I wish to express my appreciation to my present colleagues at Lafayette College for their kind and patient indulgence, especially my immediate co-workers Professors Charles M. Merrick and Charles E. Moore. I am also indebted to the staff of the Department of Industrial and Engineering Administration, Sibley School of Mechanical Engineering, Cornell University, especially to Professors J. William Gavett, Martin W. Sampson, Jr., and Andrew S. Schultz Jr., for their contributions to the author's professional development. Sincere appreciation is also extended to Mrs. Edwin J. Harte, Mrs. Robert Henthorn, and Mrs. Lou Robinson for their patience during the typing of the manuscript, to Mrs. George Berkemeyer for her capable editorial assistance, and to Mr. Robert J. Rees, president, and William E. Van Order, executive vice-president of the Everclean Rowboat Company, and to Mr. Richard O. Jones, general manager of the Mechanical Garden Trowel Company, for their kind cooperation.

EDWARD V. KRICK

Easton, Pennsylvania
October, 1961

Contents

APPENDIXES

List of Tables

Appendix Tables

1

Introduction

A MAJOR CHANGE

For those readers who have recently emerged from several semesters of college level work heavily steeped in basic science and mathematics, the applied subject matter you are about to be confronted with represents a drastic change, one which some students have considerable difficulty adjusting to satisfactorily in a relatively short period of time. It is an adjustment that must eventually take place if a person is to deal effectively with real industrial problems. It is unfortunate, however, that so much of this change in point of view must take place so abruptly.

With the hope of facilitating and accelerating this change in perspective, it seems worthwhile to discuss at the very outset some of the commonly troublesome respects in which the present subject matter differs from the work that predominates in the first several semesters of many college curriculums. Thus, a number of relevant characteristics of methods engineering and to a major degree of engineering problems in general, are to be discussed. Contrast these properties with the problems encountered in chemistry, physics, mathematics, and the like. In view of the sizeable roles played by judgment and creativeness, a drastic change is involved, from the so-called "exact" sciences to what has many of the attributes of an art.

The Number of Alternative Solutions and the Difficulty of Evaluating Them

Ordinarily there are many if not an infinite number of possible solutions to the typical engineering problem. Very rarely in the course of working out a solution to a problem does the engineer find it possible or practical to conceive of more than a very small percentage of these many alternative solutions. Exhaustion of all possibilities is rare indeed. Furthermore, any two engineers working independently on the same problem will usually conceive of substantially different sets of alternatives.

Almost invariably the engineer finds it impossible to obtain a perfect or even comprehensive evaluation of the alternative solutions to his problem. Ordinarily there are many variables involved and only a relatively short period of time in which to evaluate solutions, so that in most instances other than the most important factors bearing on the decision must be ignored or treated superficially. Furthermore, in many situations the engineer is unable to measure important variables bearing on the decision, and must treat such factors as intangibles. Finally, errors are made in measurement and prediction of the performance of various designs. Thus, much of the final evaluation of alternative solutions is of necessity left to judgment.

The fact that we ordinarily deal with only a small fraction of the many solutions available and that we must choose from among these on the basis of imperfect evaluation of the alternatives makes it impossible to justifiably refer to a solution to a practical engineering problem as *the* answer or the correct answer. Ordinarily, the most we can justifiably say is "here is *an* answer, *a* solution, which among the alternative solutions conceived of appears superior." Furthermore, the fact that other people are likely to have different ideas, and that considerable opinion is involved in the evaluation of these, means that the solution an engineer presents is not ordinarily automatically accepted. And as an engineer you will find that there is room for lack of confidence in your judgment as well as room for differences in ideas and in opinions. It is such circumstances as these that make it necessary that you do more than merely present your solution to a problem. It becomes necessary to sell, to convince others of the superiority and workability of your ideas.

Role of Creativeness in the Solution of Engineering Problems

In much of his work, an engineer must rely heavily on his creative ability in generating solutions to problems. True, he has a certain body of knowledge comprised of principles, accepted common practices, results of his experiences with previous problems, and the like, but all this is only of limited assistance in view of the uniqueness of most problems encountered. Because successive problems differ in so many respects, significant portions of the designs produced by an engineer are the result of his own inventiveness. He must rely on his own ingenuity to produce ideas in those many situations in which set solutions derived from existing principles and practice do not exist (or at least are not known to the designer). Therefore, creativeness is vital, an attribute well worth beginning to cultivate now.

The Role of Judgment

The extent to which judgment must be exercised in dealing with everyday business and engineering problems is sometimes disturbing and difficult to accept at some stages in our academic development. Actually and perhaps unfortunately judgment ordinarily plays a major role in almost every phase of an engineer's work. For example:

1. Considerable judgment is usually required in deciding on the best way to approach a particular problem, the best procedure, technique, or practice to follow. Such matters are seldom clear-cut; in fact they are often the subject of considerable disagreement among practitioners.
2. An objectionable degree of judgment is often required in the techniques and procedures themselves as well as in their selection.
3. And as mentioned earlier, considerable judgment is required in the evaluation of alternative solutions to a problem.

The extent to which judgment must be employed is inevitable. Decisions must be made relatively quickly in the dynamic business world. If a solution to a problem is to be arrived at and implemented before the problem itself ceases to exist, a liberal amount of judgment must be employed in lieu of an exhaustive quantitative investigation and evaluation of all the alternatives and factors involved.

The Manner in Which Problems are Presented

Problems will never be handed to you on a silver platter; they will not ordinarily be defined for you. Rather, the usual procedure will be to confront you with a situation, most likely the present solution to the actual problem, and you must proceed from there. Therefore, you will sometimes miss the real problem at hand (it happens to even the most experienced of engineers). This is likely to result in subsequent embarrassment which in turn sometimes leads to some form of face-saving reaction. *If* you conscientiously and effectively attempt to isolate and define the problem underlying the situation confronting you, you will minimize the chances of the real problem eluding you.

Variation

Virtually all aspects of the phenomena an engineer deals with, the environment he works in, and the techniques that he employs, display variation. The materials he uses or specifies, the people he deals with, the measurement and testing equipment he uses, the sales volume of the product he is concerned with, and so on, all display considerable variation. This variation presents no end of difficulties and means that things are never as "well behaved," never as clear-cut as you would like them to be or as you are probably accustomed to having them. This variation is also evident in industrial practices. The way things are done, the terminology used, the degree of success achieved, and the like, vary radically among companies, so that it is often difficult or impossible to generalize concerning such matters in practice.

Imperfection

No procedure, practice, or policy employed in the business world is without its flaws and disadvantages. The means of solving real world problems *and* the solutions thereby produced are almost invariably far from perfect. Furthermore the flaws are sometimes quite obvious, yet however objectionable, these procedures and techniques *are* used for the very simple and practical reason that they are the best means available at the moment for achieving the purpose at hand. Because answers and results *must* be attained, because decisions must be made, practitioners *will* use the best techniques and procedures available regardless of imperfections in them. We must not lose sight of the fact that in choosing solutions to real world problems it is *always a matter of selecting from a set of imperfect alternatives.* Imperfection very seldom deters usage.

The Indefiniteness Property

Another property of real world problems which leads to much trouble and dissatisfaction especially at first, is the indefinite nature of so many matters. There will be many instances in this course and in the business world in which you cannot get a clear-cut yes-or-no answer to a question, no clear-cut result from an investigation or experiment. Unfortunately things are seldom simply black or white. Instead of a simple yes or no, possible or not possible, economical or uneconomical, in response to an inquiry you get an answer that amounts to "MAYBE." This inconclusiveness is something you must become accustomed to, for it is a very common property of the answers to questions you pose.

The Evolution Property

In attempting to learn why some aspect of a business enterprise is organized or operated in the way it is, we often find that the present system, rather than being the result of a deliberate, carefully thought out decision process, is something that over the years just "grew." In fact there are numerous outlandish, incredible schemes, procedures, and systems to be found in practice, that owe their existence mainly to a haphazard evolution process. They were never really designed, they simply evolved, somewhat piecemeal, over a period of years. This is likely to be true of a factory's existing layout, the existing company organization, accounting procedure, wage payment scheme, and many other aspects of the operation; they just "evolved."

You should remember this when attempting to find objective explanations for certain matters or situations. You may well be seeking a profound explanation that just does not exist. Furthermore, the fact that so many situations are an outgrowth of this haphazard evolution process means that in many of these instances there are tremendous dollar-saving opportunities.

The difficulty that some students have in believing that such situations abound in industry today is related to the rather widespread misconception among undergraduate students that many if not the bulk of industrial problems are well-nigh solved. Undergraduates often have the impression that because a company employs engineers to design its products, product design problems are for the most part overcome; or that because engineers are utilized to lay out the plant, the arrangement of facilities and flow of product are therefore optimum; or that because the company employs methods engineers, work methods about the plant leave little room for improvement. This is a gross misconception (and what relative peace and contentment managers could enjoy if this state of virtual perfection existed!). The typical business enterprise has virtually an endless backlog of problems worthy of attention. There is ample opportunity for each new shining light to employ his problem-solving prowess.

Methods Engineering's Lengthy and Lively History

For many readers, methods engineering[1] will represent a rather drastic change in subject matter not only because of its applied nature, but because this is no ordinary applied subject. The field is a surprisingly controversial, troublesome, and explosive one, as indicated by its lively history dating back to the late nineteenth century. Methods engineering is directly concerned with the establishment of work methods and work loads. As a consequence, workers, union, and supervisory personnel have much at stake in this process and therefore show an active interest in the results of the methods engineer's efforts *and* even in the means he uses in arriving at work methods and loads.

Time study, which constitutes a major phase of methods engineering work, has a history that can be traced back many decades, notably to the work of Babbage who reported his use of rather elaborate time studies of manufacturing operations in a classic book entitled *Economy of Machinery and Manufacture,* published in 1833. The main impetus to formal time study much as we know it today is universally attributed to Frederick W. Taylor. His pioneering work involving the use of extensive timing procedures in the course of studying manufacturing activities is amply documented.[2] The time-study procedures developed by Taylor around the turn of the century were rapidly adopted in industrial circles and were extensively used by 1920. During this period and in the period 1920 to 1930 as well, the technique was much abused, overpromoted, and misapplied by a multitude of self-appointed experts unaffectionately referred to then and now as "efficiency experts." Unfortunately, the blunders committed primarily in this period have blotted the history of time study to the point where the field has been given a generally poor reputation. These past malpractices and abuses have helped to generate the rather antagonistic attitude the time study practitioner must ordinarily cope with on the part of union and workers.

What might at the outset appear to be a rather straightforward, routine procedure is in reality a focal point of much labor-management strife. It is difficult to imagine the number of hassles between company and union representatives, of worker grievances, of arbitration cases, and of strikes that arise from time study matters. Also surprising

[1] Although the term methods engineering is becoming more widely used, this field is still referred to by many as motion and time study.

[2] Frederick W. Taylor, *Principles of Scientific Management,* Harper and Brothers, New York, 1915.

are the amounts of lost time, heated words, measures and countermeasures, and bargaining, attributable to the same source. Thus, time study and related problems present somewhat of a headache to all concerned, especially the peace-seeking industrial relations manager.

Time study is not new, yet surprisingly enough the practices of today do not differ appreciably from those introduced by the pioneers in the field. As a consequence, tradition has a virtual stranglehold on practice today. The power of tradition, the simple fact that *it* has been done this way or that for the past 35 years, in maintaining the status quo is easily underestimated. Many time study theories, techniques, procedures, and situations owe their existence to this one factor: tradition. To look for a sound theoretical and economic rationale in such cases is futile.

The other major phase of methods engineering, classically referred to as motion study and referred to herein as methods design, has had an equally lengthy but not so lively history. The main impetus to the development and the spread of motion study throughout industry in the early part of this century is attributed to Frank Gilbreth.[3] Many of the contributions of Frank and Doctor Lillian Gilbreth to the philosophy and techniques associated with design of work methods are still in evidence. The characteristics of time study and its history outlined previously are true of motion study also, although generally to a lesser degree. Thus, the theories, procedures, and results of motion study are the subjects of controversy and cause considerable friction between labor and management; the field has an unfavorable reputation mainly as a result of past abuses and malpractices; tradition exerts a strong influence on theory and practice; labor maintains a vigorous interest in the results and in the manner in which the engineer arrives at those results. An awareness of these important characteristics of the field of methods engineering and its history will facilitate a fuller appreciation and understanding of what is to follow. The realization that these circumstances exist constitutes a more mature outlook, which is highly desirable in approaching this subject.

REFERENCES

Davidson, H. O., *Functions and Bases of Time Standards,* American Institute of Industrial Engineers, New York, 1952.

Fillipetti, G., *Industrial Management in Transition,* rev. ed., Richard D. Irwin, Inc., Homeward, Ill., 1953.

[3] Frank B. and Lillian M. Gilbreth, *Applied Motion Study,* Sturgis and Walton Co., New York, 1917.

part I

THE DESIGN PROCESS

2

Introduction to the Design Process

Characteristics of a Problem: A Simplified and Special Case

A familiarity with the general attributes of a problem and the fundamentals of problem solving will facilitate an understanding and appreciation of design and the procedure it entails, for design *is* problem solving applied to a special class of problems society traditionally relegates to the engineer.

A simple case is that of the spatial problem, characterized by two locations or points in space, the interval between which must be traversed. The problem may be one of getting from a river bank to the opposite one, or from one city to another, or from one planet to another. A characteristic of this and in fact most types of problems is the ordinarily large (often infinite) number of alternative solutions, that is, alternative methods of getting from one location to the other. There certainly are many possible modes of travel and many possible routes between two cities, some known, some unknown, some reasonable, some unreasonable, and some ridiculous. A problem always involves dealing with a number of different methods of accomplishing the desired result; in fact if there are no alternatives, known or unknown, there is no problem.

If all solutions are equally desirable, no problem exists; but this is not usually so. A problem involves more than finding just any method of getting from the one location to the other; it requires finding a preferred method, such as the least costly. If the most preferred method is obvious at the outset, then again no problem exists.

The final important problem characteristic in this simplified case is the usual requirement that an answer be provided in a certain finite period of time.

Characteristics of a Problem: The General Case

The general case includes seeking a method of getting from one physical state (form, condition, or status) to another, as well as from one location to another. Thus a problem may involve seeking a method of getting from untoasted to toasted bread, from one temperature level to another, from individual parts to completely assembled automobile, from warehouse to boxcar, as well as from one city to another or one river bank to the other. In any problem there is an originating state of affairs (input or starting point) referred to herein as "state *A*." Similarly, there is a state of affairs (output, objective, or result) the problem solver is seeking a method of achieving and which will herein be designated as "state *B*."

To get from state *A* to state *B* there are certain things that must take place. For example, to get from individual parts to complete automobile, the parts *must* be physically assembled and certain parts *must* be assembled before certain others; to accelerate a mass a force must be provided; for plant growth to take place, water, light, and certain nutrients *must* be provided. For physical reasons these things must take place or must be provided in order to accomplish the transformation from state *A* to state *B*. Usually there are certain other things that must take place or must be provided to achieve the transformation from state *A* to state *B*, not for physical reasons but because they have already been decided on by someone whose authority the problem solver must accept. For example, it has been specified by management that the automobiles must be assembled in a certain area of the plant. In the river crossing problem, it has already been specified that a bridge shall be used. In the toasting problem, it has already been decided that this shall be accomplished by heat supplied by electricity. These things that must be true of an eligible solution to the problem, whether for physical reasons or simply because of previously made decisions, will henceforth be referred to as restrictions. This is a simplified view of a restriction, but it will suffice until the concept is expanded in a later chapter.

In any problem there are alternative methods of achieving the transformation from state *A* to state *B*. The following common characteristics of these possible solutions have important effects on problem solving.

1. The number of these eligible solutions is usually very large or infinite.

2. Very seldom are all possible solutions to a practical problem obvious at the outset; in fact very seldom are all possible solutions to a problem known even after considerable search.

3. These alternative solutions are not equally desirable. A preferred solution is sought and a screening or decision process is necessary. The existence of methods of unequal preference means that it is usually worthwhile to "shop around" (search) for a number of alternative solutions before choosing one. Henceforth a basis of preference will be referred to as a criterion. In many problems of the business world, return on the investment is the primary criterion used in choosing from alternative solutions. In general, the method sought is the one that maximizes return on the investment of time, money, and other resources.

4. The relative desirability of the alternative solutions to a problem is very seldom immediately obvious. Usually investigations involving fact gathering, measurement, and computation must be made to provide satisfactory estimates of relative desirability of solutions. Even after thorough investigation, some uncertainty as to the relative desirability of alternatives remains.

The number of times that the object or person is to be transferred from state A to state B becomes a significant characteristic of a problem whenever *total* cost, the cost of arriving at, creating, and using a given alternative solution, serves as a criterion. For example, the number of times that the river is to be crossed or the number of automobiles that are to be assembled become significant. If the river is to be crossed only once at a given spot, a bridge is not the alternative minimizing total cost or total time, or maximizing return on the investment. If millions of persons are expected to cross the river at the given spot, obviously a rowboat is not the preferred method with respect to these criteria. The same reasoning can be applied to assembling automobiles, toasting bread, or any other problem in which the criterion of total cost is involved. Thus different solutions become the most desirable as to cost as the number of repetitions changes.

The desire to maximize return on the investment or to minimize *total* cost has a second important effect. The cost of the time and other resources expended in arriving at a final solution to a practical problem constitutes part of the investment required by that solution; in fact, this cost is frequently the major portion of the investment. As an engineer continues to work on a problem, the investment continues to accumulate, as indicated by curve A in Figure 1. At the same time, as the search for better alternatives continues, a point in time is reached at which additional alternatives become increasingly difficult to find and of generally diminishing profitability, as indicated by the point of inflection of curve B, Figure 1. The latter curve represents the *expected* total value of ideas generated as a function

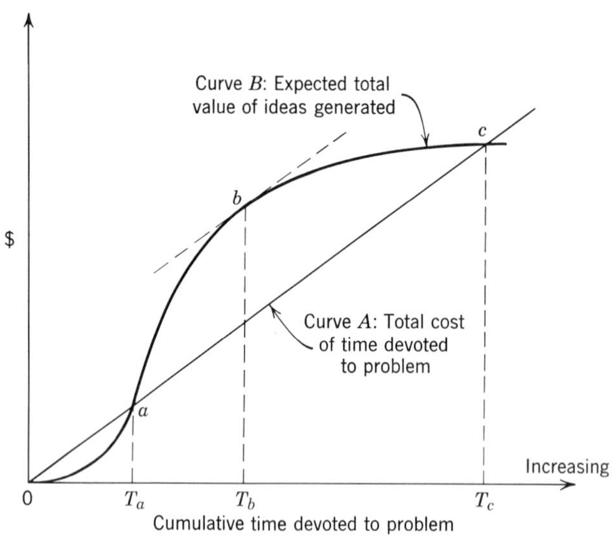

Figure 1. Relationship of cost (curve A) and value (curve B) of time devoted to a problem as a function of time expended.

of cumulative time devoted to the problem. As a consequence, as the search for improved ideas continues, a point in time is reached where it is improbable that better solutions will be found that will justify an additional investment of time and effort. At this point, T_b in Figure 1, the incremental improvement in solutions is balanced by the incremental cost of seeking better solutions. Thus a point in time is reached in the solving of a problem at which return on the investment is a maximum. Return on the investment in solution of the problem is *expected* to diminish beyond this point, because of the cost accumulated by continued attention to the problem and of the generally decreasing profitability of solutions not yet conceived of.

Thus, there is an optimum amount of money and time to be spent on solving a given problem whenever the criteria mentioned are employed. However, this optimum expenditure is difficult to estimate satisfactorily, so that the amount to be spent on a problem and a deadline for a final solution are frequently established arbitrarily.

There are two reasons why a problem is rarely solved in the sense that the term solved is used in everyday language, namely, arrival at the correct or perfect answer. First, the period of time required to achieve this would often be longer than the life of the problem itself, at least for most industrial problems. Second, it is almost invariably uneconomical to attempt to find the perfect solution because of the situation described earlier, namely, that it usually becomes profitable to divert the effort to other problems awaiting attention long before there is any hope of finding the perfect answer to a given problem. Therefore, in problem "solving" there is no intention of searching until the ideal solution is found, nor is there much if any expectation of finding it, nor is there much chance that it would be recognized as such if it were discovered. Rather, it is a matter of attempting to progress *toward* the ideal solution, continually seeking successively better solutions, until it no longer seems profitable to continue the search.

In summary, a problem has the following characteristics:

1. Two states of affairs, call them A and B, and the desire to reach state B from state A one or more times.
2. Certain things which must be done to accomplish the transformation from A to B.
3. More than one (and probably many) possible methods of accomplishing this transformation.
4. Unequal preferability of these methods.
5. A certain time period within which a solution is to be furnished.

If all alternatives and their respective relative desirabilities are known at the outset, no problem exists. Ordinarily this is far from the case. Even after considerable attention has been devoted to a problem, the problem solver must ordinarily select from a much smaller number of alternative solutions than actually exists, and he must select on the basis of incomplete knowledge of the relative desirability of each alternative.

The General Problem-Solving Procedure

Three sequential and functionally different phases may be distinguished in the basic problem-solving procedure: a definition phase, a searching phase, and a decision phase. Specifically:

1. *The definition process* is a determination of the characteristics of the problem, namely:
   a. The specifications of states *A* and *B*.
   b. The restrictions.
   c. The criteria.
   d. The number of repetitions.
   e. The time limit.

   This is a determination of what may be described in various ways as what is "given," or the limitations within which the problem solver must perform, or the things which insofar as the problem solver is concerned "must be."
2. *The searching process* involves the seeking of alternative solutions, different methods of accomplishing the transformation from state *A* to state *B*. This process is characterized by investigation, synthesis, a considerable amount of creative thought, and an element of chance.
3. *The decision process* involves an evaluation of the alternatives generated, then a choosing from among them on the basis of the criteria given. This is a process of elimination.

These phases of the problem-solving process may overlap in practice. Similarly, there will probably be some recycling in this process before a satisfactory solution is reached.

The playboy problem. Here is a sophomore, penniless and desolate, yearning to be transformed into a wealthy Florida-type playboy. How might his problem be formulated?

State A: Sophomore, on verge of expulsion; tired, hungry, and thirsty.

State B: Ex-sophomore, permanently and delightfully situated at his own Miami Beach estate, with surplus of secretaries, cooks, and maids; with sports cars of various styles; with pool, plane, golf course, and other essentials.

Criteria: Time, cost, and pain involved in reaching objectives.

Time limit for reaching a solution: By the time expulsion notice is received from the university.

Restrictions: No prison term should result.

Searching for alternative solutions should not overtax the imagination in this case.

Application of this Problem-Solving Process

The approaches to problem solving and design ordinarily advocated in the textbooks and commonly found in practice are characterized by vagueness and haphazardness. Traditional procedures have exhibited two important weaknesses.

1. Frequently the problem solver fails to get at the true nature of the problem at hand, primarily because of attraction to and over-involvement with the present solution to the problem.

2. The proposed solution too often turns out to be simply a rehash of the present solution, whereas radical changes are desirable and possible. This results from overinvolvement with the present solution and failure to actively and effectively seek other alternatives.

Evidently it is such shortcomings as these that have aroused the rather recent interest in the subject of problem-solving methods, as evidenced by articles, papers, discussions, and the promotion of formal problem-solving procedures within several large corporations. Although these procedures differ in specifics, they exhibit the same definition-search-decision pattern described earlier. The General Motors Corporation, for example, advocates the following problem-solving procedure.

1. Determine the nature of the problem to be solved or the objective that is desired.
2. Study the conditions—the causes and effects related to the problem.
3. Plan all possible solutions.
4. Evaluate these possible solutions.
5. Recommend whatever action should be taken.
6. Follow up to assure that action has been taken.
7. Check the results to insure that the problem has been eliminated and the objective attained.

About this procedure they state:

This step-by-step plan assures better results because it (*a*) places emphasis on correct problem analysis as a sound basis for investigation, (*b*) encourages a search for all possible solutions, (*c*) insures evaluation of each possible way of resolving a problem aiming at the most effective and economical solution, and (*d*) requires a check of results to be certain that the proposed corrective action has eliminated the problem or achieved the objective.[1]

The General Electric Corporation recommends the following procedure:

1. Recognition: Recognize the unsolved problems that exist around you.
2. Definition: Determine what the specific problem is that you can undertake. You must decide what portion of the problem you will solve, what assumptions you can make, and what specific results you desire. Here you make your investigation of the problem and set the specifications on the solution.

[1] By permission from R. D. McLandress, "Methods Engineering for More Effective and Economical Use of Manpower," *General Motors Engineering Journal,* General Motors Corporation, November-December, 1953.

3. Search for Methods: Apply your creative imagination to ascertain what methods may possibly be used to arrive at the desired solution.
4. Evaluation of Methods: Evaluate the various methods under consideration using such things as analysis, experiment, and testing.
5. Selection of Methods: Compare your evaluations and choose the method that seems the most desirable.
6. Preliminary Solution or Design: Make a preliminary design or solution, using the method you have chosen.
7. Interpretation of Results: Interpret the results of your preliminary solution as a check point before completing the solution or design.
8. Detailed Solution or Design: Perform the necessary follow-through action, as dictated by your interpretation of results, such as the detailed solution or design.

About this procedure they state:

"A good approach to the solution of engineering problems is of utmost importance to any successful engineer. Here is a formalized approach consisting of the eight basic steps necessary to the solution of any problem."[2]

A similar problem-solving procedure is described in the Western Electric Corporation engineering magazine.[3] These procedures are in general agreement on the importance of including a careful definition of the problem and an effective and thorough search for alternative solutions, matters so easily neglected in practice.

A survey of enlightened industrial practice indicates a general belief in the existence of a particular problem-solving or design procedure that *in the long run* yields superior results, both in the quality of solutions *and* in the cost of arriving at them. It is true that even the most inferior problem-solving approach will occasionally yield a commendable solution, for there is an element of chance involved in the generation of ideas. Furthermore, use of an optimum approach will not guarantee that all final solutions will be superior to what inferior procedures might have produced. The difference lies in the *probability* of superior problem-solving results, so that the payoff is in the *long-run* performance of the designer.

The belief that there is a preferred, superior, problem-solving procedure is supported by theorists on the subjects of problem solving, innovation, creativity, and decision making, as well as by enlightened practice. What this procedure involves in light of current knowledge is described herein. In presenting this procedure it is assumed that the problem solver is attempting to perform in an optimum manner both

[2] By permission from L. W. Warzecha, "New Course Emphasizes Unique Problem Approach," *General Electric Review,* September 1954.

[3] Robert F. Brewer and James A. Hosford, "Designing Automatic Machines," *The Western Electric Engineer,* vol. I, No. 1, January 1957.

with respect to the final solution he provides and the manner in which he arrives at the solution.

THE DESIGN PROCESS

The problems ordinarily encountered by engineers are exactly as characterized in the preceding portion of this chapter. The process of solving these problems, to be described and henceforth referred to as the design[4] process, differs only in minor respects from the general problem-solving process introduced previously. This design process concerns the fact gathering, thought processes, decision making, and other phases of activity ordinarily engaged in by a designer in the course of arriving at the solution he ultimately specifies. Therefore, the design process is the general problem-solving methodology of the designer.

The design process differs in two respects from the general problem-solving process from whence it derives. In the design process it is suggested that definition of a problem be executed in two separate steps: a broad, *detail-free* formulation of the problem followed by a relatively *detailed* analysis of it. Defining a problem in two distinct steps in this manner is strongly advocated to encourage the designer to adopt a broad perspective of the problem *before* becoming enmeshed in the details. The second respect in which the two processes differ is the addition of a specification phase, for the practical purpose of communicating the solution to permit its implementation.[5] The result is the following five-phase design procedure.

Formulation of the problem.
Analysis of the problem.
Search for alternatives.
Evaluation of alternatives.
Specification of the preferred solution.

Whether it be a bridge, production tool, automatic dish washer, power plant, ore refinery, jet engine, manufacturing plant, materials handling system, or single manufacturing operation, this same basic design procedure applies. The definition and scope of these phases are summarized and illustrated in Table 1, and discussed in detail in succeeding chapters.

[4] The term design derives from the Latin word *designare,* meaning to mark out or point the way, and is truly descriptive of the process that is taking place.

[5] Implementation, as used herein, refers to adoption, creation, and proper use of the final design.

TABLE 1. Outline of the Design Process

Phases of the Design Process	How this Process Might be Applied to an Assembly Problem
I. FORMULATION OF THE PROBLEM	
A *brief, general* description of the characteristics of the problem, *free of detail and restrictions*, and concerning at least 1. States A and B, and perhaps 2. The main criterion or criteria, 3. The volume* and 4. The time limit.	To design, within approximately 3 weeks, a method of assembling 15,000 model X watches from their component parts, in such a manner as to minimize total cost.
II. ANALYSIS OF THE PROBLEM	
A *detailed* determination of the characteristics of the problem, *including the restrictions.* This phase is primarily concerned with the specifications of states A and B, criteria and their relative weightings, and the restrictions. It is characterized by the gathering, investigation, clarification, and screening of facts concerning the above.	Analysis of this problem would entail: Determination of the specifications of the component parts and of the complete assembly; Determination of the specific criteria to be employed and their relative weightings, such as cost of installation and operation of a proposed method, employee fatigue, effort, monotony, learning time required, etc.; Determination of the restrictions on the method of assembly. For example, some component parts must be assembled before others, and it has already been decided that this assembly activity must be located in a certain area of the plant, and performed on an assembly line.
III. SEARCH FOR ALTERNATIVES	
The seeking out of alternative solutions.	The designer seeks a variety of alternative assembly procedures, workplace layouts, sequences of events, types of equipment, etc., relying on his own ideas and contributions from many other sources.

* Volume, previously referred to as number of repetitions, refers to, for example, the number of automobiles to be manufactured, the expected usage of a certain highway, the expected number of power plant customers, and the like.

TABLE 1. Outline of the Design Process (*Continued*)

Phases of the Design Process	How this Process Might be Applied to an Assembly Problem
IV. EVALUATION OF ALTERNATIVES The evaluation of alternative solutions on the basis of the established criteria, in preparation for a decision.	Here investment costs (costs of installation, capital, training, etc.) and operating costs (costs of operating equipment, of labor, etc.) are estimated for competitive proposals.
V. SPECIFICATION OF THE PREFERRED SOLUTION Delineation of the specifications *and* performance characteristics of the selected method.	In this phase the details of the selected method, (the layout, procedures, and equipment), are recorded to facilitate installation and control of that method.

Some overlap between these phases is to be expected. For example, some solutions are conceived during the period in which formulation and clarification of the problem predominate. Similarly, generation of alternatives later becomes the predominate activity, but not to the exclusion of occasional reformulation and further clarification of the problem. Partial or complete reiteration is not uncommon.

EXERCISES

1. Identify states A and B for each of the following problems, making assumptions where necessary.

(*a*) Formation of paper clips.
(*b*) Inspection of electron tubes.
(*c*) Home construction.
(*d*) Travel to the moon.
(*e*) The mountain climbing problem.

2. Describe some criteria that might well be applied in choosing between alternative solutions to each of the problems cited in 1.

3

Formulation of the Problem

Do you advocate trying to solve a problem without actually knowing what the problem is? Surely not, yet this is exactly what we frequently do, to the detriment of our long-run design performance. The purpose of the formulation phase is to maximize the chances that the problem at hand *will* be isolated and satisfactorily defined, and that this will be accomplished at the outset. The time devoted to problem formulation is at least as profitably spent as that devoted to any other phase of problem-solving activity.

A Case in Point

The management of a large organization producing and distributing feeds and fertilizers is concerned over the relatively high cost of handling and storing its products. An engineer has been assigned to the problem with the hope for a significant cost reduction. Currently the materials are bagged and stored in a large warehouse by the method diagrammed in Figure 2.

This situation will provide illustrations for the discussion which follows.

Formulation of this problem. Recall that formulation entails a *brief, general* description of the characteristics of the problem, free of detail and restrictions. It is very important that this take place at the outset of the engineer's approach, that this be a broad phrasing of the problem, and that detail, restrictions, and the present solution to the problem be avoided during this period. This phase is a crucial one, and the fact that it may be brief and consume only a minor part of the total time devoted to a problem may belie its importance.

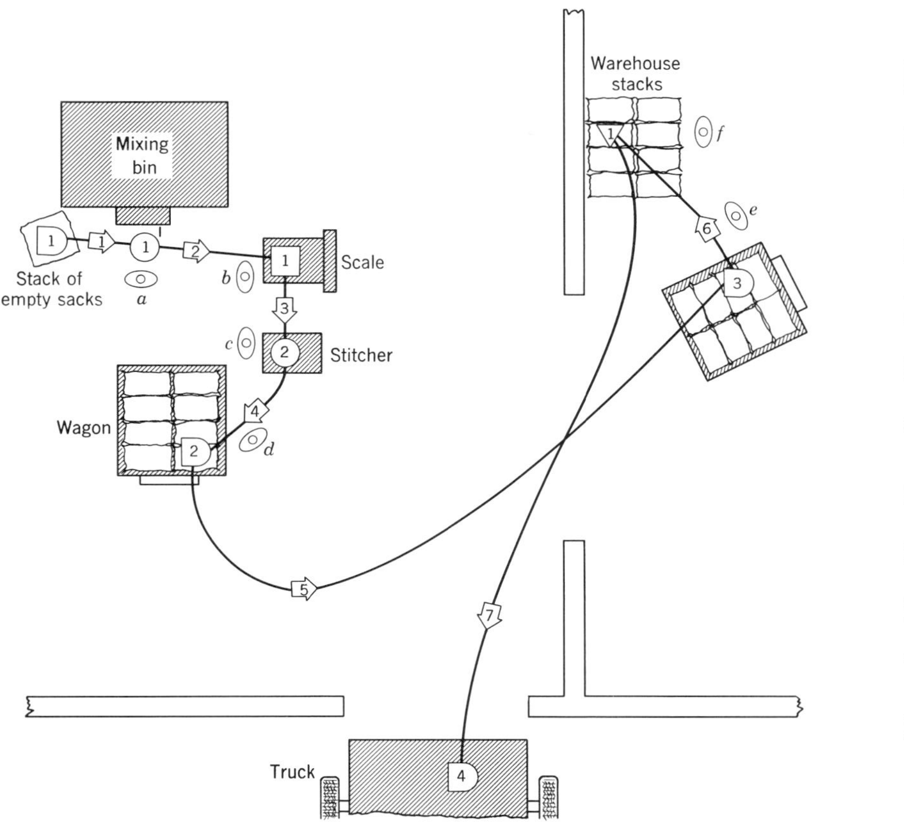

- [1] (storage symbol) Stacked sacks await filling.
- ⇨1 Man A lifts empty sacks from stack and places it under spout for filling.
- (1) Man A fills the 100-pound sack by gravity feed, manually controlling the rate of flow.
- ⇨2 Man A hands the bag to Man B.
- □1 Man B checks the weight and adds or removes material when necessary to adjust the weight to approximately 100 pounds.
- ⇨3 Man B hands the bag to Man C.
- (2) Man C folds and stitches the top of the bag.
- ⇨4 Man D takes the bag and loads it on wagon.
- ⇨5 Loaded wagon is pushed to warehouse.
- ⇨6 Bags are stacked by E and F.
- ▽1 Bags are stored, awaiting sale.
- ⇨7 Bags are loaded on awaiting truck, two or three at a time by handtruck, then delivered to consumer.

Figure 2. Flow diagram of present method of filling, storing, and loading bags of feed.

Conceptually, problem formulation entails a delineation of states A and B, the main criteria, the volume, and the time limit. A formal statement embodying all of these characteristics is seldom necessary or even common in practice. The essential and crucial feature of problem formulation is the identification of states A and B. This identification for a given problem may be phrased verbally or diagrammatically.

The problem for the feed distribution system described may be viewed in a number of different ways. For example, to find, in a period of time x, the maximum profit method of transferring approximately y tons of feed from:

1. *mixer* to *stockpiled sacks in the warehouse, by* filling, weighing, stitching, and stacking; or
2. *mixer* to *stockpiled sacks in the warehouse;* or
3. *mixer* to *sacks in the delivery truck;* or
4. *mixer* to delivery *truck;* or
5. *mixer* to the *transportation medium;* or
6. *mixer* to the *farmer's storage bins;* or
7. *storage bins of the feed ingredients* to the *farmer's storage bins;* or
8. *warehouse* to *farmer;* or
9. *producer* to *consumer.*

There are still other possible formulations of this problem, broader, narrower, and ranging between the samples given.

Formulation 1 is inadvisable in that it includes restrictions, namely, "by filling, weighing, stitching, and stacking," which should be excluded from a problem formulation. Furthermore, note that these "restrictions" are actually no more than particular characteristics of the current solution to the problem. Yet this is exactly how we would often approach a problem like this, unjustifiably accepting numerous features of the current solution as essentials, and proceeding as if they were unalterable, thus excluding many profitable changes that might otherwise have been introduced. The overcoming of this tendency is a major objective of the design process described herein.

Although the remaining formulations of the feed problem are free of restrictions, they are not equally preferable or profitable. This array of formulations and the fact that the likely consequences of each are quite different, calls attention to an important matter henceforth referred to as breadth of the problem formulation.

Breadth of the Problem Formulation

With respect to formulation of a problem, the term breadth will be used to refer to (1) the extent to which states A and B are assumed to be specified as the design process is begun, and (2) how much of the total problem the designer is addressing himself to. The completeness with which states A and B are assumed to be specified as the design process is begun is a major determinant of the number and variety of alternative solutions available to the designer. In formulations 1, 2, and 3 of the feed problem, it is assumed that the feed is in sacks in state B. In formulation 4, however, only "truck" is specified, *which opens the problem to a whole group of possibilities* not involving sacks. In formulation 5, only "transportation medium" is specified for state B, opening up additional possibilities not involving trucks. This trend toward a less specific definition of states A and B continues until only producer and consumer are specified, thus leaving the way open for a wide variety of methods of handling, modes of transportation, package types and sizes, etc. In formulating a problem, it is *suggested that the designer ordinarily assume specifications on states A and B that are as general as the economics of the situation and organizational boundaries will permit.* Failure to follow this policy will cause the designer to exclude whole realms of profitable possibilities from consideration, merely because in setting up his problem he has unjustifiably accepted certain specifications on states A and B.

Breadth of the problem formulation also refers to how much of the total problem the designer is addressing himself. For example, in formulations 1 and 2 of the feed problem, state B goes only as far as the warehouse stockpile. In formulation 3 it is extended to delivery truck, and in formulation 6 it is extended to the consumer. In formulation 7, state A is also extended. Thus, as states A and B are extended to include more of the total problem the problem formulation becomes broader. (In most instances the total problem may be considered to be: how to maximize the return on an investment.) A designer's problem might entail the design of a single production operation, a sequence of such operations, a production department, an entire factory, or an entire production system embracing all that takes place between natural resource and ultimate consumer. *In general, a problem should be formulated so as to include as much of the total problem as the economics of the situation and organizational boundaries will permit.* The more a total problem is split up into parts to be attacked independently, the more suboptimum the total solution or system is likely to be. If, in the feed problem, bagging is treated as one problem, transportation to and stacking in the warehouse as another problem, loading of trucks another, transportation to the farmer another, and unloading of trucks at the farm another, it is likely that the overall system that eventually evolves will be unnecessarily far from optimum. Treating this problem broadly is likely to yield a superior overall system involving a much lower total cost.

Henceforth, a specification the designer unjustifiably assumes is being imposed on him will be referred to as a fictitious specification. Although never told so, the designer might well automatically and unjustifiably assume state *B* to be stockpiled sacks in the warehouse, and complete the whole process of design without realizing that it is only he that has limited the problem to this breadth. The detrimental effects of fictitious specifications through exclusion of profitable alternatives and through suboptimum fractionation of problems have already been mentioned.

If the engineer assigned to this project is successful in freeing himself from the bagging limitation, he opens the possibility of "handling in bulk." If he is successful in formulating the problem so as to encompass delivery to the farmer, he opens the possibility of "bulk handling" directly to the farmer's storage bins. Evidently someone has succeeded in this fashion, for after many years of utilizing laborious handling methods, dealers are beginning to deliver feed directly to the farmer's storage bins by blowing it from large bulk trucks, which results in a substantial saving to producer and consumer.

How Broadly Can the Designer Formulate a Problem?

How broadly a designer formulates a problem is his decision to make. The problem formulation is a point of view, the manner in which the designer perceives the problem. It may be no more than a series of thoughts or some scribbled notes. This formulation is not binding. It is not inflexible. This being the case, the designer can and should formulate the problem broadly; it is his prerogative and much to his advantage in the long run to do so. He is selling himself and his employer short if he does otherwise. Thus, in the feed problem, it behooves the engineer to view the problem as one of "transferring feed from producer to consumer."

Note, however, that a broad formulation such as this is one matter;

how far the designer is justified and able to actively pursue this broad formulation through the remainder of the design process is quite a different one. Pursuance of a broad formulation often involves trying to have specifications changed, penetrating into organizational areas outside those of the designer, and so forth. These activities, if pressed too far, may well meet with resistance or curtailment.

Consider the engineer who is confronted with the problem of designing an improved method of assembling rather intricate cardboard shields that are subsequently placed around the latching mechanisms within refrigerator doors to keep the insulation in the doors from fouling the mechanisms. This shield, pictured in Figure 3*b*, is assembled from two prescored cardboard stampings, Figure 3*a*. In the assembly operation, a series of flaps are first folded over, then the two sections are stapled together as illustrated in Figure 4. Ultimately, in a later operation, the assembled shield is glued into place over the latching mechanism inside the refrigerator door, as pictured in Figure 5.

Notice that the engineer is presented this problem by showing him the present method of assembling the shields. As usual, *the problem is not formulated for him; instead he is merely confronted with the*

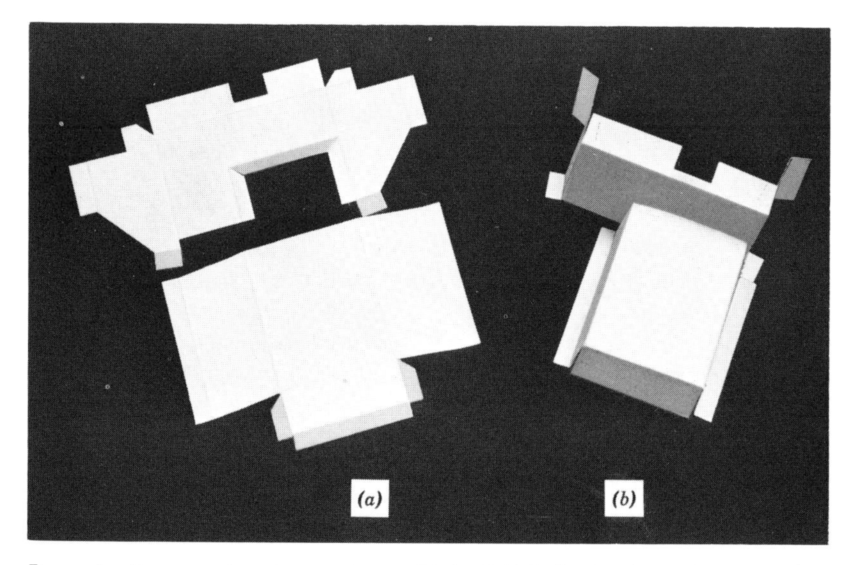

Figure 3. Component parts a and assembled shield b for keeping refrigerator door insulation from fouling latching mechanism.

Figure 4. Stapling phase of the shield assembly operation. (Courtesy General Motors do Brazil)

present solution to the actual problem. Suppose the engineer formulates this problem as that of finding the most economical method of keeping the refrigerator door latching mechanism free of insulating material, instead of that of finding the most economical method of assembling the cardboard shielding device specified. The former is a broader formulation and opens the situation to many different ways of keeping insulation out of the works, including the use of a rigid sheet-foam insulation that will not shift around inside the door as the current insulation does, applying an adhesive on the inside of the door panel to hold the currently used insulation in place, or a much less complicated and less expensive cardboard shield. These and any number of other practical alternatives are surely superior to the method given to the engineer.

If in this problem the designer chooses to pursue the broader formulation, he must attempt to obtain a revocation of the current shield's specifications and perhaps certain other specifications of the door. It may well be that the product designers or others will not allow the "specs" to be changed. This is not unlikely for a variety of reasons, one of which is that if there is the usual pressure to get a new product into production, management will be reluctant to reconsider decisions already made. It is highly desirable that the designer pursue his broader formulation at least this far. If he succeeds in getting away from the current cardboard shield, a better all-around solution will probably result, and to his credit. If he is refused, at least he has protected himself if a question is raised in the future as to "why this ridiculous shield?" Another means of self-protection in a situation like this is for the designer at the conclusion of his efforts to present several alternatives for management to choose from, among which would be an improved method of assembling the current shield as well as some of the superior methods of keeping the latching mechanism free of insulation.

Consider the engineer working on the feed problem. If he decides to pursue the broad formulation, it may be necessary for him to persuade those responsible to give up the use of individual sacks, to get into matters of sales policy, methods of warehousing, and the like. It may be that those who assigned him to the project had something else in mind. Perhaps they expected him to devote his attention to improving the method of filling, checking, stitching, and

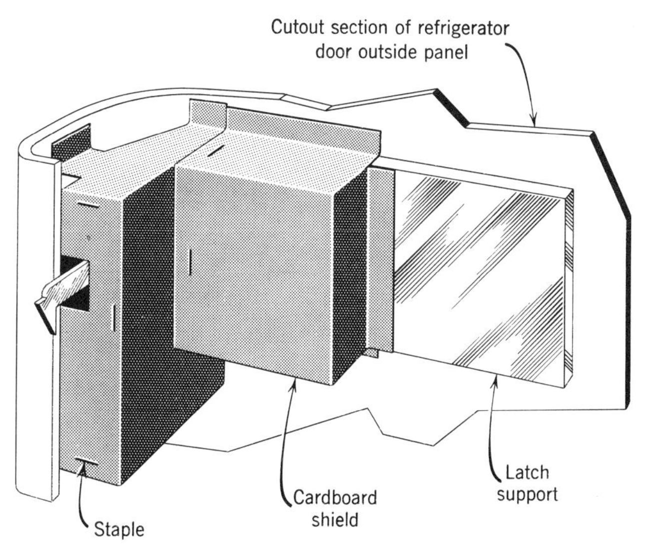

Figure 5. Shield glued into position over latching mechanism inside refrigerator door. (Courtesy General Motors do Brazil)

transporting the bags. Thus a designer may well be forced to pursue a very narrow formulation of the problem at hand, even though it would be more profitable from the point of view of the enterprise for him to do otherwise.

To what extent a designer is justified and able to *pursue* a broad formulation of a problem depends on such factors as:

1. The scope of his responsibilities. The official capacity of the designer is a primary determinant of the decisions he is authorized to question or change.
2. The economics of the situation. In general, the less important the problem is financially to the organization, the fewer possibilities that may justifiably be investigated, the fewer specifications we can afford to challenge, and the more things we must accept "as they are" because it is simply unprofitable to devote attention to them. Volume is very influential in this respect.
3. The arbitrary limit (if any) that has been placed on time and money that can be devoted to the problem.
4. Special circumstances. For example, the particular personalities involved may make it impossible to challenge previously made decisions even though organizational lines justify it.

Pitfalls at the Start of the Design Process

The feed distribution and shield assembly problems were introduced to the assigned engineers by presenting to them the respective current solutions to these problems, with the stated or implied charge of finding a better method. *Rarely is the true problem laid before the engineer;* instead he must attempt to see through the existing design and through tradition and the opinions of others, to determine for himself what the problem actually is. This matter is made difficult not only by the manner in which problems so neatly disguise themselves, but also by the rather universal practice of colleges of presenting problems to students in an unrealistically pure form so that they are unaccustomed to and unskilled in identifying a problem in "real life." These two situations make the designer especially vulnerable to the following pitfalls.

One pitfall that a designer should be especially wary of as he approaches his problem is partially or completely solving what in reality is a fictitious problem, a problem for which there is no real need. In this case the attention of the designer was never justified. Designing a component part that is actually not needed, redesigning a manufacturing operation that actually could be eliminated entirely, are examples of a fictitious problem. Failure to actively and effectively seek out and define the problem in the initial approach to the situation begs a result such as this to the subsequent discredit and embarrassment of the designer.

The second major pitfall is attacking the current solution instead of the problem itself. *The current design is simply one solution among many to the problem at hand. The current solution is not the problem, yet the designer frequently attacks the present design as if it were.* The ease with which one falls prey to this pitfall and the importance of avoiding it cannot be overemphasized. There is a subtle but nevertheless crucial difference between starting with the current solution and picking at it in an effort to eliminate the inefficiencies and inadequacies, and starting with a basic definition of the problem and synthesizing a superior solution to it via the design process. The latter procedure is a major contributor to superior design performance over the long run.

Concentrating initially on a phrasing of the problem and temporarily avoiding detail, restrictions, and present solution, seem to be the most effective means of evading these pitfalls.

The "Black Box" Point of View

This unique and useful method of viewing a design problem will be referred to as the black-box approach. The solution to a problem is visualized as a black box of as yet unknown, unspecified contents, with a specified input (state A), specified output (state B), and given criteria for evaluation of the box's performance. Figure 6 illustrates this point of view. Other examples of the input are wood pulp, soiled clothes, or electrical energy, with their respective outputs: paper, clean clothes, mechanical work.

The black-box approach facilitates proper identification of states A and B as a problem is formulated. It is a visualization gimmick that is especially helpful in taking the designer's attention away from the present solution. In order to obtain maximum benefit from this approach, it is important that the designer let this black box block out the present solution in his thinking; that he make no assumptions for the moment as to what will be or even what must be part of the ultimate solution. The frequency with which the engineer is introduced to problems primarily by being familiarized (if he is not already so) with the current solution(s) will probably never change appreciably. Thus, it behooves the young engineer to develop the ability to

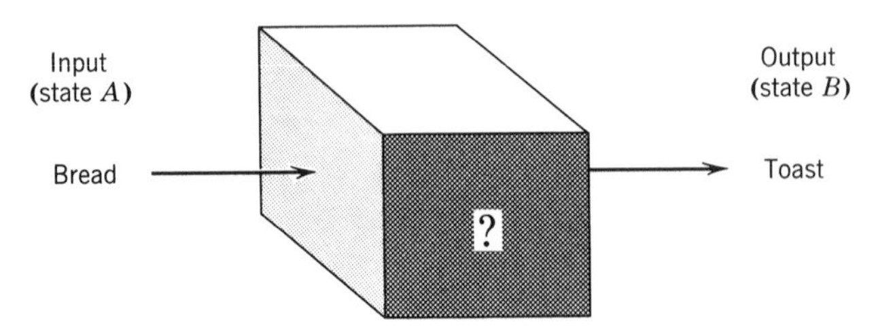

Figure 6. The "black box" formulation of a problem. Problem: To find, within a given period of time, the contents of the box that will perform the indicated conversion in a preferred fashion.

effectively isolate the problem in spite of this method of introduction. The black-box concept is one means of developing and enhancing this skill.

Summary

At the very outset of his approach, a designer should formulate the problem at hand, making sure (1) that a problem worthy of his attention exists; (2) that his view of the problem is broad; (3) that he avoids detail and restrictions for the moment; (4) that he is wary of the fictitious problem; and (5) that he does not become occupied with the present solution to the problem, nor confuse problem and solution. Ordinarily these matters deserve most of the attention devoted to this phase. Identification of the criteria, volume, and time period will ordinarily be subordinate in the formulation phase. Depending on the specific problem, they may be routine matters, requiring only superficial consideration until the analysis phase, where all characteristics of the problem are considered in more detail.

Any problem can be formulated in varying degrees of breadth. These may range from a very broad definition that maximizes the number and scope of decisions to be made, to a formulation that offers very little latitude and that minimizes the number and scope of variables to be altered. Ordinarily, the designer should formulate the problem as broadly as circumstances will permit. If by pursuing such a broad formulation the designer oversteps his organizational or economic bounds, he will almost automatically be informed of the fact by other personnel. If he fails to view the problem as broadly as warranted, there will be no automatic sign that he is selling himself and the organization short.

Note that there is no place for a description or detailed consideration of the present solution in this phase. This is an overall phrasing of the problem, made *before* becoming involved in the details of the situation. This broad view, made at the outset, maximizes the probability that the designer will define the problem satisfactorily.

4

Analysis of the Problem

A Case in Point

A certain manufacturer of washing machines is about to redesign the automatic washing machine currently marketed. The management has already decided on the maximum manufacturing cost, maximum dimensions of the unit, maximum cycle time, maximum weight, and a number of additional requirements enumerated below. They want the plans for this within six months. The forecasted sales volume is 300,000 units. An analysis of this problem, on the basis of the facts given, is as follows.

State A: A quantity of soiled clothes.

State B: The same clothes with a specified minimum amount of dirt and moisture remaining, with damage, shrinkage, and wear limited to a specified minimum degree.

Criteria: Return on the investment, or more specifically, manufacturing cost, customer appeal, cost of supplying replacement parts, life of product, ease of repair, etc.

Restrictions:

1. Since clean clothes are to be obtained from soiled ones, dirt *must* be removed.
2. Total cost-of-manufacture *cannot* exceed $125.
3. The unit *cannot* be larger than 30 inches wide, 38 inches high, or 30 inches deep.
4. The unit *cannot* weigh more than 250 pounds.
5. The cleaning process *must* be complete in 20 minutes or less.
6. The unit *must* operate on ordinary 60 cycle, 110 volts A.C.
7. The unit *must* be operable by the ordinary housewife with no more than a few simple instructions required.
8. The unit *must* be foolproof against blunders in operation.
9. The unit *must* be foolproof against accidents caused by exposed parts.

10. The unit *must* satisfactorily handle all current washable textile materials, and those that might be developed in the foreseeable future.

11. The unit *must* come in six different (specified) colors.

Volume: Approximately 300,000 units, averaging approximately 12,000 per month.

Time period: Design to be completed within approximately six months.

Expansion of the Restriction Concept

A restriction was defined earlier as a characteristic or feature that *must* be true of any eligible solution to the problem. Insofar as the designer is concerned, this is a solution characteristic that is inescapable if a solution is to be acceptable.

All restrictions arise, directly or indirectly, from previously established specifications. Indirectly they arise because once states *A* and *B* have been decided on, certain features of all acceptable solutions are automatically established. That is, once these states are specified, there are certain things that for physical reasons must take place to accomplish the transformation. Once states *A* and *B* are specified as soiled clothes and clean clothes, this automatically fixes a character of all acceptable solutions, namely, that they *must* remove dirt. In these matters the designer has no choice, or a limited range of choices, because of the very structure and behavior of the universe (the speed of light, the composition of matter, the relationship between temperature and pressure, the nature and behavior of the living organism, the force of gravity, etc.). If a conventional wheel is to be satisfactorily mounted on an automobile, the designer has no choice; the wheel must for physical reasons be placed before the lugnuts. Thus, the specification of states *A* and *B* automatically determines certain necessary characteristics of an acceptable solution.

Directly, a restriction is established when someone whose authority the designer must accept specifies a characteristic that must be true of an acceptable solution. Illustrations of this type of restriction are the requirements 2 through 11, laid down for the method of cleaning clothes in the problem introduced above. These are characteristics that by command must be incorporated in an alternative solution if it is to be eligible for consideration.

Note that specification of states *A* and *B* determines *what* must take place to accomplish the transformation, as illustrated by restriction 1 in the washing machine problem. The restrictions established directly by imposition relate to *how* the transformation must take place, as illustrated by restrictions 2 through 11 in the washing machine problem.

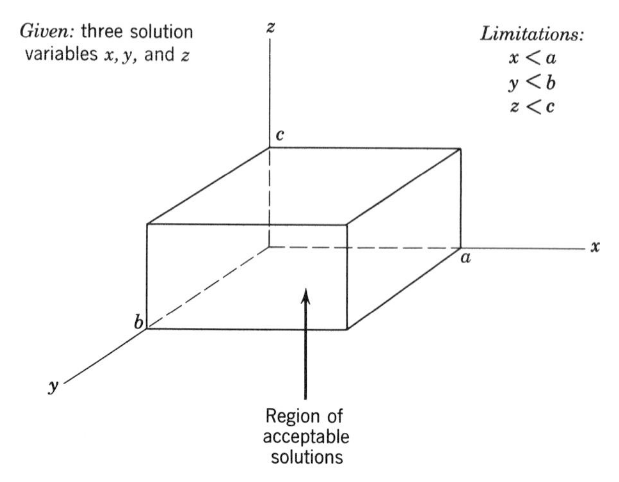

Figure 7. Problem with three solution variables with limits on each, pictured in three-dimensional space.

The Concepts of Solution Variables and Solution Space

In a design problem, alternative solutions differ in many respects. In a product design problem, for example, solutions differ in size, weight, shape, type of material, color, and many other attributes. The respects in which alternative solutions to a problem may differ will be referred to as the solution variables. The final solution to a problem consists of a specified value for each of these variables—a certain size, certain weight, certain shape, etc. The situation may be viewed as one of N variables in N dimensional space. For a problem with three continuous solution variables—x, y, and z—the situation is readily pictured in three dimensions, as in Figure 7. The range of values the designer may specify for a given variable is often limited, as in the washing machine problem, which means that there are boundaries on this multidimensional space that define a space from within which the designer must select his final solution. This space is appropriately referred to as the region of acceptable solutions. In the problem with three solution variables postulated above, suppose that x may not be greater than a, y greater than b, or z greater than c. These limitations establish a region of acceptable solutions, shown in Figure 7. The problem involves selection, from within this bounded space, of a solution that is optimum, superior, or satisfactory, with respect to the criteria selected.

These limitations on solution variables are restrictions. Some potential variables are fixed a priori, and may not be altered by the designer. For example, in the washing machine problem the power source has already been fixed at 110 volts, 60 cycles. Other variables are alterable by the designer, but only within an imposed range. For example, in the clothes cleaning problem it has already been specified that the final product shall weigh not more than 250 pounds. This restricts the variable to the 0- to 250-pound range.

Types of Restrictions

1. Real. They *have* been imposed, they are quite genuine, there is no assumption or imagination involved. Of these
   (*a*) some are accepted,
   (*b*) some are deviated from, because it is necessary and/or profitable to do so.
2. Fictitious. They are unnecessarily and probably unwittingly assumed by the designer.

Real restrictions—accepted. It is likely that the designer will accept and conform to at least some of the real restrictions imposed on him. Such will be the case if:

1. The restriction is an optimum or nearly optimum decision, or appears so at least.

2. The restriction is apparently a suboptimum decision, but not enough to warrant the time and effort required to challenge it. The benefits that might accrue *if* a revocation of the specification were achieved do not seem justified by the cost of preparing and presenting a case for change. For example, it may appear to the washing machine designer that to produce units of six different colors is hardly economically justified, but the apparent cost of convincing the sales department of this discourages any action on his part. It seems more economical to conform.

3. The restriction appears sufficiently suboptimum to justify deviation, but the designer recognizes at the outset that it is futile to even suggest doing other than specified. Strong and sometimes rather arbitrary personalities of certain key personnel associated with the matter often create this situation.

4. For personal reasons, the designer elects not to deviate even though it would be profitable to do otherwise.

5. The designer has attempted to have the restriction revoked and failed.

Real restrictions—not conformed to. Not all initially imposed restrictions are irrevocable. Nor are they always compatible. Sometimes the designer finds that the restrictions are such that he cannot arrive at a solution that satisfies them all, or that can satisfy them without prohibitive expense. Thus, of the real restrictions imposed, it is likely that some will be deviated from because they represent suboptimum or incompatible decisions, which make deviation profitable or even unavoidable.

It is naive and unwise to assume that all specifications given for states *A* and *B* (which indirectly determine restrictions) and specifications given for solutions (which become restrictions) are the best decisions that could have been made and thus are optimum with respect to the overall criterion. In fact, almost all decisions are suboptimum to some degree, which is not so surprising to one who considers:

1. The element of chance involved in gathering alternative solutions, and the fact that most decision makers are under time pressure.
2. The magnitude of the role played by judgment in making real-life decisions; mistakes are made; not all future implications and consequences are foreseen.
3. The degree to which an overall problem is commonly subdivided into many relatively independent subproblems.
4. That decisions are certainly not always made on a completely objective and rational basis.
5. That an engineer is sometimes given a restriction that is more an expression of hopefulness than a "must."

In view of these circumstances it is inevitable that a designer will occasionally encounter a set of restrictions that he cannot satisfy, at least with a single solution.[1]

Much more likely than the foregoing situation is the one in which all restrictions can be satisfied, but only at an unnecessary and perhaps prohibitive expense. A designer frequently encounters a suboptimum specification established by someone preceding him in the decision-making hierarchy, and realizes that it is unnecessarily costly to adhere to this restriction. For example, in the clothes cleaning problem, it is possible to design a method that will satisfactorily handle all known and expected washable fabrics, but the expense involved in achieving this feature does not seem justified by the resulting increase in sales revenue. In this type of situation, which is far from uncommon, the designer must appraise the situation, apply his best judgment, and

[1] In terms of the solution space, there are two (or more) bounded areas that do not overlap, such that there is no common solution.

choose from a number of alternative courses of action. He must decide whether under the circumstances the restriction in question can be ignored, whether it can be violated to a limited degree, whether he should accept it and proceed, or whether it is profitable to seek and possible to obtain a revocation of the specification, and if so, how far he should go in attempting to achieve a change. These and probable other choices facing the designer in this type of situation contribute substantially to the judgment, frustration, and human relations problems associated with design work.

The foregoing situations raise the question of the revocability of previously established specifications. Ordinarily, not all of the specifications imposed on a designer are final and irrevocable. Many are tentative, subject to change upon agreement between the designer and other parties involved. A methods engineer working on an assembly problem might well find the design of the component parts and finished assembly (states A and B) to be completely fixed, so that any time devoted to redesigning the parts to facilitate assembly is wasted. More often, however, he will find that there is some flexibility in the product design, and that if he conceives of a product-design change justified by savings in assembly costs, this change will be adopted. The "rigidity" of imposed restrictions ranges from completely fixed (by reason of pure arbitrariness, resistance to change, politics, aesthetics, personality clashes, as well as for objective reasons) to completely removable at the discretion of the designer. In between these extremes there is a continuum of indefiniteness in which a majority of these specifications lie. In this "zone of indefiniteness" the designer is not sure as to the removability of a restriction, and only after varying degrees of inquisition, consultation, compromise, and bargaining is he subsequently able to learn if it is revocable. Certainly he should not assume that every imposed specification is final and unalterable. Certainly many very profitable design changes have come about through a designer effectively challenging the specifications previously established by another decision maker in the organization.

The statements concerning incompatible, suboptimum, and revocable restrictions also apply to the criteria and time limit imposed on the designer. It is not inconceivable, for example, that a designer be given certain specifications and criteria that are virtually impossible to satisfy by the deadline given.

Fictitious restrictions. The term fictitious restriction will be used herein to refer to a "restriction" unnecessarily assumed by the designer. This assumption has probably been made unconsciously and is both undesirable and unjustified. In most cases the fictitious restriction is not an explicit decision made by the designer through conscious deliberation. It is a matter of automatically and unjustifiably *acting as if* a certain thing "must be," or as if a certain alternative cannot be used. Using the feed problem introduced earlier for illustration, many would proceed with this problem *as if* the feed must be handled in sacks even though no one has said that this must be so.

Fictitious restrictions are ordinarily not explicit, which accounts for their elusiveness. In fact, if they are explicitly stated, their fictitious and oftimes absurd nature usually becomes obvious. For example, referring again to the feed problem, suppose we state some of the characteristics of the current solution as restrictions, which they actually are not. Feed *must* be placed in sacks, the sacks *must* be transferred to the warehouse by wagon, the sacks *must* be handled individually. All we have done is substitute the words "must be" for "is" or "are," which is in effect what we often do in accepting this type of restriction.

There are two types of fictitious restrictions prevalent. One is nothing more than some feature of the current or past solutions to the problem, unjustifiably treated as a restriction, as illustrated in the previous paragraph. The other type involves the automatic and unjustified exclusion of solutions to a problem, but in this case not as an outgrowth of the current solution to that problem. In this instance exclusion results because the designer unwittingly and unfortunately defaults on whole areas of feasible solutions, so that they go completely untouched. To illustrate, suppose we state some additional artificial restrictions for the feed problem: a belt conveyor cannot be used, pallets and fork trucks cannot be used, an automatic filling and weighing machine cannot be used, bulk handling cannot be used. It is likely that many, in the process of seeking alternative solutions to this problem, would act *as if* some or all of these actually were restrictions, even though when explicitly stated as such they may seem absurd.

It is remarkably easy to fall prey to fictitious restrictions; evading them is difficult. Yet they warrant considerable attention, for success in isolating them opens the problem to many superior solutions. Some suggested means of avoiding fictitious restrictions arising out of the present and past solutions are:

1. Careful formulation and analysis of the problem.
2. Remaining constantly aware of the tendency to accept such "restrictions," and being alert for their presence.
3. Using the "black box" approach suggested earlier. The problem

Work Measurament

Krick, Edward V.

Methods Engineering; design + measurement of work methods

T56 .K7

Karger, Delmar W.

T60 .W6 K3 1966

~~Mackenzie, R. Alec~~

Oakes, William S.

Developing work standards

T60.2 .D48 1985

is formulated with all solutions temporarily replaced by the so-called black box. It is a gimmick for conceptually blotting out the present solution and its ancestors. Starting with only the black box and thus from scratch, and proceeding from there to a synthesis of the genuine restrictions on the basis of logic and the *facts* of the situation, is an excellent means of avoiding this type of fictitious restriction.

4. Paraphrasing various features of the current solution *as if* they were restrictions, as attempted earlier for the feed problem. This is a mental exercise that is effective in calling attention to fictitious restrictions.

Fictitious restrictions of the other type, those arising by "default," may best be avoided by actively and effectively seeking out alternative solutions, as discussed in the following chapter.

Analysis of a Problem

The analysis of a problem consists of a relatively detailed phrasing of the characteristics of the problem, including restrictions. It differs from the previous phase with respect to degree of detail and recognition of restrictions. Since the time limit and volume data often involve little detail, this phase is primarily concerned with states A and B, the restrictions, and the criteria. And in fact, only when one or several of the customary criteria are to receive unusual weighting is it necessary to give detailed consideration to the criteria in this phase. A criterion that is of special importance and to be given more than usual weight in the final decision, affects the types of alternative solutions that will be emphasized in the "search for alternatives." For example, high quality is to be a special sales feature of a new model shotgun, and thus a criterion of extra importance in design of the gun. As a consequence, the designer will be considering different materials, mechanisms, features, and finishes, than he would otherwise. Actually, in many instances the criteria are implicit and routine in nature.

This phase of the design process involves considerable "fact gathering," especially with respect to the characteristics of states A and B and the restrictions. It also involves a very thorough screening of these "facts," so that when the designer has completed his analysis of the problem he will have stripped the problem down to the bare essentials by isolating the genuine limits within which he must work. In other words, he will have determined what is given, what is fixed, what insofar as the designer is concerned "must be." What is more important is that he will have determined all that he is free to and expected to specify as he proceeds. He will have determined what variables he is authorized to alter—the areas of possibility from which he is free to select. The objective in giving considerable attention to restrictions is not merely to learn what cannot be done, but rather to bring to the fore what is fixed so that the designer may proceed on the assumption that the variables remaining are his to alter. The ultimate purpose then is to find out in what respects he is *not* restricted.

At the conclusion of this phase the designer should have freed himself to the maximum extent authorized and economically justified, from previously established specifications. The designer should have objectively scrutinized the soundness of specifications imposed on him, investigated the rigidity of unsound decisions, and attempted to obtain revocation of same if it seemed economically and organizationally feasible. In addition, he should have made a special effort to free himself from fictitious restrictions that are no more than particular characteristics of the current solution to the problem. The overcoming of this tendency to accept what "is" as "must be" warrants considerable attention. The danger is too much self-restriction, with the probable result that promising solutions are unjustifiably and unfortunately excluded from consideration.

Gathering Facts

The fact gathering referred to earlier concerns the actual collection of information pertinent to the characteristics of the problem. In addition to the determination of specifications and the like, this process may involve learning the nature, behavior, and interaction of certain variables in or affecting states *A* and *B*, the criteria, volume, restrictions, or potential solutions. Examples of such variables are:

1. Relevant to a warehouse design problem, expected variation in size and time of arrival of incoming shipments.
2. Relevant to a dam design problem, expected variation in rainfall.
3. Relationship between quality of product and consumer demand.
4. Short- and long-term variation in demand for electric power in a community.
5. Relationship between thickness of insulation and rate of heat loss.

Thus, the process of gathering facts may well entail more than consulting records, personnel, handbooks, and other information sources; it may also require observing phenomena, analyzing historical data, experimentation, sampling human reaction and opinion, and similar forms of information collection.

In obtaining information relative to volume, it is important to learn of volume growth and decline trends, seasonal and other types of cycles, and inherent random variations. Volume is seldom a static matter. It is also important to determine the expected "life" of the problem in question. Is the product being designed expected to be in demand for six months? Five years? Indefinitely? Thus, the designer must obtain volume projections into the future, from persons the most qualified to make such forecasts.

This fact-gathering process may also include the collection of information relative to the current solution(s) to the problem, subject to the precautions to be mentioned.

Where Does the Current Solution to a Problem Fit In?

In a majority of cases there is a currently utilized solution to the problem at hand. In most problems the designer is familiar with that solution, sometimes too familiar. If he is not acquainted with it, in all likelihood it will be shown to him as he is being assigned the problem. So, regardless of whether he prefers it or not, for better or for worse, the designer usually knows the currently used method, and so what we commonly refer to as design is more often redesign.

For some purposes, this familiarity with the present solution is necessary or desirable. For example:

1. In appraising a problem to determine if redesign efforts are warranted, it is necessary to estimate the savings potential in that problem. To estimate the inferiority of the current method with respect to what could probably be designed and installed at a reasonable cost, it is necessary to become familiar with the current solution. However, this does not necessitate an extreme intimacy with the current method; a cursory survey seems adequate.

2. It is often profitable to know the current solution so as to benefit from its parts that are already satisfactorily worked out, and for which no redesign efforts are justified. Again, however, no intimate familiarity, no documentation, is ordinarily necessary to accomplish this.

3. The current design is a solution to the problem at hand, and in fact it is often a competitive solution for it ordinarily has at least one advantage: zero installation cost. Furthermore, the currently utilized method frequently has a number of obvious and relatively

easily corrected inefficiencies that make it more competitive, yet with little or no investment requirement. So the current method should usually be known and duly considered along with other alternatives. However, the appropriate time for this is when alternatives are being assembled and evaluated, *not* in the problem definition phase.

4. Some of the performance characteristics, like operating cost of the existing solution, must ultimately be determined for purposes of comparison with a proposed new design, to determine the advisability of changeover; also to convince others of this if the change is advisable. Again, however, this does not require detailed analysis of the physical aspects of the existing solution, nor is this cost information necessary until the evaluation phase.

Thus, there are two reasons for wanting to know the current solution to a problem before or at the outset of the design process, but in general, for these purposes no refined analysis or description seems necessary at this stage. Why should the designer document and analyze the present solution in detail, especially when as in so many instances it is obviously grossly inferior, while simultaneously hoping and expecting to subsequently replace it with a basically different and superior method? It is not so much the wasted time that matters; it is the difficulty it causes the designer in his attempt to free his thinking from the present method and to conceive of methods that differ significantly from it. The common consequence is that the designer indulges only in "picking" at trivia in the present method, and at the conclusion of his efforts has produced nothing more than a rehash of the old solution. Digging up facts about and looking hard at the present solution gives rise to what might be more appropriately described as a rehash or rework rather than a design process. Engineering supervisors become concerned over a lack of creativeness in their engineers, arising from the fact that the engineers have "lived" with basically the same product design, or plant layout, or manufacturing methods, etc., so long that they find it difficult to conceive of anything much different. So why should the designer make the situation even worse when embarking on a project by immediately becoming further enmeshed in the details of the present method in the course of making a refined analysis of it?

To summarize, the designer already is, or soon will be familiar with the present solution at the time he starts the design process. This seems fairly certain. However, the designer should avoid becoming any more involved with the existing solution than is necessary for appraisal of the savings potential of the problem and for culling worthwhile features from that solution. And more important, the designer should cultivate the ability to free his thinking from the present solution, to see through it, and properly identify the problem.

Summary

When the emphasis on formulation and analysis is complete, the designer should have:

1. Determined the specifications of states A and B to which he will adhere.
2. Determined the restrictions to which he will adhere.
3. Maximized and determined the areas of possibility open to him.
4. Determined and appropriately analyzed and scrutinized the remaining characteristics of the problem: criteria, time limit, and volume.

He should *not* have:

1. Immediately begun trying to think of improvements in the present method upon being confronted with same, thus completely bypassing a definition of the problem.
2. Spent much time scrutinizing and analyzing the present solution(s) to the problem.

He may have:

1. Attempted a formulation of the problem and then after delving into details in his analysis, reformulated the problem on the basis of something learned in the process, and perhaps repeated this a number of times.
2. Found that one or several of the specifications given to him cannot be adhered to, or are unprofitable to adhere to.

EXERCISES

1. One portion of a large oil company warehouse is devoted to the storage, packing, and shipping of road maps to service stations. The current method is illustrated by the flow diagram on page 46. The stored maps are removed from their cases (200 to a case, 8 bundles of 25 maps each) and a moderate supply for each state is stacked on the open shelves. The packer, with order slip in hand, fills the customer's order by picking the requested type and quantity of maps from the shelves and assembling these in a carton on the bench. When the order is assembled, he slides the carton to the sealing and labeling station and performs these operations. Then he slides the carton to the next position in order to weigh the shipment and add the required

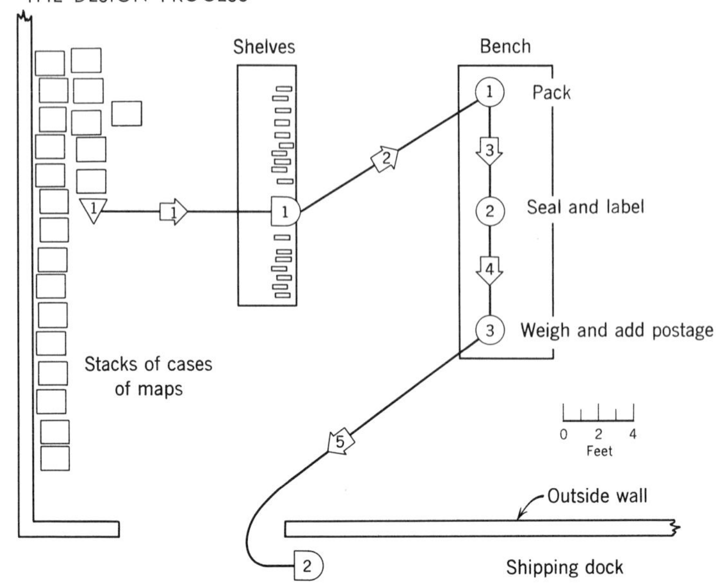

Present layout for assembling and packing map orders.

postage. He then carries the completed order to the shipping dock. Under this method, the typical order requires an average of 10 minutes for completion, starting as the packer picks up the customer's order and ending as he returns from placing the ready-to-ship carton on the shipping dock. The wage rate is $1.54 per hour. There are approximately thirty-five orders to be filled per day.

Seventeen different maps must be stocked, so that an order may involve anywhere from one to all seventeen types. An analysis of the orders received over a period of time indicates the following:

Per Cent of Orders Received	Number of Different Types of Maps Requested in the Order
3	1
3	2
4	3
7	4
14	5
16	6
13	7
9	8
7	9
3	10
3	11
1	12

Per Cent of Orders Received	Number of Different Types of Maps Requested in the Order
2	13
1	14
2	15
1	16
11	17 and above.

A service station may requisition any quantity of any type of map, but not more than a total of 500 maps in any one order. A quantity analysis of orders received over a period of time produced the following results.

Total Number of Maps Requested in Order	Per Cent of Orders Received
0–50	7
51–100	11
101–150	14
151–200	19
201–250	13
251–300	10
301–350	6
351–400	3
401–450	4
451–500	2
501 and up	11

As an engineer assigned to improve this procedure, what would you propose? Provide a step-by-step description of the procedure you recommend and a plan-view sketch of the work area.

2. A certain cosmetic company has an inventory of 200,000 glass facial cream jars on hand. Producer, trade name, and price are already painted on the jars. After this supply was obtained, the management decided to raise the price above that already marked on the jar. In addition, they have decided that the current inventory of jars will be used, and that the price marked on the jar should be sanded off so that there is no chance that a customer can learn that the price change has taken place. An engineer has been assigned the task of designing the method of removing the former price and painting on the new one. He has one week to do so.

Write out your formulation and analysis of the problem.

5

The Search for Alternatives

This phase of the design process is a concerted attempt to find alternative solutions which comply with the restrictions imposed. No doubt in the course of appraising, formulating, and analyzing the problem, the designer has conceived of or stumbled upon alternative solutions. However, it is now that the designer concentrates his efforts on the matter of generating solutions. In previous and subsequent phases of the design process, generation of solutions is a by-product; in this phase it is the objective.

Specifically, this phase involves the accumulation of alternative solutions and partial solutions from many sources. These sources include textbooks, handbooks, and the like, conversations with various individuals, the designer's own background knowledge, solutions to similar problems, and past solutions to the current problem.

Actually, this phase seldom culminates in the generation of a set of complete, mutually exclusive solutions. Rather, the output will probably consist primarily of partial solutions—ideas that concern only one or several of the steps or variables that a complete solution must eventually encompass. Some of these partial solutions can be meaningfully combined, others can not. For example, at the culmination of this phase in the design of a method of assembling television sets, the designer will probably have generated a variety of alternatives with respect to each of the method variables—e.g. order of assembly, number of workstations, arrangement of stations, method of supporting and moving the subassembly, method of supplying component parts—some of which are compatible, others of which are not. Or, in the case of a toaster design problem, the designer will have generated a variety of alternatives for basic shape of the unit, material, method of loading and unloading, etc. In the following phase of the design process the designer will evaluate the alternatives in each category, progressively weed out the inferior ones, and eventually synthesize a complete solution that represents the most favorable meaningful combination of partial solutions. Thus, the assembly method designer will subsequently evaluate alternative sequences of assembly, numbers of workstations, arrangements of stations, etc., and combinations of these, until the best overall solution is found.

Alternatives are seldom specified in detail at this stage. It is not ordinarily necessary or advisable to do so because many alternatives can be satisfactorily evaluated while still in a relatively crude state of specification. Therefore, many alternatives are specified in skeletal fashion, to be specified in more detail later if their worth justifies it.

Determinants of Creativeness

Because the engineer's own ideas are a major determinant of his success in design work, and ultimately of his advancement opportunities, the emphasis in the remaining portion of this chapter is on the generation of ideas. Creativeness is used herein to refer to the quantity, quality, and diversity of ideas a person generates. The objective is to maximize the designer's creativeness.

The primary determinants of a designer's creativeness appear to be:

1. His knowledge: the information he has available to draw on in generation of ideas.
2. The effort he exerts: how actively he seeks ideas, the degree to which he applies himself.
3. His aptitude: the inherited qualities which contribute to creativeness.
4. The method he uses: the particular manner in which he goes about generating ideas. For example, the type of thought process, the thought jogging aids, the problem-solving process, the sources he consults, and so on.
5. Chance: of the large number of alternative solutions to a problem, the particular ones a person will conceive of depends considerably on chance. For example, what he conceives of depends on the particular chain of ideas he happens to pursue, on things he happens to know, and even on the things he happens to do, see, and hear during this period.

In the author's opinion, aptitude usually has a negligible effect on creativity in contrast to the method of generating ideas. The author believes that a person who has a so-called low aptitude for creativeness, which of course is beyond his control, may ordinarily compensate for this "deficiency" by the method he follows in seeking ideas, which *is* under his control.

An Analogy to the Process of Generating Ideas

A physical analogy to the mental process of searching for ideas will be used to assist in describing how this search ordinarily progresses and some of the common pitfalls and difficulties involved. The "points in space" in Figure 8 are representative of the large, perhaps infinite number of feasible alternative solutions or partial solutions to a given problem, existing in a multidimensional solution space. Assume that the distance separating any two points is indicative of the similarity of the ideas represented. Adjacent points represent similar ideas. We would hope that the designer could start somewhere in this space and move steadily to progressively better solutions until a time limit or perfection ends the process. However, limitations of the human mind ordinarily prohibit such an effective performance. Instead, we find the searching process displays objectionable degrees of regression, bias, and chance, and that it yields a relatively high percentage of worthless and inferior solutions, in a manner to be described.

Starting from the present solution, the point S_p in Figure 8, a problem solver generally proceeds from one point (idea) to another in a manner indicated by the arrowed path in the figure. To just what solution he will proceed, when dwelling on a given idea, depends considerably on chance. Notice that the jumps tend to be relatively small, and that the ideas tend to cluster about the current solution. This failure to deviate significantly from the current solution seems to be a path of least resistance. Large jumps to remote locations, to radically different ideas, seem to be difficult and relatively rare. Often, if we succeed in conceiving of a radically different idea that impresses us, it tends to assume the attractive powers formerly held by the present solution, resulting in a subsequent cluster of ideas about it.

There are boundaries on the space the designer will select ideas from. In Figure 8 three types of boundaries are shown.

1. Genuine restrictions (some alternatives have been ruled out of bounds to the designer);

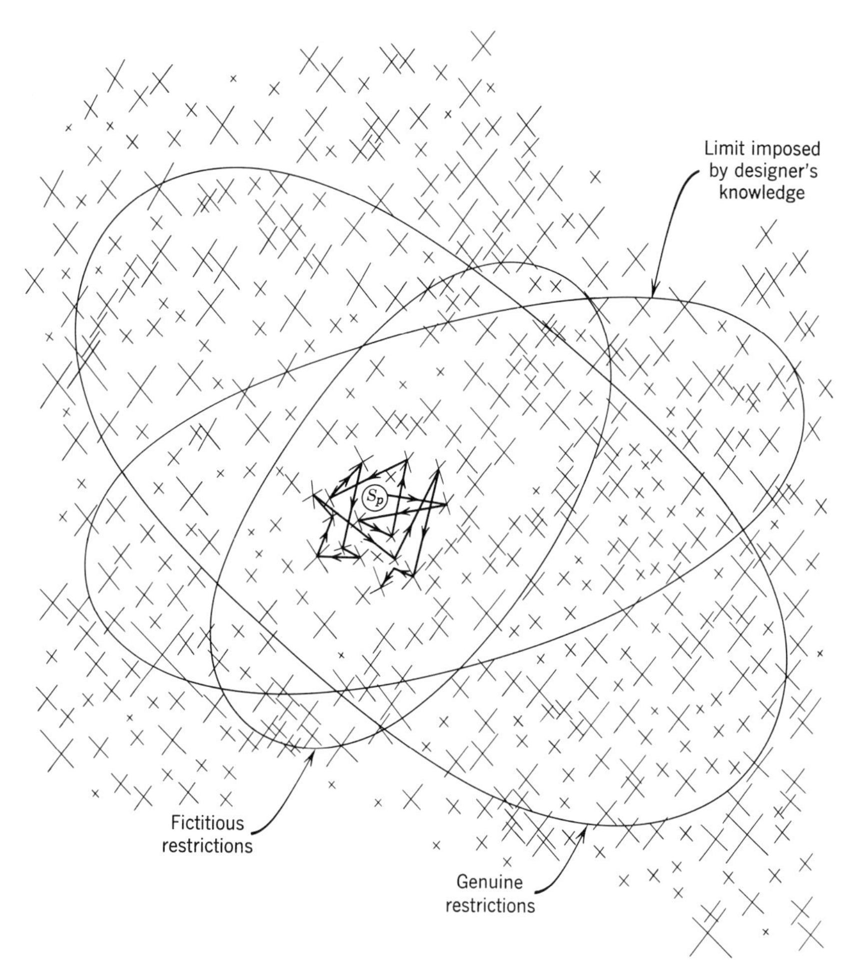

Figure 8. Solution space with an illustration of the manner in which a problem solver is prone to sample it. X's represent feasible solutions to a given problem. Similarity of solutions is inversely proportional to separation distance.

2. Limits imposed by the person's knowledge (ideas a person generates must come from his mental store of potential solutions and partial solutions, and this knowledge ordinarily embraces only a small fraction of the possible alternatives);
3. Fictitious restrictions (some alternatives are unjustifiably and automatically ruled out by the designer himself).

These boundaries are depicted as having different degrees of restrictiveness. It is thought that for a majority of situations the real restrictions are the least limiting on the designer, the fictitious restrictions the most limiting. Although there are these boundaries on the areas of possibility the designer is authorized to penetrate, it is highly doubtful that he will effectively sample even the space within the fictitious limits, let alone the space he is entitled to sample. Furthermore, the fictitious boundaries frequently cause the acceptable solution space to be searched in a biased manner, leaving major feasible solution areas untouched.

The mentioned tendency for ideas to cluster about the present solution is the result of several factors, among which are the following.

1. The failure to put forth some mental effort and actively seek different ideas.

2. A failing in formulation of the problem, such that the designer is in effect looking for nothing more than relatively small modifications of the current method, rather than for a variety of basically different solutions to the actual problem.

3. The biasing or blinding effect of the present solution. The designer may have correctly formulated his problem and may be trying, but because he has "lived" with the existing situation for so long it is very difficult for him to conceive of anything much different. This is an influential factor in stifling creativity, especially when the designer has long been associated with and is intimately familiar with the present system.

4. The somewhat natural tendency to be overly conservative, coupled with the virtually automatic assumption that large initial investments in solutions are undesirable, forbidden, or impossible.

Maximizing the Number, Quality, and Diversity of Alternative Solutions to a Given Problem

There are a number of very effective measures a designer can employ to enhance his creativeness. A wealth of literature on this subject is available, including a number of books devoted entirely to the topic of creativity.[1] A summary of the measures that a designer can take to maximize the number, quality, and variety of alternative solutions he can generate for a given problem is as follows.

[1] See for instance: Alex F. Osborn, *Applied Imagination,* Charles Scribner's Sons, New York, 1953; or Eugene Von Fange, *Professional Creativity,* Prentice-Hall, Inc., Englewood Cliffs, N.J., 1959.

1. The designer should maximize the number and variety of alternatives from which he may choose. In terms of the analogy, he should maximize the space from which he may sample, by pushing back the limiting boundaries as far as practical. This may be accomplished by: *a.* making a deliberate and conscientious attempt as the problem is analyzed, to eliminate fictitious restrictions and genuine restrictions insofar as economically and organizationally justified; *b.* broadening his knowledge, if only for the benefit of the particular problem in question. Certainly the designer should already be well grounded in the principles and practices fundamental to his particular engineering specialty. For the methods designer, this grounding would include principles and established good practices in the areas of work procedure, workplace layout, and equipment design. In addition to these fundamentals, the designer might well arm himself with additional knowledge peculiar to the particular problem with which he is concerned. In effect, he is enlarging his store of bits of information from which he can synthesize ideas.
2. Following this, the designer should sample the alternatives in the region of acceptable solutions as effectively as possible to obtain as large and as diverse a group of ideas as the time will allow. In terms of the foregoing analogy, the designer should ordinarily search the space available to him by sampling in many different areas. This requires that he make large jumps to remote points, rather than allow his ideas to cluster as described earlier.[2]

Ideally, this search of solution space should be directed, systematic, and devoid of any element of chance, but of course this is impossible. The search of solution space will always be at least partly random because of the very nature of the human mind and the process of generating ideas. However, the designer should attempt to introduce as much direction and system into his search as possible.

Volume and criteria offer bases for direction of this search for alternatives. Recall that the volume (usage) involved affects the areas of possibility likely to produce superior solutions (recall the row boat versus the bridge), and thus the general areas that should be searched. Specially weighted criteria have a similar effect (recall the high-quality shotgun). It is primarily for this reason—for direction of the search of solution space—that volume and criteria are given consideration during the problem analysis.

System is introduced into this search mainly through proper organization of the designer's thoughts, inquiries, and investigations, so that a wide range of basically different solutions are brought under consideration. In addition, several of the creativity-enhancing measures to be presented serve to add system to this search.

Regardless of how much system, direction, and randomness characterize this search, the important matter is that the designer penetrate into the

[2] This is not advisable if the approximate nature of the optimum solution is known, but that this should be known before searching seems unlikely.

worthwhile areas of possibility that he is entitled to call upon for ideas. There are a number of suggestions the designer should consider as ways of improving the effectiveness of this search process. Some of them are thought joggers, others are cautions against certain detrimental tendencies, and still others are methods of facilitating a systematic search. Typical of these measures are the following suggestions.

a. Exert the necessary effort. The designer must concentrate and vigorously apply himself, for creativeness is not achieved without a good measure of mental effort.

b. Make liberal use of a questioning attitude. Frequent use of the simple question "WHY?" can do wonders in uncovering profitable alternatives.

c. Try a systematic approach. For example, systematic application of different questions, systematic alteration of various variables, systematic examination of analogous situations, or systematic substitution, inversion, rearrangement, or combination of ideas.

d. Try check lists of good ideas, or good questions, that previous experience has shown are frequently worthy of consideration. This is a reminder device. It is a thought-jogging aid intended to "upset" a person's chain of thought and direct his thinking into different channels. In terms of the space analogy introduced, use of a check list helps to force "big jumps" to distant points and thus assists in combating the clustering tendency. This same purpose is true of a number of these measures intended to improve creativeness.

e. Seek many alternatives. Let the immediate objective be to accumulate as many alternatives as possible in the time allowed.

f. Consult others. Actively seek information and suggestions from engineers, salesmen, customers, supervisors, inspectors, set-up men, and others. Such conversations not only have the direct effect of increasing the designer's knowledge, but they also have an important effect through jogging his thoughts. Such conversations have a third beneficial effect. They facilitate the future acceptance of the engineer's proposals on the part of the persons who were consulted and thereby offered the opportunity to contribute.

g. Attempt to divorce your thinking from the existing solution. This is a direct attack on the clustering tendency. It is not an easy matter to free the designer's thinking from the present solution, which more often than not acts as a formidable thought block. Yet with some mental disciplining, different and very worthwhile chains of thought can be started. These same statements apply to that appealing idea the designer gets, which often offers as effective a block to further creative thought as the present solution did previously.

h. Try the group approach. This method, popularly known as brainstorming, involves a group of persons assembled for the stated purpose of generating solutions to a problem. Ideas are called out and recorded on a blackboard or chart so that they may be viewed by all. Volume of ideas is encouraged. Any form of evaluation during this period is strongly discouraged. All ideas are to be contributed regardless of how ridiculous they may seem at the moment. Apparently the variety of ideas generated by a number of persons working together is greater than would be obtained if this same number of persons worked independently in generating solutions. This is to be expected for two reasons. First, the limits imposed on each individual by his own knowledge are extended because the spheres of knowledge of different individuals do not completely overlap. Secondly, the rapid-fire flow of ideas is constantly "upsetting" each individual's chain of thought. This technique is very effective as a device for directing thoughts into many different channels.

i. *Avoid conservatism.* Do not shy away from radically different ideas. In terms of the analogy, when we are successful in making a big jump, the tendency is to immediately backtrack, to shy away from it. It seems natural to favor ideas and proposals that are time tested and that we are reasonably confident will work satisfactorily. There is a tendency sometimes to be overly conservative in this respect and choke off ideas for which the risk is certainly not unreasonable. By its very nature, originality tends to bear with it some additional element of risk, but history has certainly borne out the long-run profitability of originality.

j. *Avoid premature rejection.* Do not be hasty in rejection of ideas; in fact, postpone evaluation until later. Forget the cost of new ideas for the time being. As one author has put it, "do not mix ideation with evaluation." What seems to be a ridiculous idea at one moment might a short while later be modified by another idea and become a very worthy and workable alternative. It pays to remain open minded.

k. *Avoid premature satisfaction.* Do not be tempted to settle for the first "good" idea or the first one that appears to be an appreciable improvement over the current solution, when actually there is economic justification for further searching. It seems relatively easy to be blinded by this first "brilliant" idea. Related to this is the case in which a designer has spent a great deal of time on details of a new design, then learns or thinks of a much superior solution that renders these details useless. Under the circumstances, the designer might be unjustifiably and subconsciously attracted to the solution over which he has previously labored and be inclined to unwittingly resist the newer and better idea. Since this is probably an unintentional reaction, an awareness of the tendency may be helpful in avoiding it.

l. *Refer to analogous problems* for ideas.

m. *Work in an atmosphere conducive to creative thought.* It is helpful to have quiet physical surroundings where the designer is free to think for prolonged periods without interruption.

n. *Remain conscious of the limitations of the mind in this process of idea getting.* If the designer is constantly aware of his tendency to impose artificial restrictions, to be conservative, to prematurely evaluate, etc., then he has already made an important step toward overcoming creativity stifling tendencies.

There are many other such aids intended to enhance creativeness, for example: concentrating on the ridiculous, trying it yourself, putting the problem aside for awhile, and recording all ideas.

The foregoing discussion relates primarily to what a designer can do to assist himself in the solution of any given problem in the short run. There are a variety of measures directed to improvement of the designer's creativeness over the long run that he should seriously consider. Some of these are discussed in a later chapter dealing with the personnel responsibilities of the engineering administrator. Such measures include the encouragement of originality and the cultivation of open-mindedness of engineers by their superiors, and keeping up to date with technical developments in the field.

Summary

This phase entails a search, partly random, partly systematic, partly directed on the basis of restrictions, volume, and criteria. This is a search primarily of the designer's store of knowledge, for progressively better solutions to a problem. Conceptually, the search should terminate when the incremental cost of search balances the incremental probable improvement in solutions. In practice it is difficult to decide just when to terminate this phase.

The effectiveness of this search is enhanced by adherence to the design process described herein, for a number of its features are directed to this matter of facilitating and encouraging more and better alternatives. In this particular phase, it is suggested that the designer first attempt to maximize the number and variety of eligible alternatives, then sample these as effectively as possible by avoiding certain undesirable tendencies and by utilizing the searching aids described.

REFERENCES

"Brainstorming—Better Way to Solve Plant Problems," *Factory,* vol. 114, no. 5, May 1956.

"Creative Thinking Training," *Factory,* vol. 115, no. 11, November 1957.

Osborn, Alex F., *Applied Imagination,* Charles Scribner's Sons, New York, 1953.

Osborn, Alex F., *Your Creative Power,* Charles Scribner's Sons, New York, 1950.

Pearson, D. S., *Creativeness for Engineers,* Edward Brothers, Ann Arbor, Michigan, 1958.

Von Fange, Eugene, *Professional Creativity,* Prentice-Hall, Englewood Cliffs, New Jersey, 1959.

6

Evaluation of Alternatives, Specification of a Solution, and Remainder of the Design Cycle

In the previous phase the designer has generated a number of solutions and partial solutions to the problem at hand. The search phase may appropriately be described as an expansion process, expansion of the number and variety of alternative solutions from which to choose, as pictured in Figure 9. What is needed at the termination of this expansion phase is a process of elimination involving evaluation and comparison, combining and recombining, that will reduce this collection of ideas to the best single solution. The expansion process, therefore, is logically followed by a process of reduction (Figure 9). This reduction process begins with a relatively crude means of evaluation, perhaps by pure judgment, of the alternatives still in a rough state of specification. As some are rejected, remaining solutions are specified in additional detail; then there is further evaluation by a more discriminating procedure, resulting in the elimination of still more possibilities. This multistage screening process, characterized by increasing refinement in the evaluation and specification of alternatives as the number is gradually reduced, is continued until the best solution emerges. Much of this process may actually involve partial solutions, so that combining and recombining of these partial solutions is probably going on concurrently.

At the culmination of this phase the "final" solution evolves and remains to be specified in detail sufficient to permit installation. On occasion, the designer is not expected or permitted to make the final choice. Under these circumstances he presents a limited number of alternative solutions, complete with associated performance and cost data, to whoever makes the decision.

It is difficult to generalize concerning evaluation procedure; however,

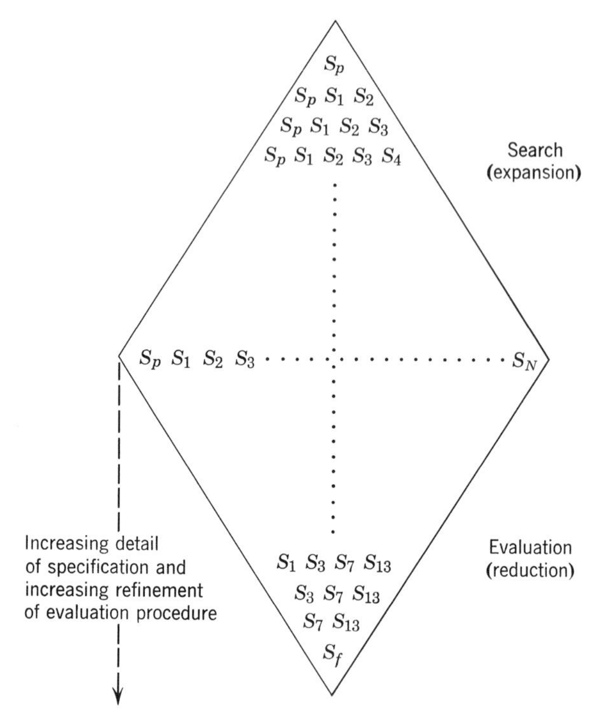

Figure 9. Diagrammatic formulation of the search and evaluation phases of the design process, starting with the present solution (S_p) and terminating with the final solution (S_f).

in instances where evaluation is other than purely subjective, where it is formal, refined, and quantitative in nature, a general evaluation procedure is detectable. This process involves selection of the criteria, prediction of the "effectiveness" of each alternative with respect to each criterion, conversion of these estimates to monetary terms insofar as possible, and comparison of the alternatives in meaningful fashion to facilitate a decision, followed by selection of the preferred alternative. These steps will be described in some detail.

Selection of Criteria

The basis for this evaluation is provided by the criteria identified, at least in general terms, as a part of the analysis of the problem. In many engineering problems the primary criterion is return on the investment, which in its broadest sense refers to the benefits expected from the alternative solution in question relative to the cost of creating that solution. In product design, this requires an appraisal of the benefits in the form of sales revenues and service to the public expected from each alternative design, and an appraisal of the total cost of producing each. In design of a bridge, dam, or highway, this criterion requires an appraisal of the benefit each alternative offers to the public and the cost of constructing each. In work methods design, this criterion requires estimation of the relevant[1] cost of operating each alternative method as well as the cost of creating each. The latter includes equipment and material purchase costs, cost of setup, and lost-production costs.

Although return on the investment may be the most common primary criterion, it is naive to assume that maximum return on the investment is always sought in practice. A variety of other less objective criteria are commonly used. For example, there is the elusive yet realistic "happiness criterion" in which the designer's objective is to select the alternative solution most likely to keep peace among the parties involved, to be least offensive to everyone involved, to be accepted by all those concerned. In many instances this is the major criterion, and in a good many more cases it is *a* criterion along with others.

Prediction of Performance and Conversion to Monetary Terms

A major task facing the designer in evaluation is to quantitatively predict the performance of each alternative with respect to each of the important criteria involved. For example, the designer of a work method should predict the performance time required by each alternative, as well as the effort required, the skill demanded, the fatigue caused, the flexibility offered, the maintenance required, and so on, then reduce these to dollar costs. A majority of these predictions must be made while the method is still in the conceptual stage, because actual experimentation is seldom economically feasible. Especially under these conditions the future performance of an alternative and the cost estimates derived therefrom cannot be predicted with certainty. The elimination of error in measurements, forecasts, and the ensuing decisions is impossible even under the most favorable conditions.

Inevitable error in measurement and prediction is not the only cause of uncertainty in evaluation of alternative designs. Ordinarily there

[1] Costs unaffected by a decision are customarily excluded from consideration in the course of making that decision.

are so many criteria and interdependencies of same, that some of lesser importance must be ignored in evaluation because of limitations of time and money. Furthermore, there are usually a number of unquantifiable criteria involved, for which no quantitative measure is possible or economically feasible. For example, the customer appeal of alternative product designs defies quantitative prediction. Still other criteria are quantifiable, the energy required by alternative work methods for example, yet cannot satisfactorily be converted to monetary terms. As a consequence there are usually numerous criteria to be considered in the final decision that are not in monetary or even numerical terms. Thus judgment ordinarily plays a role in the evaluation of and choosing from alternatives, often a major role.

Comparison of Alternatives

To facilitate a decision, it is desirable to aggregate the costs and unquantifiable criteria in some form which will make possible a convenient and meaningful comparison of the alternatives. Several typical methods of achieving this are by estimating the total annual cost of each alternative, estimating the capital recovery period (payback period), and estimating the rate of return of the investment required. Simplified versions of these common methods of quoting the relevant costs in a comparison of alternatives will be illustrated by means of the following example. A firm is considering two types of paint spray equipment to replace the current method of applying the external finish to its products. Cost data pertinent to the decision are as follows:

Alternative	Initial Cost (IC)	Annual Operating Cost (OC)	Expected Useful Life of Equipment (N)
P (present method)	0	$38,000	5 years
A	$16,000	31,000	7 years
B	11,000	34,000	6 years

Total annual cost method of comparison. To compare the foregoing alternatives in this manner, it is necessary to convert the initial investment to an annual basis, then add this to the annual operating cost to obtain the total annual cost (TAC). Thus $TAC = OC + IC/N$. For the alternative paint spray methods:

$$(TAC)_P = \$38{,}000 + \frac{\$0}{5} = \$38{,}000 \text{ per year}$$

$$(TAC)_A = \$31{,}000 + \frac{\$16{,}000}{7} = \$33{,}286 \text{ per year}$$

$$(TAC)_B = \$34{,}000 + \frac{\$11{,}000}{6} = \$35{,}833 \text{ per year}$$

Under this method of comparison, alternative A offers the lowest total annual cost, which takes into account both the annual operating cost and the initial investment prorated to a yearly basis.

Capital recovery period method of comparison. By this method, calculations are made to estimate the period of time required for the cumulative savings in operating costs offered by a proposal to equal the initial cost of that proposal, or in other words, to recover or pay back the initial investment. The savings here are the difference in operating costs of the present and proposed methods. Thus, an estimate of the capital recovery period (CRP) is

$$\frac{(IC)_a}{(OC)_p - (OC)_a}$$

where $(OC)_p$ is the operating cost of the present method and $(OC)_a$ is the operating cost of the alternative under consideration. For this example then:

$$(CRP)_A = \frac{\$16{,}000}{\$38{,}000 - \$31{,}000} = 2.3 \text{ years}$$

$$(CRP)_B = \frac{\$11{,}000}{\$38{,}000 - \$34{,}000} = 2.8 \text{ years}$$

Thus, proposal A offers to recover its own initial cost in a shorter period than proposal B. In using this method of comparison the decision maker must of course decide whether or not the 2.3-year capital recovery period of proposal A makes the investment worthwhile in the light of other possible uses of capital. Some companies have a policy on capital recovery periods, sometimes ultra conservative and inflexible. This policy states a certain capital recovery period that a proposed investment must equal or do better than, in order to be approved. A one-year maximum seems common, yet is certainly not always justified. Under this one-year rule, a proposal that does not offer to recover the initial cost in one year or less is automatically rejected.

The capital recovery period method of stating relative profitability is commonly used and is quite meaningful to the ordinary decision maker.

Rate-of-return-of-the-investment method of comparison. Under this procedure an estimate is made of the percentage of the initial investment that will be recovered per year through the resulting savings in operating costs. The annual rate of return of the investment (R/I) then is

$$\frac{(OC)_p - (OC)_a}{(IC)_a}$$

For this example:

$$(R/I)_A = \frac{\$38{,}000 - \$31{,}000}{\$16{,}000} \times 100 = 45\% \text{ per year}$$

$$(R/I)_B = \frac{\$38{,}000 - \$34{,}000}{\$11{,}000} \times 100 = 36\% \text{ per year}$$

Thus for alternative A it is expected that the initial investment will be paid back at a rate of 45 per cent of that investment per year, offering a 9 per cent advantage over alternative B.

Note that these are simplified versions of the total annual cost, capital recovery period, and investment rate-of-return methods of comparison. The effects of interest, taxes, and other important factors have been ignored. For more refined descriptions of these procedures, consult a standard textbook on engineering economy.

Evaluation procedure in design work is highly variable. Uniqueness of the engineering specialties and of the problems within any specialty limits the extent to which generalization can be made. One very safe generalization however, is that judgment ordinarily is heavily relied on in evaluation of alternative designs.

SPECIFICATION OF THE PREFERRED SOLUTION

The specification phase of the design process involves a delineation of the attributes *and* performance characteristics of the selected design. The primary purpose of this phase is to communicate the solution to a number of different parties, such as:

1. The persons responsible for approval of the solution.
2. The persons charged with physical creation of the solution.
3. The persons responsible for administration of the solution in use, such as the role a foreman serves for a new method of manufacture.
4. The persons responsible for maintenance of the solution, such as the personnel responsible for servicing a product after it is on the market.

5. Anyone else who sometime in the future has need for the detailed specifications of the solution.

The fact that it is not likely to be the designer that is serving in these capacities is an indication of the careful attention the designer should devote to this communicative function. The designer must record his solution clearly and in sufficient detail to permit intelligent decisions concerning it and to permit successful implementation. The importance of the designer's ability to communicate ideas effectively and convincingly cannot be overemphasized.

THE DESIGN CYCLE

The designer's task seldom terminates with specification of a solution. His responsibility ordinarily extends to gaining acceptance of his design, overseeing its installation and use, observing and evaluating the design in use, and deciding (or contributing to the decision) as to when redesign is advisable. These functions form a complete cycle which is diagrammed in Figure 10. The post-specification functions are described below.

Implementation of the design. To assure insofar as possible that the final solution will be successfully put to use, it is vital that the

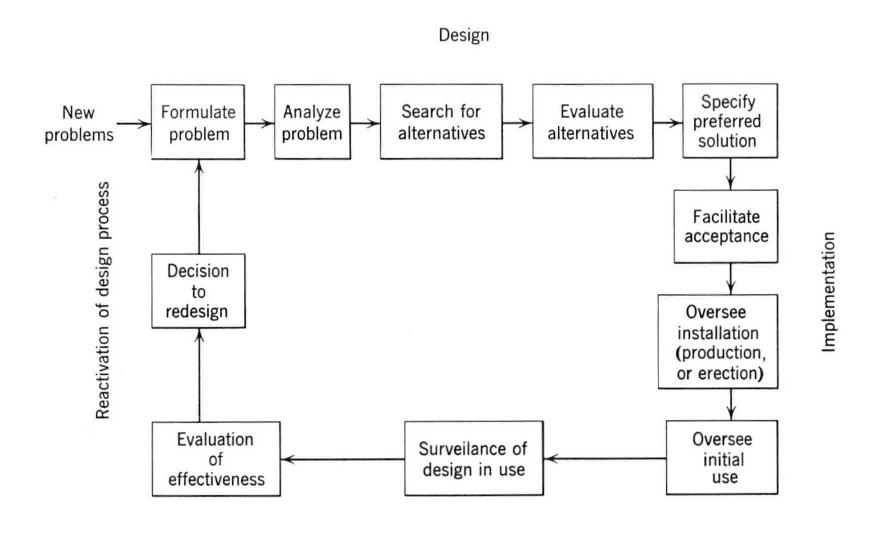

Figure 10. The design cycle.

designer facilitate acceptance of, oversee the creation of, and oversee initial utilization of his proposed design. As indicated in Figure 10, these measures constitute the implementation phase of the design cycle.

Certainly the designer hopes to and in fact must gain acceptance of his proposal on the part of executive and operating personnel. Almost without exception, engineering is a staff function in the organization, which means that the engineer acts in the capacity of an advisor or consultant to nonengineering personnel. In general, the engineer has no authority to issue an order to other than his own engineering subordinates. Outside of this he is authorized organizationally only to *advise*. Thus it is possible and not uncommon to devote considerable engineering hours to the solution of a design problem, only to have it flatly rejected by those who have the power of approval.

Gaining the desired approval requires that the design be more than technically and economically acceptable to the appropriate people. Approval often requires also that the proposal not jeopardize the personal best interests of those holding a veto power. Thus, the advisory role of the engineer, along with the fact that differences of opinion can arise since an element of judgment is always involved in making design decisions, and along with the fact that people often allow personal motives to influence their decisions, combine to make it imperative that the designer give careful attention to this matter of gaining acceptance.

Young engineers often start out in their profession with the mistaken impression that if their proposals are technically and economically superior they are automatically adopted. They underestimate the need for thorough and effective presentation of their proposals, for *selling* others on the worthiness of their ideas, for a certain amount of "realistic compromise" with respect to some features of their proposed designs, and for caution and careful planning to minimize "resistance to change." The latter is an important matter given further discussion later in this chapter.

Once his design has been accepted by the appropriate personnel, it becomes important for the designer to diligently oversee the creation (erection, production, or installation, whichever the case may be) of his design. Ordinarily it is both necessary and desirable that the designer serve in an advisory capacity as his design is brought to actuality, for example as the bridge is erected, as the refrigerator is produced, or as the communication system is installed. This is to insure adherence to specifications and to detect and remedy details that are incorrect, omitted, or unwisely chosen. It is both common and advisable that a certain amount of design modification take place as the proposal is created, and desirable that this be performed or supervised by the original designer.

Follow-up. Any design, whether product, structure, or production method, should be subject to the periodic scrutiny of its designer. This observation permits continued assurance that the design is being used as intended and offers an opportunity to evaluate the effectiveness of the design in use.

Evaluation of the results attained through a particular design offers an excellent opportunity for the designer to learn and thus to improve his future design performance. This is a most profitable but apparently a readily neglected phase of the design cycle. Rare is the designer who cannot benefit by observing his creation in use "in the field" over a significant period of time. For example, note the wealth of useful information available to the product designer from the population of consumers, retailers, wholesalers, salesmen, and others that have an intimate association with his creations in actual use. Yet opinions and symptoms indicate that this wealth of useful feedback generally goes relatively untapped.

Adequate attention to this follow-up evaluation of results achieved is one of the true marks of a good engineer.

Reactivation of the design process. Periodic evaluation of effectiveness also provides a basis for deciding when it is profitable to devote time and money to redesign. No solution to a practical problem remains superior indefinitely, better solutions are discovered, new demands arise, new materials and tools are developed, conditions change, and physical depreciation occurs, so that a point is reached in the life of a design at which it is profitable to seek a new and better solution to that problem. An engineering department can intelligently decide when to engage in redesign only if the current solutions to problems within its realm are periodically reviewed and appraised. These functions are easily neglected yet they are important responsibilities of an engineer.

The design cycle is complete when, after the solution to a problem has been devised and put to use, it is decided that this design is obsolete, and the process of designing a superior solution is again begun. Completion of this cycle is not the only basis for initiation of the design process. Sometimes it is a new problem, a new need. The introduction of the transistor, for example, created a problem of process design that manufacturing engineers had not faced before. (Actually, situations that are not truly redesign are relatively rare.)

Resistance to Change

Resistance to change may be defined as an unwillingness to accept a change where the unwillingness is based on something other than demerits of the change itself. The frequency with which such resistance is encountered makes this a problem of primary importance to an engineer, for most of his time is devoted to originating (potential) changes.

Many of us, as we start out in the business world, require a considerable period of time to become accustomed to the fact that worthy ideas are not automatically accepted, that it is often necessary to do a thorough selling job to gain acceptance, and that even then, a surprising number of obviously profitable ideas fall by the wayside. In fact, the number of times we fail to gain acceptance sometimes becomes difficult to understand and rather discouraging. This is not quite so difficult to understand, however, if we are aware of the numerous potential causes of resistance to change. Fortunately, an awareness of these causes and an anticipation of them assist appreciably in minimizing the frequency and severity of this reaction to change.

Ability to achieve acceptance of his ideas is important to an engineer's rate of advancement. He is not hired merely for the technical knowledge he can supply. He is expected to produce by putting that knowledge to use through creativeness. But even being creative is not sufficient; he must also have the ability to achieve acceptance and effective use of his creations. A deficiency in any of these respects jeopardizes an engineer's advancement opportunities.

A surprising amount of an engineer's time is consumed in dealing with fellow employees. His conduct during such "dealings" strongly affects his ultimate success in getting these people to accept his proposals. Yet the engineer is notoriously unskilled in the area of human relations, and it is his frequent clumsiness in introducing changes that appears primarily responsible for this reputation. Therefore, it appears desirable to learn now what some of the pitfalls are and how to minimize the difficulties resulting from the introduction of changes.

Many worthy engineering proposals are discarded on grounds other than their technical content. Perhaps of even greater concern, however, is the high percentage of proposals that are accepted and installed and that eventually prove notably unsatisfactory, not for technical reasons but because the people affected by the new methods have resisted and responded in such a way as to render these proposals failures. Not only can a new system meet with lack of cooperation and indifference when it is installed, it can and often does encounter deliberate attempts to contribute to its failure or to cast it in an unfavorable light. Thus, resistance to change presents a formidable problem in getting ideas adopted *and* in obtaining the cooperation of persons affected by new procedures after adoption.

Basic causes of resistance to change. Knowledge of the basic causes of resistance to change is useful in planning the introduction of changes to minimize resistance, and in diagnosing and remedying situations in which resistance has been encountered after changes have been introduced.

There is an all-important fact concerning "business life" to be recognized in dealing with resistance to change problems. It is that almost every individual in an organization has some personal motives that are in conflict with the overall objectives of the enterprise. The more common personal objectives of interest here are:

1. To attain *advancement.*
2. To be of some *importance* in the eyes of associates, superiors, family and friends.
3. To be *liked* by fellow employees.
4. To earn more *money.*
5. To receive some *satisfaction* from the work involved.
6. To *participate* in making decisions concerning personal welfare.
7. To attain *security* with respect to employment, position, salary, etc.[2]

The important point is that a person is *likely* to resist a change, in spite of the fact that it is capable of increasing company profits, if it is in conflict with the foregoing personal objectives. What reaction should we expect from a person if a proposed change lessens his chances for advancement, or undermines his importance, or makes him unpopular with associates or subordinates, or lessens his chances of earning more money, or jeopardizes his current wage level, or reduces the amount of satisfaction he receives from his work, or threatens his security of employment? Is he likely to be receptive to a change he has had no opportunity to contribute to or express his opinion on previously? Is he likely to approve a change that offers to cause him the embarrassment of appearing negligent or ignorant because he be-

[2] These are not the only objectives sought, nor does every individual hold all of these objectives, nor do different individuals weight them equally.

lieves it is he that should have thought of the idea, perhaps long ago? Ordinarily, the individual is willing to make little or no personal sacrifice for the sake of the welfare of the business. It is unrealistic to think that everyone acts at all times in the best interests of the enterprise employing them. Consequently, there is a surprising amount of "suboptimization" going on in the administration and operation of any enterprise.

There are numerous specific causes of resistance to change that stem from this practically universal conflict between personal and organizational objectives. A comprehensive check list of these common specific causes of resistance to change is presented in Appendix B. An outgrowth of this analysis of the causes of resistance to change is a series of recommendations on the minimization of such resistance, also included in Appendix B.

It is important that attention be given to this matter of resistance to change from the moment a design project is begun. This matter should not be neglected until the designer finds himself in the position of having completed an excellent solution to a problem, a solution that the appropriate personnel will not accept. Nor should he exclude consideration of this matter as he works out a problem, until he is ready to propose his solution, then stop to plan his "acceptance gaining strategy," for it may already be too late. Instead, he should be laying the groundwork for acceptance *from the moment the project is begun.* He should give careful consideration from the outset, to his behavior, his manner of dealing with various personnel, the features of his proposal, etc., with the objective of minimizing the chances of subsequent rejection of his ideas. Thus attention to this problem of resistance to change should be, insofar as possible, in the form of *planning* to minimize the likelihood of its occurrence, as opposed to the *diagnosing* of unfortunate situations to learn why rejection has occurred and what might be done, if anything, to reverse these situations.

part II

INTRODUCTION TO METHODS ENGINEERING

7

Methods Engineering: An Overall View

Engineering is primarily concerned with application of analytical methods, principles of physical and social sciences, and the creative process, to the problem of *converting* our raw materials and other resources to forms that satisfy the needs of mankind. The process involved in solving this conversion problem is ordinarily referred to as design.

The problems dealt with by the various engineering specialties lend themselves very appropriately to the problem characterization and problem-solving procedure introduced herein, for each concerns itself with a transformation from one state of affairs to another. For example, the mechanical engineer is primarily concerned with the transformation of energy in its natural state to energy in a readily usable form. Broadly speaking, state A is energy in the form of coal, oil, the atom, the sun, and the like. State B is the useful energy, usually in mechanical form, such as the output of the steam or internal combustion engine, the turbine, the jet engine, and so forth.

Industrial Engineering

The industrial engineer is primarily concerned with the transformation of materials to a different and more useful state, with respect to form, location, or time. It is his responsibility to design the medium (method) that will achieve this conversion in a preferred fashion, as for example, in a manner that maximizes return on the owners' investment.[1] This medium might be an automobile factory, a textile mill, a foundry, or any other type of production facility. In this case state A

[1] Maximizing return on the investment like the ideal solution is something that the designer works toward but an objective he is not ordinarily expected to actually achieve. Furthermore, return on the investment is not always the primary criterion, to be sure.

is unprocessed material, state B is the processed material. The former might be steel and other materials, the latter a typewriter, or the former might be grapes, the latter, wine.

The industrial engineer, then, specializes in design of production media. The medium with which he is concerned is a complex aggregate of men, machines, materials, and communication networks, such that a thorough and intelligent design of its physical composition and coordinating systems becomes essential if the objectives of owners and creators are to be realized.

This medium of production, whatever product it may create, is actually an organism—a highly complex, highly integrated whole composed of many interdependent, mutually reliant parts—quite similar in fact to the living organism in a variety of respects. Each has a physical facility, each has a nervous system, each is comprised of many interdependent parts, each has the ability to learn, and each is constantly striving to adapt itself to an everchanging environment.

The manufacturing organism's physical facility (plant) is composed of buildings, equipment, and people. It is this component of the organism that does the actual producing; however, it is incapable of acting constructively without a second component, a system to plan, actuate, coordinate, and control activity. Without a system analogous to the living body's nervous system, the manufacturing organism is totally helpless. The business enterprise, like a living organism, is constantly affected by internal and environmental changes and disturbances, to which the organism must be constantly adjusting if it is to obtain its objectives. Both types of organism must have a system that will *sense* changes and disturbances in or affecting behavior, that will *decide* what compensating action to take, that will instigate this *action,* and that will perform this sense-decide-act cycle as a continuum ad infinitum, as pictured below. With such a system, the

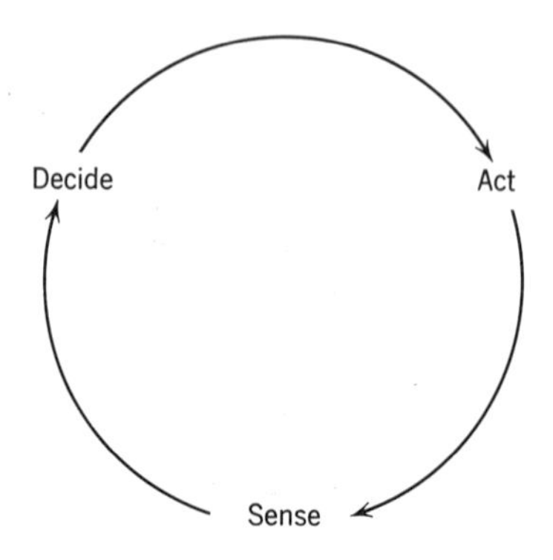

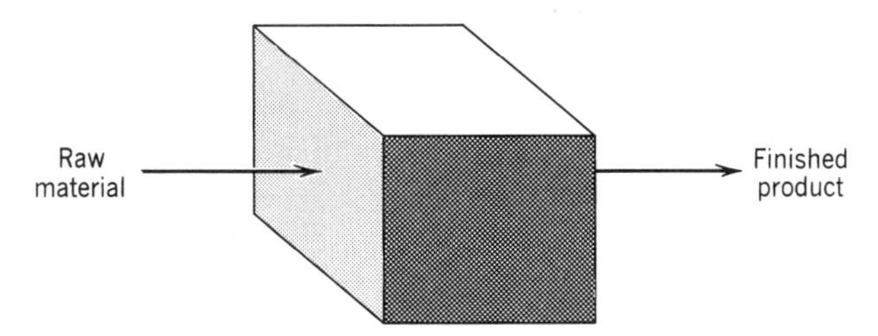

Figure 11. "Black box" formulation of the industrial engineering problem. What must the contents of this box consist of in order that satisfaction of its owners will be maximized?

living animal senses the dangers that jeopardize its existence and acts accordingly. The business enterprise must do likewise with the developments that threaten its existence, as with the maneuvers of its competitors for example. In actual practice, the behavior of the manufacturing plant is governed by a number of operating systems. There are systems for planning and controlling profit, movement of materials, utilization of equipment, product quality, inventory level, and so forth. To illustrate, the business decides what course of action it must take in order to maximize profits. Then it acts on the basis of the plans made. In the face of inevitable disturbance and imperfect plans, the company senses the need for changes in course of action, then decides what compensating action to take, takes this action, and so on.

The industrial engineer is responsible for design of the plant *and* of the systems necessary to operate it successfully. This production medium might be viewed as a black box of as yet unspecified contents, with a specified input of materials and other resources and an output of the same materials in a new and more useful form, as pictured in Figure 11. The industrial engineer's task is to specify the contents of this black box. The result of his efforts is a primary determinant of the profits investors will realize on their investment, so to be sure, they are vitally interested in the thoroughness and effectiveness of the industrial engineer's performance.

A design problem. The management of a certain steel company has decided to set up a plant to manufacture corrugated steel drainage pipe for use along highways, railroad beds, and residential driveways. To design a manufacturing facility that will meet their requirements, a rather lengthy specification process is necessary. It will be informative to trace this process starting after management's decision to invest in this enterprise and terminating with the specifications for the complete plant.

Market analysis and product design. After the decision has been made to enter this business, a team composed of product engineers, industrial engineers, and market analysts is assigned the task of estimating the market potential, of determining what types and sizes of pipe to produce, and estimating the quantity of each type and size that should be manufactured. (The results of their efforts will be indicated later.) Note that there is a strong interdependency between these decisions, as well as with most of the decisions that follow.

The role of industrial engineering in this decision-making process. Industrial engineering has participated in making and has the right to appeal the preceding decisions with respect to the product design and the scale of production; however, its major responsibility ordinarily begins after these decisions have been made. The industrial engineer's prime problem is to find the method of producing the specified pipe from flat sheets of corrugated steel in the volume given, that will maximize return on the investment. This problem may be visualized as follows:

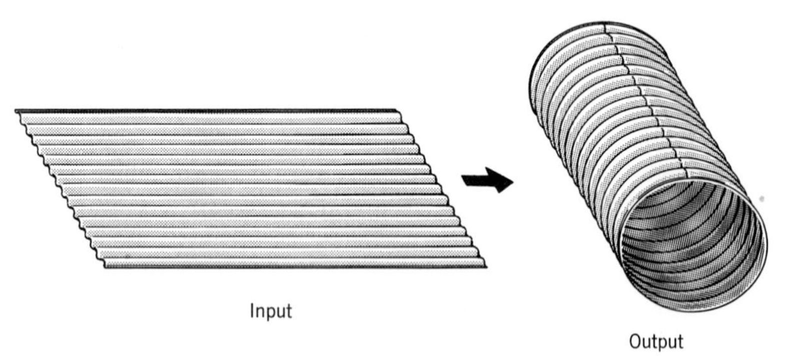

This is a large-scale, complex design problem which in practice is handled in several stages, each one of which is dealt with by a specialist within the industrial engineering field. These stages and the specialist dealing with each are now introduced as evolution of the complete plant design is described.

Specification of the basic method of manufacture—manufacturing engineering. Once the pipe has been designed and production volume decided on, the emphasis turns to specification of the *basic* manufacturing steps required to create this pipe. In practice this phase of industrial engineering is commonly referred to as manufacturing engineering (production engineering). The problem facing the manufacturing engineer may be analyzed as follows:

State B: Corrugated drainage pipe in 12-, 24-, and 48-inch diameters, in standard lengths of 8, 12, and 16 feet. These are made up from basic 2-foot, 2-inch sections of pipe. All securing of seams and joints is accomplished by riveted overlaps of 2 inches.

State A: 2-foot, 2-inch by 13-foot sheets of 16-gage corrugated, galvanized steel.

Volume: The capacity of the plant should be 750,000 linear feet per year, distributed in unequal proportions (not given here) over various diameter-length combinations, with certain expected seasonal fluctuations, and with a gradual build-up to capacity volume over the next 5 years.

Criterion: The primary criterion is return on the investment.

Restrictions: To transform the flat sheets into the specified sections of pipe, the following must take place:

1. The 2-foot, 2-inch flat sheets must be sheared to the proper length (Figure 12*a*).
2. These sheets must be formed (Figure 12*b*).
3. Holes must be punched at prescribed intervals along the seam and around the ends (Figure 12*c*).
4. The seam must be secured by overlapping and riveting (Figure 12*d*).
5. The 2-foot, 2-inch sections must be joined to make up the desired 8-, 12-, or 16-foot sections, by overlapping and riveting the appropriate number of sections together (Figure 12*e*).

The charge of the manufacturing engineer is to specify the basic process for accomplishing the foregoing steps. In the course of doing this he will decide in collaboration with the methods engineer, which of the required operations will be accomplished by machine, by man, and by man and machine combined. He will specify the machines, tools, jigs, and fixtures to be used. The results of the manufacturing engineer's decisions are usually spelled out on a document commonly referred to as an operation sheet, which serves as the official reference with respect to the basic method of fabrication and assembly of a product or product component. The resulting operation sheet for the case at hand is shown in Figure 13. The three basic operations described on the operation sheet, "cut to size," "roll," and "assemble, punch, and rivet," are shown in Figures 14, 15, and 16 respectively.

Specification of the flow of materials—plant layout and materials handling. Once the basic process has been specified, the emphasis turns to specification of the physical path the material will follow through the plant. This phase of industrial engineering, ordinarily referred to as "plant layout and materials handling," is usually

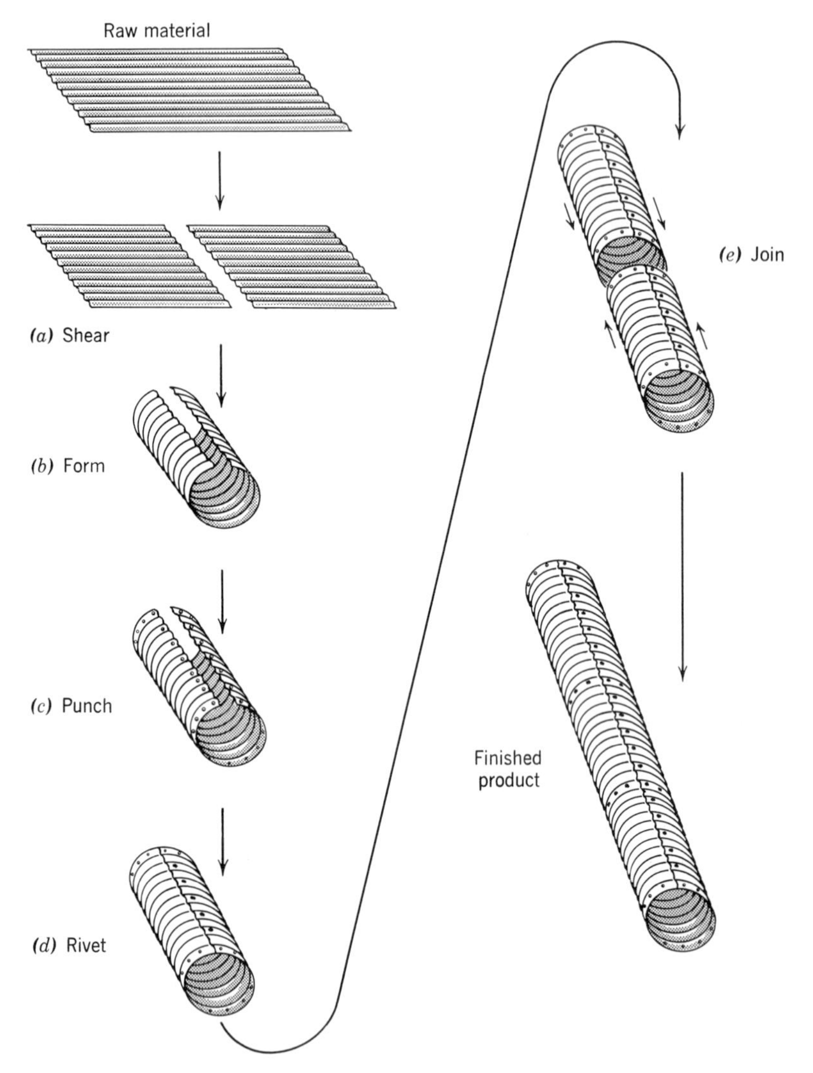

Figure 12. Steps necessary to fabricate and assemble the specified pipe from flat sheets.

handled by industrial engineers who specialize in this phase of the total problem. In this instance the problem might be analyzed as follows:

State A: Sheets, as specified earlier.
State B: Varying diameter and length of pipe, as specified earlier.
Volume: As given earlier.

Operation Sheet

Part: Corrugated Culvert Pipe | Date: 6/19

Mfg. Engr.: R. King | Approved: C. Coburn

Operation No.	Operation Name	Machine and Auxiliary Equipment	Expected Hourly Production
1	Cut to size	Bertsch culvert shear	
2	Roll	Bertsch rolling mill	
3	Assemble, punch and rivet	Bertsch stake riveter and punch	

Figure 13. Operation sheet for the manufacture of corrugated culvert pipe. (Expected hourly production is subsequently filled in by the methods engineer.)

Figure 14. "Cut to size"—first operation in the manufacture of corrugated culvert pipe. (Courtesy of Empire State Culvert Corporation, Groton, New York)

Figure 15. "Roll"—second operation in the manufacture of corrugated culvert pipe. (Courtesy Empire State Culvert Corporation)

Figure 16. "Assemble, punch, and rivet"—third operation in the manufacture of corrugated culvert pipe. (Courtesy Empire State Culvert Corporation)

Criterion: As given earlier.

Restrictions: In this instance the designer must adhere to the restrictions arising for physical reasons to obtain the specified pipe from the flat sheets, restrictions that the manufacturing engineer likewise faced, *plus* the restrictions imposed by the manufacturing engineer's decisions. The restrictions then are:

1. The sheets must be cut to size by a Bertsch shear.
2. These sheets must be formed by a Bertsch rolling mill.
3. The holes must be punched along the seam and around the ends by a Bertsch riveter-punch.
4. The seam must be riveted by the same machine.
5. Successive 2-foot sections must be joined together to obtain the desired lengths, on the same machine.
6. The material must be transported from the warehouse, between operations, and to the finished-pipe storage location.

The charge of the plant layout and materials handling engineer is to specify the physical arrangement of facilities and the methods of transporting material along the path it is to follow. In this problem the location of the sheets, the shears, the rolling mills, the riveters, and the finished pipe stockpile are to be specified, as well as the method of handling the work-in-process between these locations. The results of this decision phase are presented in Figures 17 and 18.

Specification of the work methods—methods engineering. The final and most detailed stage in the evolution of the completely designed production facility concerns integration of the human operator into the process in a manner that is consistent with the objective of the enterprise. This is ordinarily performed by a specialist called a methods engineer. It is his responsibility to design in detail the particular work method for each of those activities in the process that require use of the human. He specifies the *procedure* each machine operator is to follow, the *layout* of materials and tools at the work-station, and the *nature of the equipment* insofar as the operator is involved. For this particular problem, he will specify the method the operator should use in handling and positioning uncut and cut sheets at the shear (see Figure 14). He will specify how material should be handled and the machine loaded and unloaded during the rolling operation (see Figure 15), as well as for the assembly of pipe and the method of punching and riveting (see Figure 16).

In general, the methods engineer's problem is to find that method of producing the specified pipe which maximizes return on the investment, but under all the restrictions imposed through decisions made by specialists preceding him. After the decisions concerning the basic process, plant layout, and materials handling, have been made, the task that remains and becomes the responsibility of the methods engineer is design of the productive process where man is concerned.

After these decisions have been made, the production facility has been completely specified. This is only the physical component of the complete organism that is necessary. The production planning and control system, quality control system, and others that are necessary to successful operation of the plant are yet to be specified. Some of these operating systems would actually be designed in parallel with the design activity just described, for there is considerable inter-

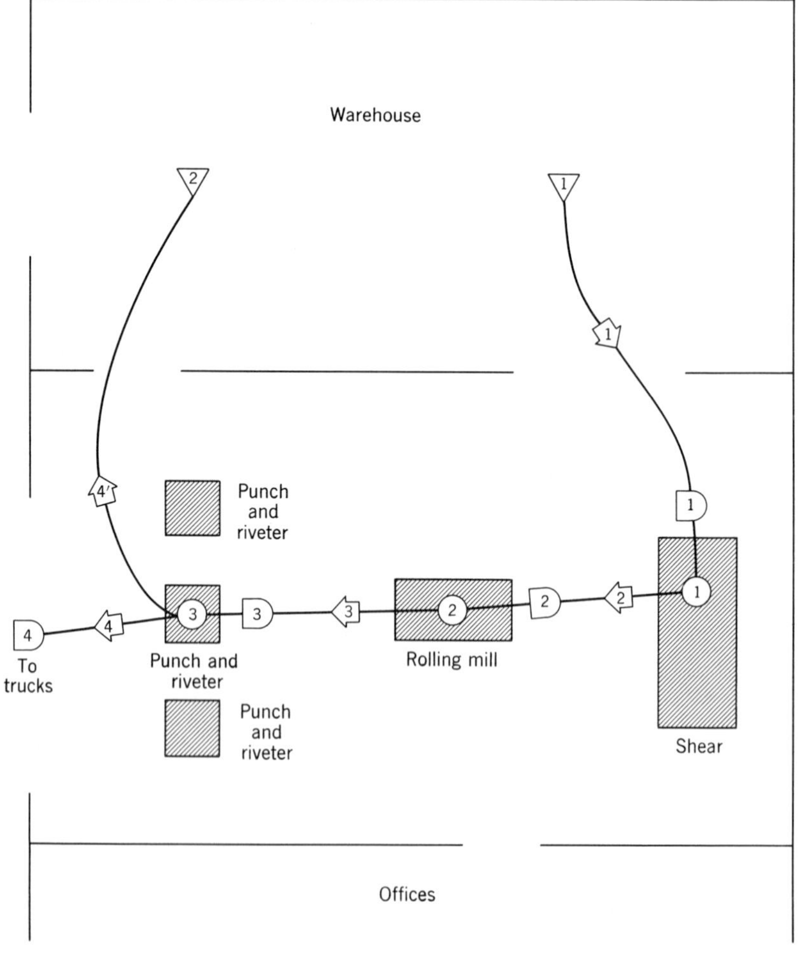

Figure 17. Flow diagram for pipe manufacturing process. (Companion to flow chart shown in Figure 18.)

Flow chart

Manufacture of corrugated culvert pipe

Symbol	Description	Distance
▽1	Raw stock in warehouse	
⇨1	Wagon to shear	90 feet
D1	Wait at shear	
○1	Cut to size (Figure 14)	
⇨2	Wagon to mill	25 feet
D2	Wait for forming	
○2	Form (Figure 15)	
⇨3	By rolling on floor	30 feet
D3	Wait for assembly	
○3	Assemble, punch, and rivet (Figure 16)	
⇨4	Hand carry to dock	30 feet
	or	
⇨4′	Hand carry to warehouse	90 feet
D4	Wait for truck	
▽2	Finished goods inventory	

Figure 18. Flow chart for pipe manufacturing process. (Companion to flow diagram shown in Figure 17.)

dependency between decisions concerning the two components.

Notice how successive decisions in the foregoing specification process impose additional restrictions on succeeding design specialties. Usually, the right to appeal and obtain alteration of previously made decisions exists.

The preceding discussion of industrial engineering and its subspecialties indicates the general role of methods engineering in the engineering hierarchy of a manufacturing organization. An elaboration on the role, objectives, and content of methods engineering follows.

METHODS ENGINEERING

Methods engineering is concerned with integration of the human being into a productive process. Alternatively, it may be described as design of the productive process insofar as the human being is involved. The task is one of deciding *where* the human being will be used in the process of converting raw material to finished product, *and* of deciding *how* man can most effectively perform the tasks assigned him.

As an illustration, consider the assembly of electron tubes. First, the decision is made as to what function, if any, the human operator will serve in the assembly process. Second, for those tasks for which man is to be utilized, the methods engineer will specify the work *procedure* the assembly operator should follow, the *layout* of tools, materials, and equipment at the workstation, and the *equipment* (tools, controls, etc.) that the operator will be associated with. He will do similarly for the human's role in machining operations, where his primary concern is with the nature and location of controls and with preferred operating procedure. He will do likewise for inspection, packaging, materials handling, maintenance, repair, clerical, cleaning and numerous other types of activity in which the human has a part. In addition to a wide variety of types of activity, a methods engineer deals with operations covering a wide range of performance times, volumes, degrees of mechanization, levels of skill, types of working conditions, and degrees of repetitiveness.

The view of methods engineering advocated here includes concern with the human's role anywhere in the organization, from manager to laborer. Traditionally, however, the methods engineer has concentrated his efforts on predominantly manual activity as opposed to the activity that is predominantly mental in nature (which characterizes the work of supervisors, executives, and engineers). In the material to follow, the emphasis will adhere to tradition and likewise be on manual activity even though *conceptually* our interest is a broader one. It behooves the profession to follow suit and develop a conceptual and eventually an active interest in nonmanual activity in view of the current cost reduction potential therein, and of the increasing cost, complexity, and criticalness of this type of task.

Importance of Methods Engineering

Human beings play a crucial role in the profitable operation of a manufacturing organism. Management is vitally interested in effective performance of its personnel and justifiably so. The cost of labor continues to rise. Not only are wages higher, but the cost to procure, indoctrinate, and train a factory laborer now averages $600, which is especially significant in view of the fact that factory turnover averages about 35 per cent per year nationally. In addition, labor plays a very important role in achieving effective utilization of equipment. For a $1.50 an hour operator to work inefficiently is one thing, that his $20,000 machine is likewise being used inefficiently is another and more important matter. Country-wide, the investment in facilities per worker was $15,000 in 1960, triple what it was in 1940, and increasing at an accelerating rate. Thus, because of its natural interest in profits and costs, management attaches considerable importance to the methods engineer's role in achieving higher productivity[2] of man and machine.

In spite of popular misconceptions, the human is still indispensable in the operation of a manufacturing plant, and in the vast majority of instances he still competes rather effectively with the machine. Facilities are needed to move, apply force, manipulate, position, and so forth, and man is still difficult to equal in many of these respects. He still has much to do with the movement and processing of materials through the plant. Similarly, facilities are needed to make decisions, to plan, to reason, to direct activities. Machines are as yet virtually noncompetitive with man in these thought-requiring functions. Then some type of facility is needed to note or "sense" deviation from schedules and specifications, variation in conditions, and so on, so that corrective action can be taken. In this respect too, it is difficult for machine to compete with man. A similar situation exists with the communication medium that is necessary between the decision-making, action, and error-sensing functions. In view of these comparisons of the relative competitiveness of man and machine for various functions to be fulfilled in a productive process, it is evident that the human being still is, and will be for a long time to come, a major component with which to cope.

Furthermore, as mechanization and automation increase and the human being appears less frequently as an integral component in the overall process, the functions the human does serve as a decision maker, sensor of irregularity (monitor), temporary replacement component, troubleshooter, and the like, will require the utmost in skill, speed, vigilance, and mistake-free performance. To be sure, the company that has a $6,500,000 "automated" production system is vitally interested in keeping this facility operating a maximum percentage of the time. (Some managements are doing well to keep their systems in operation 75 per cent of the time. There is a tremendous savings potential in that 25 per cent down-time.) Thus the

[2] Definitions and interpretations of "productivity" are numerous and diverse. In this text it will be viewed as output per man-hour.

human being is not likely to become any less of a determinant of company profits. Actually he is becoming an increasingly critical "link" in the total system so that *more* attention must be given to integrating and utilizing him with maximum effectiveness.

Time Study and Time Standards

Design culminates in the specification of a solution to the engineering problem at hand. A portion of this specification is a quantitative statement of the expected performance characteristics of the solution given. In the problem of bridge design, this includes such things as expected life of the span, expected maintenance costs, maximum load, and so forth. In methods design this requires statement of, among other things, the production time the specified method is expected to require; for example the amount of time to shear a plate for the corrugated drainage pipe, to roll a sheet, to punch and to rivet. Determination of the production time a proposed method of operation is expected to require is a very important and special matter in the eyes of most managements. In fact, it is such an important part of the methods engineer's responsibility that it is given a special name and considerable attention. This phase of methods engineering, the process of estimating the performance time for some phase of productive activity, is generally referred to as work measurement (time study) and the resulting time estimate is commonly referred to as the time standard for the activity in question. This time standard is important to the management of a manufacturing enterprise for purposes of scheduling, bidding and pricing, wage payment, anticipating facility needs, and so forth.

Summary

The position and scope of methods engineering is indicated diagrammatically in Figure 19. Note that there are two phases to the methods engineering function. The first, methods design, is the process of designing the work method. It is truly a design activity. The second phase—time study—is an outgrowth of the first since it is specification of a particular and especially important performance characteristic of the final design, namely, the production time. Time study has become a somewhat separate and specialized procedure because of the importance of its product, the time standard, to the management of a manufacturing enterprise. Do not underestimate man-

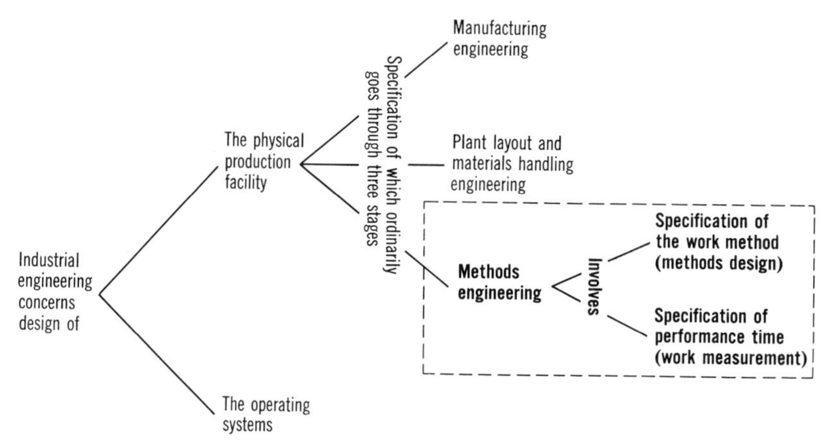

Figure 19. Diagrammatic summary of the organizational position and scope of methods engineering.

agement's interest in production time both with respect to minimizing it and knowing what it is going to be.

In Part III methods design is emphasized. Work measurement is treated in Part IV.

part III

METHODS DESIGN

8

Introduction to Methods Design; Formulation and Analysis of Methods Design Problems

As with any engineering specialty, methods design involves application of the design process along with a body of knowledge and a group of techniques peculiar to that specialty. The body of knowledge in this case is primarily a collection of generalities relating to effective integration of the human into a productive process. The special techniques are for analysis, measurement, and communication of the manual task. The overall view of methods design is summarized in Table 2. This and the next several chapters elaborate on the contents of Table 2, describing how the design process is applied and augmented in the course of designing the work method.

TABLE 2. Summary of the Approach to a Methods Design Problem

The Methods Designer's Approach Is Built Around the Design Process	Supplemented By
Formulation	
Analysis	Certain analytical techniques, unique to this engineering specialty.
Search	Principles concerning the relative capabilities of man and machine with respect to various production tasks, and principles concerning preferred methods (layout, equipment, procedure) of using man in the production process.
Evaluation	Special techniques and procedures for evaluation of alternative work methods.
Specification	Special media, language, and associated symbology, for description and communication of work methods.

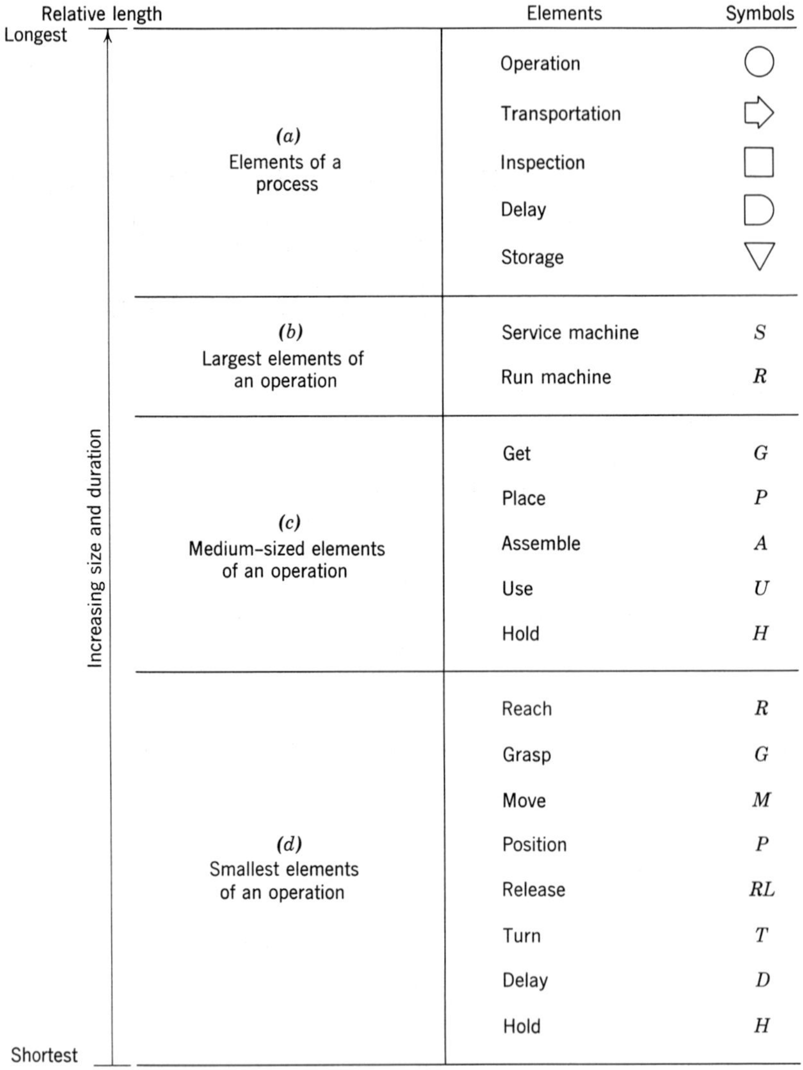

Relative length		Elements	Symbols
Longest	(a) Elements of a process	Operation	○
		Transportation	⇨
		Inspection	□
		Delay	D
		Storage	▽
	(b) Largest elements of an operation	Service machine	*S*
		Run machine	*R*
Increasing size and duration	(c) Medium-sized elements of an operation	Get	*G*
		Place	*P*
		Assemble	*A*
		Use	*U*
		Hold	*H*
	(d) Smallest elements of an operation	Reach	*R*
		Grasp	*G*
		Move	*M*
		Position	*P*
		Release	*RL*
		Turn	*T*
		Delay	*D*
Shortest		Hold	*H*

Figure 20. Samples of various sets of work elements, arrayed by size.

THE SPECIAL LANGUAGE AND SYMBOLOGY EMPLOYED BY A METHODS ENGINEER

The specification phase in particular and the methods design process in general are supplemented by a special and reasonably standardized language with associated short-hand symbols for purposes of description and communication of work methods. This language involves several standard sets of "elements" from which it is possible to rather rapidly and effectively describe a sequence of productive activity. These sets of elements are summarized in Figure 20. As indicated in the figure, these elements range in size from major subdivisions of the overall process at one end of the scale to individual motions of the hands and fingers at the other extreme.

Elements of a Process

The American Society of Mechanical Engineers (ASME) has established a standard set of process elements and symbols, those shown in Figure 20*a*. These elements and what they represent are as follows:

○ Operation. A sequence of activities or events taking place at a machine or workstation, during which an object is intentionally altered in one or more of its characteristics. Examples are fill bag with feed, sew bag closed, shear corrugated metal sheet, roll metal sheet, as shown in Figures 2 and 17 on pages 23 and 80.

⇨ Transportation. Movement of an object from one location to another, exclusive of the movement that is an integral part of an operation or inspection. Thus, transportations ordinarily occur between operations, inspections, delays, and storages. Examples are movement of bags of feed between the weighing and stitching operations, between the stitching operation and the warehouse, between warehouse and delivery truck, and so on.

□ Inspection. Comparison of a characteristic of an object against a standard for quality or quantity. Checking the weight of the bags of feed is an inspection.

D Delay. A delay occurs when at the termination of an operation, transportation, inspection, or storage, the next planned element does not begin immediately. Numerous examples of delays appear in Figures 2 and 17.

▽ Storage. Retention of an object in a state and location, removal from which requires authorization. Bags of feed in the warehouse are in a state of storage, as are the flat sheets of corrugated steel and the finished pipe while in the warehouse (Figures 2 and 17).

Largest Elements of an Operation

An operation or inspection can be analyzed in terms of certain standard sets of elements analogous to the ASME process elements introduced previously. The largest size elements into which an operation can be analyzed are "service machine" and "run machine," Figure 20*b*. Often the service element consists of removing finished material from the machine, loading the machine with new material, and starting the machine. The run machine element ordinarily is that period during which the machine processes the material. Similar sets of elements that are commonly encountered and that belong in this same large-element category are:

Unload machine (*U*), load machine (*L*), and run machine (*R*).
Set-up (or make ready), do, and put away.

Medium-Sized Elements of an Operation

The most useful set of elements in this size category are referred to as "gets and places." Analyzing an operation in terms of these elements is called a "get and place analysis." The elements ordinarily used in this type of analysis, summarized in Figure 20*c*, are defined as follows:

Get (*G*). The act of reaching to and securing control of an object, for example reaching to and grasping pencil in pocket.

Place (*P*). The act of moving an object into its intended position, for example moving pencil from pocket and positioning to paper in preparation for writing.

Use (*U*). The act of employing a tool, instrument, or the like, in accomplishing a useful purpose, for example writing with pencil.

Assemble (*A*). The act of joining two objects in the intended manner, for example affixing hinge to door, stamp to envelope, nut to bolt. (If joining the two objects can be accomplished simply by dropping one in or on the other, this act would be classified as a Place.)

Hold (*H*). The act of supporting an object with the hand while the other hand is preparing to or is performing work on that object, for example applying pressure to paper with hand while other hand writes.

Smallest Operation Elements: Motion Analysis

There are several popular sets of elements in the motion category. An analysis of an operation in terms of individual motions of the operator is referred to as a motion analysis or "microscopic analysis."

The set of "microscopic" elements recommended herein is presented in Table 3. These motions relate to the "get and place" elements in the following manner.

Get	Reach Grasp
Place	Move Position Release
Assemble Use	In a microscopic analysis Assemble and Use must be described in terms of the above motions.

TABLE 3. Elements for a "Microscopic" Analysis of Manual Activity

Motion	Symbol	Definition	Example
Reach	R	Movement of hand or finger without load, ending when hand is about one inch from target.	Move hand to pencil in pocket preparatory to grasping it.
Grasp	G	Movement involved in securing control of an object with fingers.	Position fingers about pencil in pocket and apply pressure.
Move	M	Movement of hand or fingers under load, ending when object is approximately one inch from target.	Transport pencil to paper.
Position	P	Movements involved in aligning, orienting, and engaging one object with another.	Move point of pencil to exact point at which writing is to begin.
Release	RL	Relinquish control of an object previously grasped in fingers.	Release pressure of fingers on pencil after return to pocket.
Turn	T	Movement requiring rotation of the forearm about its long axis.	Motion required to turn a door knob.
Delay	D	Hesitation of hand while awaiting termination of some act or event.	Hesitation of left hand while right hand reached to pocket for pencil.
Hold	H	Act of supporting an object with the hand while work is performed on object.	Support bolt in left hand while right hand assembles nut to it.

Traditionally, a set of seventeen motions referred to as "therbligs" has served as the basis for microscopic operation analysis. However, the detail and number of motions involved in the therblig system, plus the fact that no predetermined performance time values are available for therbligs, makes this set less useful than the series of motions introduced in Table 3. Tables of performance time values are available for the recommended set of motions. These times are provided in the form of a special system called Methods-Time Measurement, popularly referred to as MTM. This system is described in Appendix A. Henceforth the reach-grasp-move-position-release-turn-delay-hold set of motions will be referred to as the MTM elements.

In general, the "get and place" elements seem detailed enough for analysis of an operation. As long as the purpose is purely descriptive and qualitative, "gets and places" are sufficiently small and relatively rapid to apply. They provide most of the benefits that a finer, more detailed, more time consuming analysis might provide. The only apparent justification for analyzing an activity in terms of the more detailed microscopic elements is if use is to be made of the predetermined time values offered with some sets of motions. Thus the MTM motions are recommended over the "therbligs."

SPECIAL DESCRIPTIVE AND COMMUNICATION MEDIA EMPLOYED BY THE METHODS ENGINEER

For purposes of specification, information recording and presentation, visualization, explanation, improvement of method, and the like, the methods design process is supplemented by a number of special techniques for description and communication of work methods. Most of these aids utilize the special language and symbols introduced earlier. A number of them are a diagram of some type, offering a condensed, simplified, picture of the system or procedure being described.

Diagrammatic Aids to the Methods Designer

Those description and communication aids employed in methods design work that are based on a diagram of some form are appropriately designated as the diagrammatic aids. These diagrammatic techniques include the flow diagram, precedence diagram, operation process chart, man-machine chart, simo chart, and trip frequency diagram.

The flow diagram. A flow diagram includes a plan view of the work area under concern, a line diagram indicating the path followed by the object under study, and the ASME process analysis symbols superimposed on this line diagram to indicate what happens to the object as it passes through the process. Figures 2 and 17 (pages 23 and 80) provide illustrations of the flow diagram. This aid is especially useful in providing a compact, overall view of an existing or proposed process.

The precedence diagram. In most production systems there are certain tasks or elements of work that must precede certain others, as is true of the task of dressing where some items of clothing must be donned before certain others. A useful means of summarizing these precedence requirements is the precedence diagram, an illustration of which is shown in Figure 21 for the process that ordinarily transpires between the time a man arises from bed and the time he arrives at work. This diagram indicates that the person must arise before all other activities, must wash and shave before dressing, must dress before getting into the car, must eat before brushing teeth, but may eat and brush teeth any time between arising and getting into the car, and so on. Another illustration of the precedence diagram is shown in Figure 110, Chapter 22. This diagram is particularly useful as a means of isolating and summarizing the precedence restrictions that must be adhered to in the course of altering a sequence of events or tasks, as must be done in attempting to balance work assignments along a production line or in a work crew.

Operation process chart. The operation process chart ordinarily shows the materials entering a process, the operations performed, and the order of assembly, in the form illustrated in Figure 22. Note that

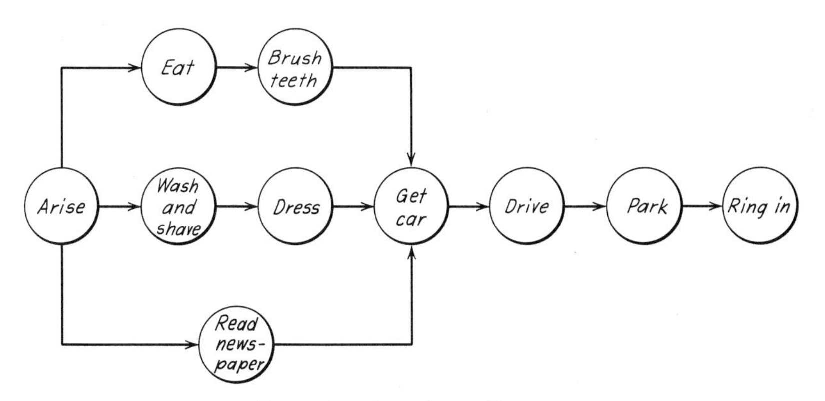

Figure 21. Precedence diagram.

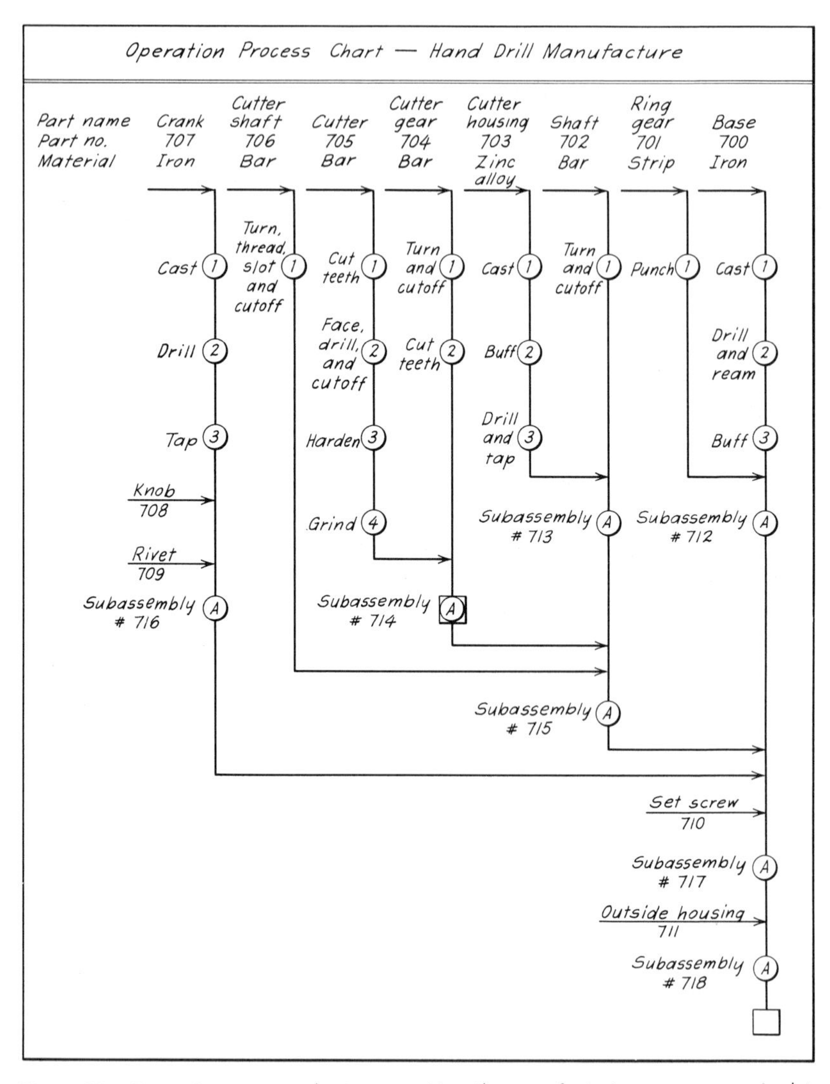

Figure 22. Operation process chart summarizing the manufacturing process required to produce a pencil sharpener.

the materials upon which operations are performed are listed across the top, operations and inspections are indicated by the appropriate ASME symbols, and purchased parts[1] are shown at the point of use.

[1] Parts purchased from an outside supplier and assembled directly without further operations by the company involved are commonly called purchased parts.

In general, materials entering are indicated by horizontal lines; processing of these materials is shown vertically.

The operation process chart has a variety of uses, the foremost of which is providing a compact overall view of the whole system of operations involved in the manufacture of a product. The comprehensive view provided by this diagram can be especially appreciated by the engineer who is attempting to get this overall picture as he sits confronted by a large stack of operation sheets. These sheets tell the whole story insofar as details of the basic method of manufacture are concerned, but the overall system is very difficult to envision from them without an aid of the type provided by an operation process chart. As a result of the overall perspective it offers and of the fact that construction of such a diagram rapidly and effectively familiarizes the maker with the whole manufacturing process and the product, the operation process chart is a valuable aid in plant layout work. It is useful also to the manufacturing engineer as he is specifying the basic manufacturing system. It is useful to the scheduler who must take into account the sequence of assembly as clearly pictured in the diagram, in his scheduling of arrival dates for purchased materials, completion dates for manufactured parts, and setup of various subassembly operations. It is useful also as an educational aid, such as in the indoctrination of new technical personnel or training salesmen and servicemen.

Multiple-activity charts. This type of chart graphically depicts the relationship of two or more simultaneous sequences of activity on a time scale. A chart of this type describing the activity of a man and of the machine(s) he is tending is a special form of multiple-activity chart commonly referred to as a *man-machine chart.* This chart is illustrated in Figure 23*a* for an operator servicing one machine, in Figure 23*b* for an operator tending two such machines, and in Figure 23*c* for an operator tending three of the same machines. The probable physical arrangement of the machines and the path followed by the operator as he completes his servicing cycle is indicated for each alternative. Notice that the chart describes one complete cycle of activity. As long as the atypical starting-up and shutting-down cycles are avoided in construction of a man-machine chart, a starting point may be arbitrarily selected and the succeeding activities charted until that point in the pattern of events is reached again and the cycle begins to repeat itself.

Figure 23 illustrates the primary use of the man-machine chart, which is to depict the consequences of varying machine assignments prior to deciding on the preferred number of machines for one man

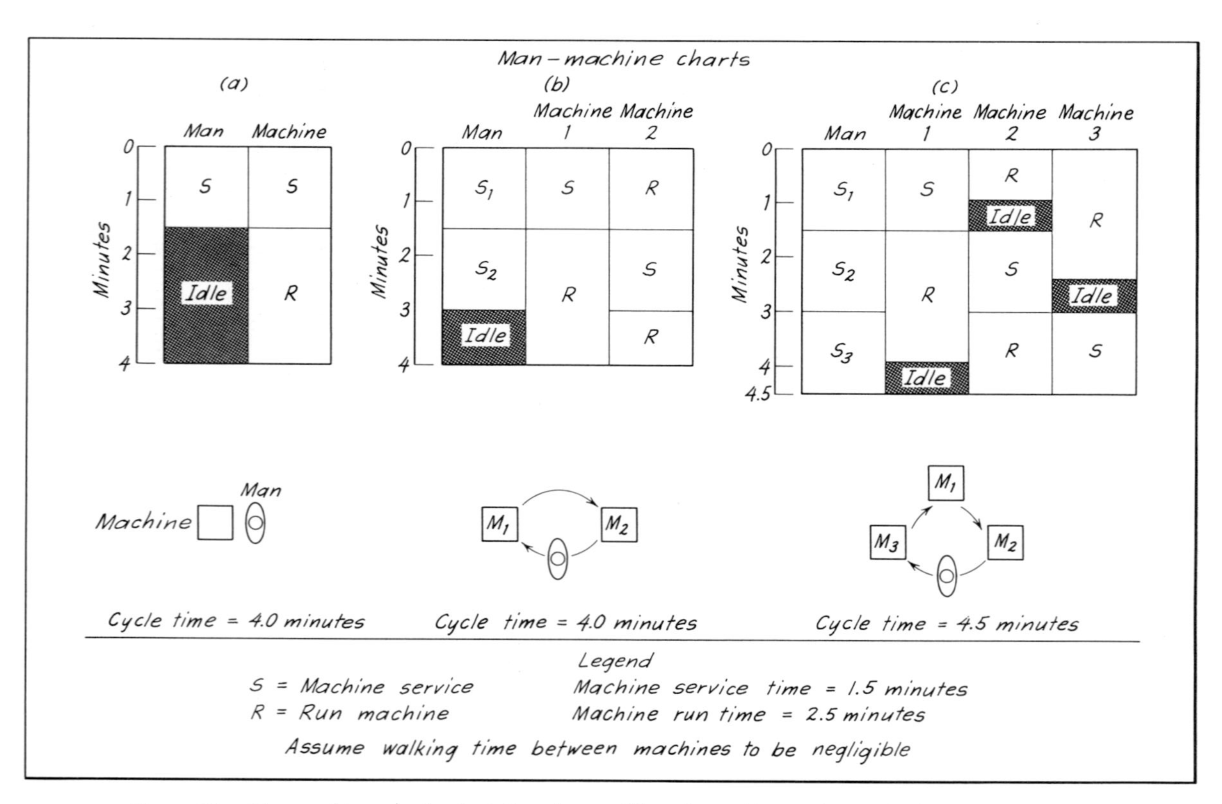

Figure 23. Man-machine charts depicting three different machine assignments to one operator.

to operate. To make multiple-machine assignments as pictured in Figures 23*b* and 23*c*, it is essential that the machine be free of any need for observation or attention during the run machine portion of the cycle and that there be no dire consequences if the man should be delayed and not be at the machine when it completes its run. A number of common pieces of processing equipment satisfy these requirements.

An assumption that is ordinarily made in the construction and use of such charts is that the service and machine run times are essentially constant from cycle to cycle. For service times this assumption is not ordinarily met in practice and occasionally this may lead to erroneous conclusions. The effects of variation in service time on the multiple man-machine system are discussed in a later chapter.

Another special version of the multiple-activity chart depicts the concurrent activities of a worker's hands during an operation. When applied in this manner the technique is traditionally referred to as a *simo chart.* A simo chart, illustrated in Figure 24, is ordinarily constructed via the following procedure.

1. A motion picture is made of the operation under study, with an ordinary spring-powered camera with a high-speed clock placed at the workstation so that time and the operator's actions are simultaneously recorded on film. If a constant-speed camera is used, no clock is necessary; the camera itself acts as an acceptable timing device. (In fact most spring-powered cameras of at least medium quality probably have a sufficiently stable speed characteristic so as to render the clock unnecessary.)
2. The film is viewed picture by picture in order to determine the motions performed and their performance times.
3. These times are then plotted as indicated in Figure 24.

The primary usefulness of the simo chart is in indicating possible improvements in the operation under study by revealing unnecessary idleness of the hands, superior sequences of motions, and the like.

The simo chart is not extensively used in practice and rightly so. It is expensive and time consuming to construct and therefore should be used only if the less expensive means of arriving at improvements in the operation under study have been applied first. If after applying these approaches it seems that additional reductions in performance time might be eked out, and if the operation under design is a very costly one as a result of high production volume and similar circumstances, use of this ultra fine technique may be justified. Only under

Simo chart						
Operation: Assemble Tab Shaft						
Part: NA 37124			Operator: R. Rees			
Department: Assembly			Date: 7/10			
Analysis: Smalley			Film No.: 16-48			
Left-Hand Description	Time in Minutes	Symbol	Time Scale	Symbol	Time in Minutes	Right-Hand Description
To shaft	0.007	R		D	0.007	
Shaft	0.016	G	0.010 / 0.020	R	0.016	To key
	0.006	D	0.030	G	0.006	Key
To assembly point	0.014	M	0.040	M	0.014	To shaft
Support assembly		H	0.050	P	0.009	To shaft
				RL	0.002	
			0.060	R	0.009	To collar
			0.070	G	0.007	Collar
			0.080	M	0.010	To assembly
			0.090	P	0.009	To assembly
				RL	0.003	
			0.100	R	0.008	To screwdriver

Figure 24. Illustration of the simo chart.

high-volume circumstances is it likely that the small reductions in performance time attainable will yield a substantial enough saving in labor cost to justify the relatively large expense. Although the simo chart finds very limited use in practical methods improvement work, it is claimed to be a useful training aid as a means of increasing the motion economy consciousness of trainees.

The multiple-activity chart idea can be applied to the concurrent activities of two or more men as well as to the concurrent activities of man and machine or of left and right hand. For example, in attempting to balance work assignments and properly relate the timing of tasks performed by members of a work crew, a multiple-activity chart is likely to be a useful aid. This includes the diesel locomotive repair crew, the highway concrete-laying crew, the surgeon and his assistants, the tree-planting crew, pilot and copilot, or any similar situation where distribution and timing of tasks is important. Figure 25 shows a multiple-activity chart for a three-man fire engine crew.

(An especially useful aid in studying crew activities is the motion picture, which permits simultaneous recording of numerous concurrent activities. These pictures might be taken at the standard silent motion picture speed of sixteen pictures per second, or preferably at a much slower and therefore less costly speed. The slower speed is quite satisfactory in most instances. Depending on the activity being studied, one picture a second, one every five seconds, or one every minute might suffice. This slow speed technique is generally referred to as time-lapse photography. One writer in the field has coined the term "memomotion technique" for this procedure.[2] Special drive units to obtain these slow filming speeds from the popular makes of 16 mm. cameras are commercially available.)

The multiple-activity chart idea may also be applied to the problem of balancing a production line. Actually the production line and crew situations are very similar. The basic difference is that in the production line the work moves along from operator to operator, whereas in a crew the work moves little or not at all and the operators move about it. There is the problem of balancing the work assignments and satisfying the precedence restrictions in both instances. Figure 26 shows a multiple-activity chart for a five-man bicycle final assembly line. This chart assists in balancing a production line such as this.

The trip frequency diagram. In many processes dealt with by the methods engineer there is no fixed sequence of events. Instead of the repeating cycle of elements true of cases cited so far, the pattern of elements often varies, depending on the task being performed and the chance interrelationship of activities. The latter is true of

[2] Marvin E. Mundel, *Motion and Time Study,* 3rd ed., Prentice-Hall, Englewood Cliffs, New Jersey, 1960.

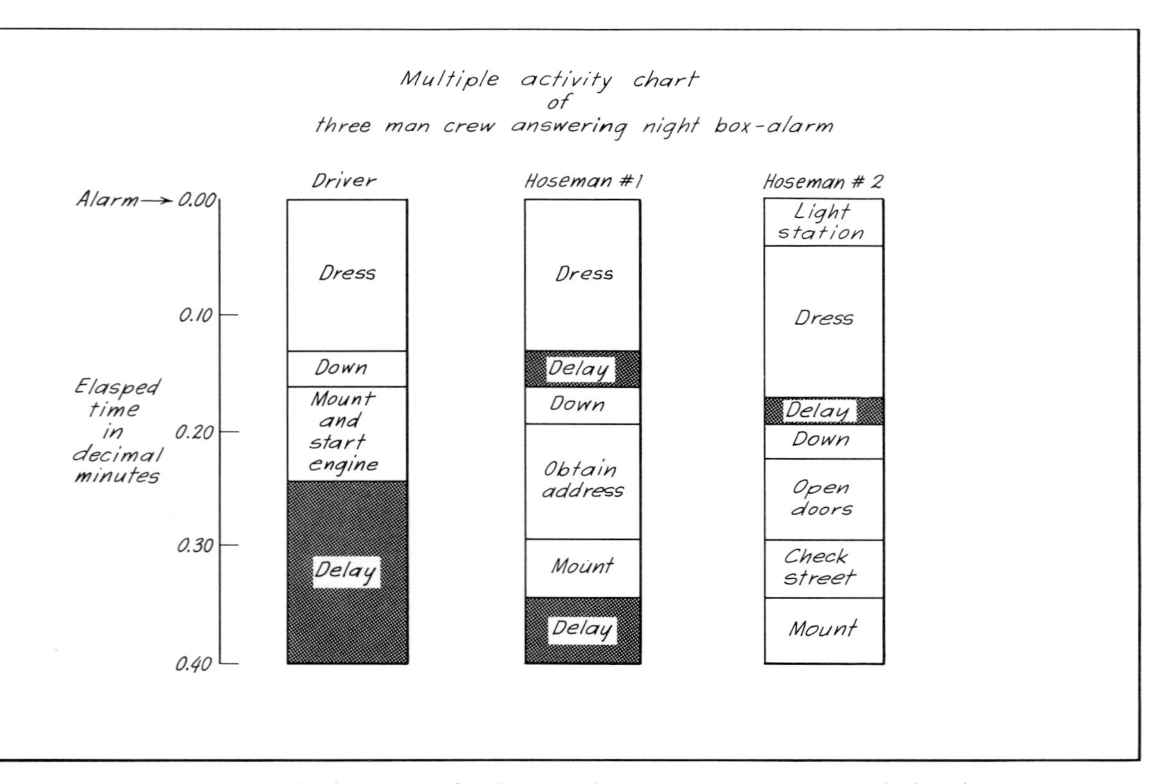

Figure 25. Multiple-activity chart for three-man fire engine crew answering night box-alarm.

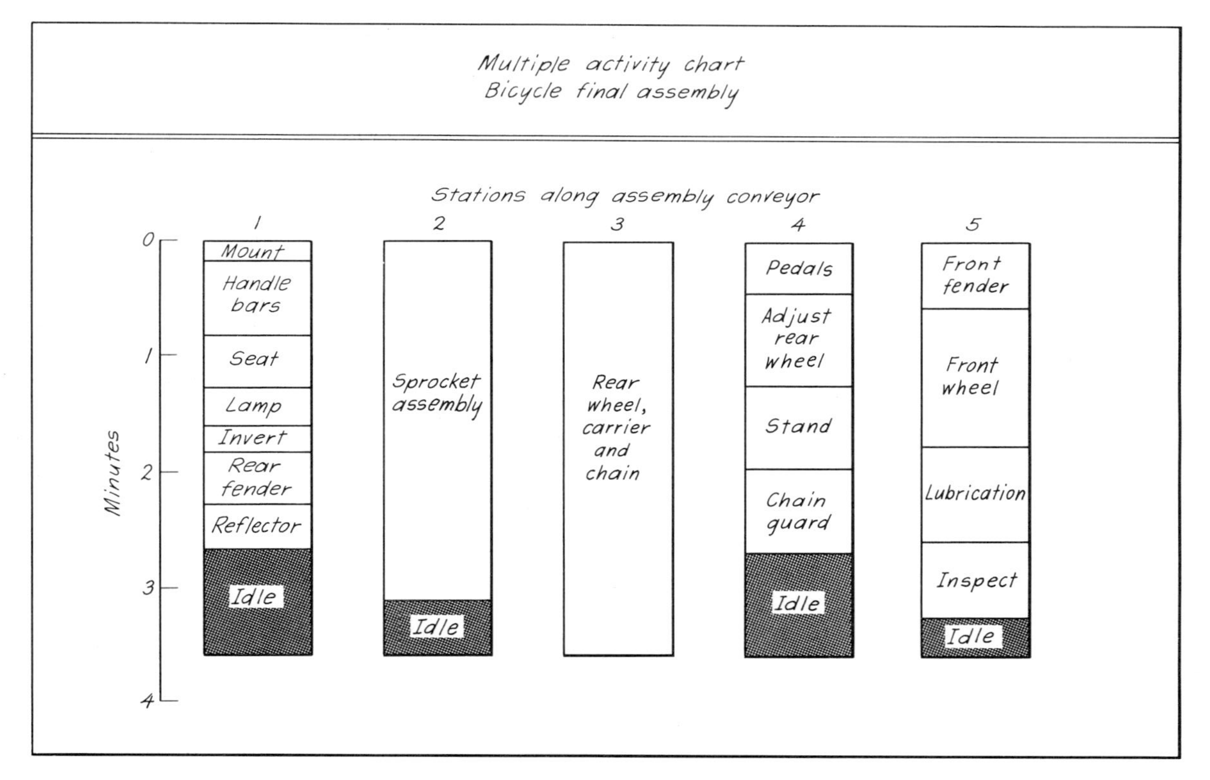

Figure 26. Multiple-activity chart for five-man bicycle assembly line.

processes found in the kitchen, the job shop, the dry cleaning establishment, the repair shop, the tool room, the office, the bakery, and numerous other situations. In each of these instances a flow diagram could be constructed for some particular item being processed. However, many of the items—devices being repaired for example—would follow substantially different flow patterns. To base the work layout and procedure on any one item would yield a very inefficient method for many of the others handled by the same process. Therefore, the process should be arranged so as to be optimum for the *aggregate* of items or products handled by the process, as for the aggregate of the menus processed in a kitchen.

One means of analyzing the flow pattern in this type of situation to obtain an estimate of the total travel is to construct a trip frequency diagram. This diagram, illustrated in Figure 27, is obtained by observing the process over a period of time and recording the trips made between work centers, as shown. To arrive at a satisfactory layout for the kitchen, a substantial number of different meals and menus must be likewise observed and the ultimate layout must be based on the aggregate of the observations. Assuming that the ultimate goal is to minimize the total distance traveled in such a case, the components of the system should be arranged so that the following sum is a minimum:

$$\text{Total distance traveled} = f_{1,2}d_{1,2} + f_{1,3}d_{1,3} + f_{1,4}d_{1,4} + \cdots f_{n,m}d_{n,m}$$

where $d_{1,2}$ = the distance between components (e.g. work centers) 1 and 2, and $f_{1,2}$ = the relative frequency of trips made between these two components.

A more convenient means of recording the same information is the trip frequency chart illustrated in Figure 28.

Other Descriptive Aids to the Methods Designer

The descriptive techniques introduced so far are based on a diagram of some form. There are several nondiagrammatic descriptive techniques commonly used, serving similar purposes. Among these are the trip frequency chart, the flow chart, and the left-hand right-hand chart.

The trip frequency chart. The travel pattern for one day of activity in the office of a small business firm has been recorded by way of the trip frequency chart shown in Figure 28. This chart, which may be used as a companion to or as an alternative to the trip frequency

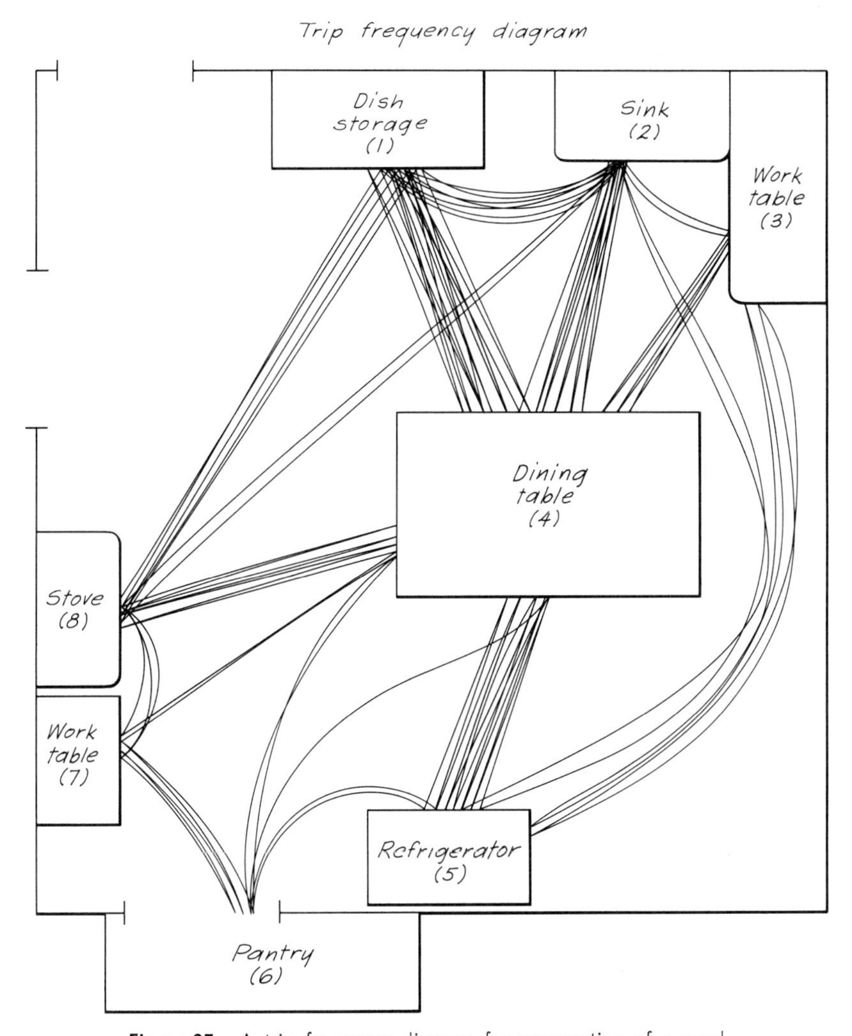

Figure 27. A trip frequency diagram for preparation of a meal.

diagram, has numerous applications to a wide variety of problems. Its applicability expands as the increase in indirect labor tasks continues, as exemplified by the sharp and continued rise in the number of persons employed in office work, maintenance and repair activity, and the service enterprises.

The flow chart. The flow chart, illustrated in its simplest form in Figure 18, page 81, is a traditional and widely promoted process

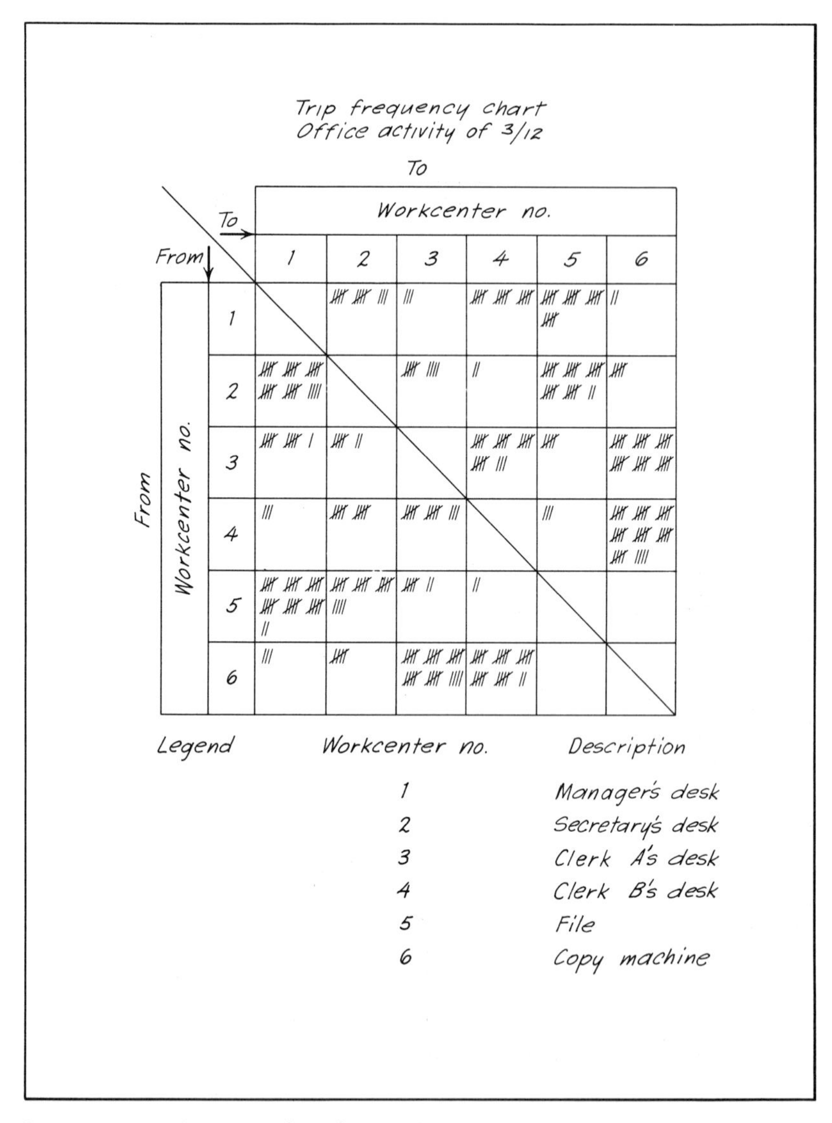

Figure 28. Trip frequency chart for one-day period of observation in small business office.

analysis technique. A version more commonly encountered is pictured in Figure 29, page 109. The flow chart is a tabular summary of the information shown on a flow diagram. It describes in terms of ASME process elements and symbols what happens to material as it passes through the process, and optionally, how various steps are accomplished. Other types of information may be added to suit the particular purposes and nature of the project.

The purposes of the flow chart are similar to those of the flow diagram. Since it lacks the physical layout feature, a flow chart does not provide the overall perspective offered by a flow diagram. In general, if one or the other is to be used, the flow diagram is preferable.

The left-hand right-hand chart. This chart is actually a simo chart minus the graphic feature. In the illustration of a left-hand right-hand chart in Figure 30, "gets and places" are used, however larger sized elements are equally adaptable to the technique. Smaller elements are not recommended *if* the analysis is only to be a qualitative one, as in the case illustrated in Figure 30. If the intent is to take advantage of the predetermined time values provided with a given set of microscopic elements in order to synthesize the time for a task, construction of a left-hand right-hand chart with motions as elements is not only justified but essential. This chart is useful as a means of communicating work procedure to those who must pass judgment on, implement, or use the procedure selected by the methods engineer. It is particularly useful as an operator training aid.

General Uses of these Descriptive Aids

1. To facilitate comprehension and understanding of the overall nature or behavior of the system under study. The details of the manufacturing process can be obtained from official records or by actual observation, but an overall view of the "whole" is not easy to obtain, especially when the system does not yet exist. In most instances the only means of obtaining a compact, overall, simplified, perspective of the whole system or procedure is by way of a diagrammatic visual aid such as the flow diagram, operation process chart, and similar techniques. This diagrammatic approach is very useful to the engineer, and should be cultivated and extended to many types of situations. For example, it is admirably suited to the analysis of a complex communication system involving considerable flow of paper work and complicated distribution of information. When used for this purpose, the technique is referred to as a forms distribution diagram (Figure 123). The same basic idea is extended to and almost essential for the analysis of company organization structure, giving rise to the common organization chart. The electrical engineer makes common use of the same idea in his

FLOW CHART

SUMMARY

	Present No.	Present Time	Proposed No.	Proposed Time	Difference No.	Difference Time
○ Operations	2					
⇨ Transportations	5					
□ Inspections	1					
D Delays	7					
▽ Storages	2					
Distance traveled	367 Ft.		Ft.		Ft.	

Process *Form #2M table leg*

☐ Man or ☒ Material *2" x 2" Maple*

Chart begins *Stockroom*

Chart ends *Stockroom*

Charted by *M.W.S.* Date *3/9*

Details of (Present / ~~Proposed~~) method	Operation	Transport.	Inspection	Delay	Storage	Distance in feet	Quantity	Time	Possibilities: Eliminate	Combine	Change: Seque.	Place	Person	Improve	Notes
1 *Stockroom*	○	⇨	□	D	▽										
2 *Hand truck*	○	⇨	□	D	▽	90									
3 *On floor*	○	⇨	□	D	▽										
4 *By hand*	○	⇨	□	D	▽	50									
5 *On table near saw*	○	⇨	□	D	▽										
6 *Cut to length*	○	⇨	□	D	▽										

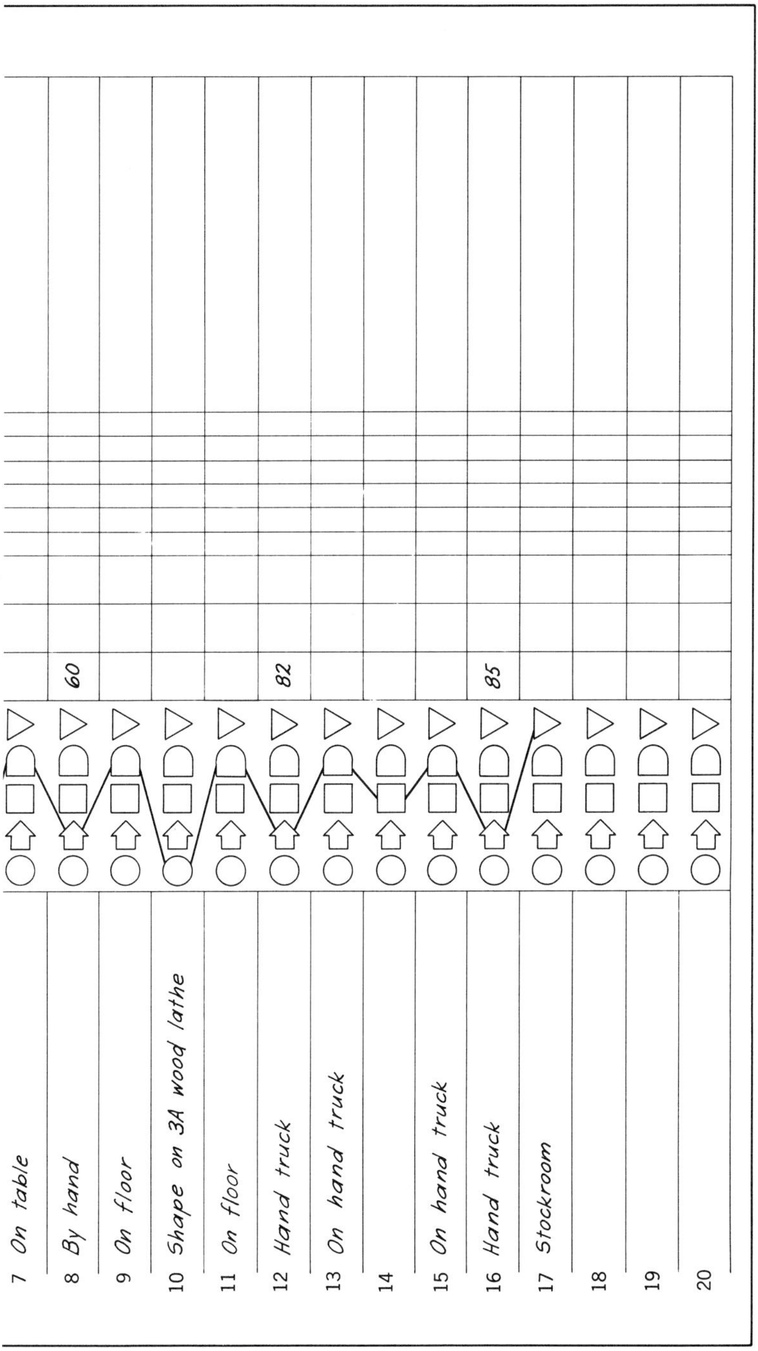

No.	Details	Symbols	Distance
7	On table	○ ⇨ □ D ▽	
8	By hand	○ ⇨ □ D ▽	60
9	On floor	○ ⇨ □ D ▽	
10	Shape on 3A wood lathe	○ ⇨ □ D ▽	
11	On floor	○ ⇨ □ D ▽	
12	Hand truck	○ ⇨ □ D ▽	82
13	On hand truck	○ ⇨ □ D ▽	
14		○ ⇨ □ D ▽	
15	On hand truck	○ ⇨ □ D ▽	
16	Hand truck	○ ⇨ □ D ▽	85
17	Stockroom	○ ⇨ □ D ▽	
18		○ ⇨ □ D ▽	
19		○ ⇨ □ D ▽	
20		○ ⇨ □ D ▽	

Figure 29. Popular format for preparation of a flow chart.

Left-hand right-hand chart

Part: *Nut, bolt and washer assembly*

Operation: *Assemble*

Date: *7/31* Analyst: *C. Honness*

Use reverse side for sketches

Left hand		Right hand	
Description	Symbol	Symbol	Description
Bolt	*G*	*G*	*Washer*
Bolt in washer	*P*	*P*	*Washer on bolt*
Assembly	*H*	*G*	*Second washer*
	↓	*P*	*Second washer*
	↓	*G*	*Nut*
	↓	*P*	*Nut*
	↓	*A*	*Nut*
	D	*P*	*Down chute*

Figure 30. Left-hand right-hand chart for a simple assembly operation.

schematic diagram and block diagram for a complex piece of equipment. The football coach uses the same scheme to explain plays to the team—an excellent illustration of the purpose and usefulness of this diagrammatic approach. Can you imagine the coach attempting to introduce a football play without the familiar play diagram, using only verbal or written descriptions of players' assignments? It is very difficult to visualize the overall picture from a series of successive detailed explanations of the individual parts, as would be the problem in attempting to visualize the overall manufacturing process from a stack of operation sheets. The diagrammatic approach then is worthy of considerable development and application, especially in view of the increasing complexity of machines and systems.

2. To communicate a procedure, as is required in specification of the solution to a methods design problem. For example, the left-hand right-hand chart is well suited for specifying the procedure to be followed by the worker in performance of his job.

3. To provide standard and convenient means of recording information about a process or operation.

4. To provide a "quick access" source of useful information, such as the operation process chart can well provide in plant layout work.

5. To assist in training and indoctrinating personnel and visitors unfamiliar with the process involved. (And in fact, the maker of one of these diagrams finds that the process of construction is an excellent familiarizing procedure.)

6. To call attention to and facilitate the search for improvements in method.

7. To assist in analysis of a manufacturing problem, as does the precedence diagram.

FORMULATION AND ANALYSIS OF METHODS DESIGN PROBLEMS

A methods design problem should be formulated and analyzed as advocated in Chapters 3 and 4. As a consequence of the decision-making hierarchy usually found in a manufacturing enterprise, the methods designer must ordinarily operate under numerous restrictions imposed by the engineering specialties that precede him in that hierarchy. To illustrate, in an assembly operation the methods designer's problem might be formulated thus: to find the most economical method of getting component parts into an assembled form. In his subsequent analysis of this problem the designer finds that many restrictions are imposed, for example:

1. By the product designer in the form of specifications of the component parts and the finished assembly.

2. By the plant layout engineer in the form of a decision as to the location at which the assembly operation is to be set up.

3. By the materials handling engineer in the form of decisions as to how the parts will be delivered to the workstation and how the assemblies will be removed.

In a metal machining operation, the methods designer's problem might be formulated as: to find the most economical method of transforming raw stock to finished piece. In this case restrictions are usually imposed:

1. By the product designer in the form of specifications of the raw stock and finished piece.
2. By the manufacturing engineer in the form of a decision as to what machine and tools will be used.
3. By the plant layout engineer.
4. By the materials handling engineer.

In other types of operations the situation is similar; many restrictions are imposed. In view of this, and of what was said earlier in Chapter 4 about the desirability and possibility of having some restrictions revoked, we begin to understand why product, manufacturing, plant layout, materials handling, and methods engineers are frequently at odds with one another. For example, the product designer is notorious for designing parts and products that are overly costly, sometimes nightmarish to fabricate and assemble. As a consequence, manufacturing and methods engineers are frequently protesting to him over unforeseen implications of his specifications. Also, it becomes obvious that it is highly desirable to facilitate overlap and liaison between these engineering specialties, in order to minimize the number of suboptimum decisions that arise from the specialization involved.

Even with the best liaison efforts however, the methods designer is likely to uncover numerous product specifications that could feasibly be changed with resulting economies in manufacture. This was true of the problem of assembling shielding devices for refrigerator door latching mechanisms. In this and similar cases, the methods designer finds it worthwhile to attempt to alter states A and B. Subject to the limitations mentioned in Chapter 4, the methods designer should investigate the soundness of a product's specifications and the profitability and feasibility of deviating from them.

Criteria

Most engineers, in principle at least, operate under the general criterion return on the investment. Of course in practice designers many times stray from this criterion. Furthermore, it is usually necessary to translate this overall return-on-the-investment criterion into a number of more specific subcriteria to guide the designer's thinking and to permit a satisfactory evaluation of alternatives. In aggregate, these subcriteria determine return on the investment. In the case of work methods design some of these subcriteria are:

1. *Original investment* required by the method in question, which depends on such factors as:
   a. The tools and other equipment required.
   b. Lost production during installation.
   c. Labor required to make the installation.
   d. Learning time required.
   e. Others.
2. *Operating cost* of the method in question, which depends on such factors as:
   a. Production time required.
   b. Cost of labor.
   c. Power required.
   d. Maintenance required.
   e. Flexibility of the method to changes in the desired rate of production, to changes in design of the product, or insofar as accommodating different products. For example, what are the consequences if demand for the product changes from 100 to 150 units per hour? Or what happens if the shape of the product changes?
   f. Fatigue.
   g. Monotony.
   h. Effort required.
   i. Safety.
   j. Employee satisfaction.
   k. Others.

The designer must take into account investment *and* operation cost. It is virtually meaningless to quote savings in operating cost that a proposed method is expected to achieve and not to estimate and quote the expected cost of achieving these savings.

In a majority of problems the criteria a designer is to operate under are implicit, automatic, and routine. In many instances only superficial consideration of the matter is necessary at this stage of the design process. On occasion however, a particular criterion must be given more than ordinary weighting. For example, in some instances safety will be of special concern, or flexibility, or space required. Of course this should be known *before* search for alternatives begins, for it means special emphasis will be put on these factors in generation as well as in evaluation of alternatives.

Volume of Production

The expected production volume for the operation or process under consideration influences the designer's course of action in several ways. First, volume is a primary determinant of the length of time the designer can economically devote to a given problem. Obviously an engineer will not under ordinary circumstances devote the same amount of time to an operation involving 500 units per year as he will to a similar one involving 500,000 units per year. In the former he will specify the method after a relatively brief and informal treatment of the problem. In the latter, his treatment of the problem will be relatively elaborate and may require weeks or months. The length of time the designer is justified in devoting to a given problem is ordinarily estimated primarily by judgment.

Second, the volume is a primary determinant of the capital that can justifiably be invested in a solution to the problem. Ordinarily, on a low-volume operation very little can justifiably be invested in machines, tools, and other equipment, whereas the reverse is true for a high-volume operation of the same type.

Third, the volume ordinarily has some bearing on the areas of possibility that should be searched for alternatives. For example, for a low-volume job the designer is not likely to generate alternatives representing high degrees of mechanization or automation. Similarly, if the volume is high enough to warrant automation, the designer will not be generating manual alternatives. Note that on low-volume jobs, it will generally be changes in workplace layout and work procedure that will offer the most likely and profitable improvements as opposed to additions of equipment. Therefore, in searching for alternatives in low-volume jobs, the designer should concentrate on layout and procedure. The designer should also emphasize flexibility in his search for alternatives, for if a method can accommodate more than one product, the volume has in effect been increased. For example, the designer might be considering a powered wrench for an operation, but the volume on the job does not seem to warrant the investment. However, if that same power wrench is adaptable to other operations about the plant, which is quite likely, the volume to be considered in deciding on the economy of the tool is now greater and the investment may well be justified.

In his analysis of the problem, the methods designer should not overlook the importance of obtaining the best forecast available of the future volume, expected trends and variations in same, and expected "life" of the problem itself. *No substantial amount of time and effort should be expended on a problem until these matters have been investigated.*

The Role of the Diagrammatic Techniques and Other Aids

The special diagrammatic techniques and associated symbolic language introduced earlier are sometimes useful in analysis of methods problems, but not as often and not as useful as textbooks in this field have traditionally led readers to believe. It is important to distinguish between what is basically a means of analyzing a problem or some characteristic of it as exemplified by the precedence diagram (a diagrammatic representation of precedence restrictions), and what is primarily a means of analyzing a solution to a problem, as exemplified by the flow chart. Rather unfortunately, aids of the latter type have traditionally been presented and promoted in the guise of problem-analysis techniques. Analysis[3] of a problem and analysis of a solution to that problem are two different procedures with different objectives and consequences. The former, essential to and typical of sound engineering practice, is a breakdown of the problem into its basic characteristics, namely: states A and B, restrictions, etc. Analysis of a solution to the problem is a breakdown of an existing or proposed method into constituent parts, conventionally by flow diagram, flow chart, left-hand right-hand chart, and the like. Furthermore, analysis of the present method or any other solution is not a necessary or even desirable part of analysis of the problem. Making a flow chart or a left-hand right-hand chart as the *first step* of the approach in redesigning an operation or process, especially if it is in lieu of definition of the problem itself, is likely to have a very detrimental effect on the designer's capacity to generate new and superior solutions. *This does not detract from the usefulness of these descriptive devices as communication and visualization media* and as *solution*-analysis techniques. It is in problem analysis that their usefulness is limited.

Much the same can be said about the use of motion pictures. The usefulness of going to considerable expense to photograph the current method in great detail is highly questionable, as is construction of a simo chart from such films. Why go to this trouble to make an expensive recording of a method that is probably grossly inferior and should be displaced almost in its entirety by a basically different and much improved method? These statements apply to misapplication of the techniques. They are not condemnations of the aids themselves.

[3] The dictionary definition of analysis is: a breakdown of the whole into its constituent parts.

When appropriately applied they serve a number of important purposes.

Summary

Although matters of emphasis and some of the details are different, formulation and analysis in this case are basically the same as they are for any design problem. Some precautions have been emphasized to counter the temptation to start the approach to a redesign problem with an immediate and detailed attack on the present solution (a tendency that seems especially strong when the present method is obviously and grossly inferior and offers tempting "bait"), and to counter traditional suggestions in the literature of this field that this very thing be done.

EXERCISES

1. The Delaware Tool Company is planning to introduce a hand drill to its line of products. Parts drawings have been prepared by the product designer and the manufacturing engineer has prepared a parts list and operation sheets for all parts of the drill to be manufactured by the Delaware Company. The manufacturing engineer feels that a diagram depicting the

Part No. 306
Hand drill assembly drawing

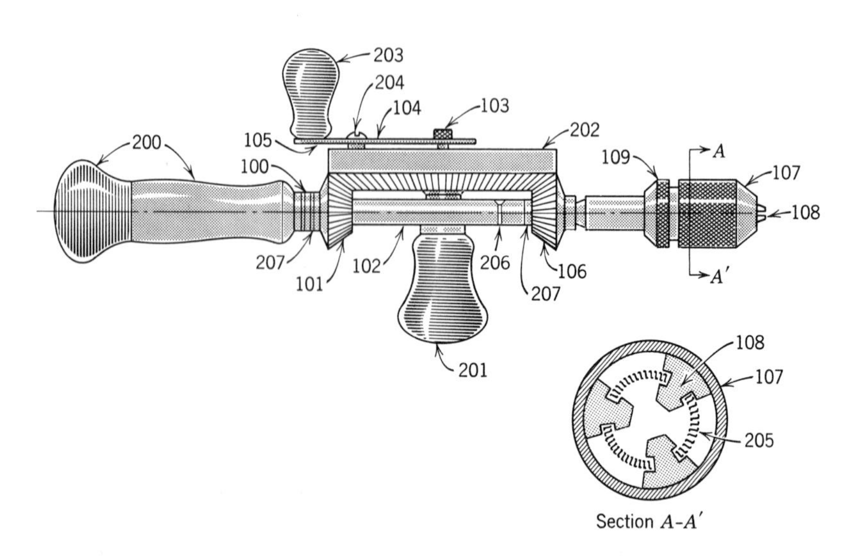

Section A-A'

overall hand drill manufacturing process would be of considerable assistance to him, to the plant layout engineer, to production scheduling personnel, and to others in the organization that must understand the sequencing, interrelationship, and timing of operations in this system. Prepare a diagrammatic aid that would best serve this purpose. The necessary supporting documents are given.

Parts List

Part Name—Hand Drill
Part Number—306

Part Name	Part Number	Quantity Required Per Final Assembly	Raw Material
Handle and ring assembly	200	1	
Handle pin	100	1	Bar stock
Upper driven gear	101	1	Bar stock
Main shaft	102	1	Bar stock
Driving gear bearing pin	103	1	Bar stock
Bearing pin handle and ring assembly	201	1	
Driving gear	202	1	
Driving gear handle	203	1	
Driving gear handle arm	104	1	Strip stock
Driving gear handle pin	105	1	Bar stock
Screw	204	1	
Lower driven gear	106	1	Bar stock
Jaw casing	107	1	Bar stock
Jaws	108	3	Bar stock
Springs	205	3	
Set screw	206	1	
Washers	207	2	
Chuck base	109	1	Bar stock

Note: 200 series are purchased parts.

Operation Sheets

Part Numbers—100 to 109 of Assembly #306

Part	Operation	Machine	Std. Time (Minutes)
Handle pin (#100)	Turn and cut off	Auto. screw	0.02
Upper driven gear (#101)	1. Turn and cut off 2. Cut teeth	Auto. screw Gear cutter	0.20 3.00
Main shaft (#102)	1. Turn, drill, and cut off 2. Drill 3. Drill	Auto. screw Drill press Drill press	0.04 0.03 0.05
Driving gear bearing pin (#103)	Turn, thread, and cut off	Auto. screw	0.03
Driving gear handle arm (#104)	Punch	Punch press	0.01
Driving gear handle pin (#105)	Turn and cut off	Auto. screw	0.02
Lower driven gear (#106)	1. Turn and cut off 2. Cut teeth 3. Thread	Auto. screw Gear cutter Threader	0.25 3.00 0.20
Jaw casing (#107)	Turn, drill, tap, knurl, and cut off	Auto. screw	0.30
Jaws (#108)	1. Turn and cut off 2. Drill	Auto. screw Drill press	0.02 0.04
Chuck base (#109)	Turn, drill, tap, knurl, and cut off	Auto. screw	0.30

Assembly Operation Sheet

Part Number 306—Hand Drill Assembly

Subassembly	Operation	Machine
#300	Assemble upper driven gear (#101), handle pin (#100), handle and ring assembly (#200), and washer (#207) to main shaft (#102).	Bench
#301	Assemble lower driven gear (#106), washer (#207), and set screw (#206) to #300.	Bench
#302	Assemble 3 springs (#205), 3 jaws (#108), jaw casing (#107), and base (#109) to #301.	Bench
#303	Assemble driving gear bearing pin (#103) into #302.	Bench
#304	Assembly bearing pin handle and ring assembly (#201) to #303.	Bench
#305	Assemble driving gear handle pin (#105), driving gear handle arm (#104), and driving gear handle (#203), then rivet.	Riveter
#306	Assemble #305, screw (#204), and driving gear (#202) to #304.	Bench
	Inspect final assembly.	Bench

2. Prepare a precedence diagram for the process of filling a cigarette lighter. Use such elements as remove cap, remove plug, fill, etc.

3. Prepare a multiple-activity chart for three apartment-mates, for the activities transpiring between arising in the morning and leaving for the first class. Attempt to phase the activities so that total time for the three will be minimized. *Assumptions:* That in addition to dressing, each person *will* eat and each *will* wash and shave; also that each bed *will* be made.

4. Prepare a multiple-activity chart for the operation of a mimeograph machine, assuming that the operator must also wash the used stencils. Is it feasible for one person to operate more than one machine simultaneously? Prepare a multiple activity chart to support your answer.

5. A high volume component of a calculating machine must be operated on by a standard milling machine with automatic feed. Two successive milling operations are required, both utilizing the same type of machine. In both instances, the operator may leave the machine during the machining portion of the cycle. The volume of parts required throughout the year is 1000 pieces per day. The operating characteristics of the two operations are indicated on the following chart.

	Operation 1	Operation 2
Unload	0.12 minute	0.11 minute
Load	0.19 minute	0.15 minute
Mill	0.38 minute	0.47 minute

On the average these machines are producing approximately 85 per cent of the 8-hour work day.

What arrangement of men and machines would you recommend? Support your answer with appropriate multiple-activity charts. Machines may be located in whatever fashion is convenient for the setup proposed.

6. Prepare a trip frequency diagram and trip frequency chart for one of the following situations:

(*a*) A period of activity in a maintenance and repair shop.
(*b*) Preparation of a meal.
(*c*) A period of activity in a business office.

7. Prepare a left-hand right-hand chart description of the motions involved in loading a stapler. Assume that the stapler is on the desk top before you, that the staples are in a box in the desk drawer, that the sequence to be described begins with the hands resting on the desk and ends when the hands have returned to that position. Analyze this activity in terms of "gets and places" and then in terms of MTM elements (i.e., reach, move, grasp, etc.).

8. A motion picture has been taken of a press operation at a constant speed of 1000 frames (pictures) per minute. A representative cycle of film of this operation has been viewed frame by frame and the number of frames (and thus the time in 0.001 minute) for each motion has been recorded. The results of this analysis are given in the following table. From the data given prepare a simo chart for this operation.

Results of Frame-by-Frame Analysis of Press Operation

Left Hand			Right Hand		
Description	Symbol	Time in Minutes	Time in Minutes	Symbol	Description
To part in die	*R*	0.006	0.008	*R*	Part in pan
Part	*G*	0.008	0.005	*D*	
	D	0.007	0.008	*G*	Part in pan
Part to chute	*M*	0.010	0.010	*M*	Part to die
	RL	0.002	0.018	*P*	Part in die
	D	0.016	0.002	*RL*	Part in die
To LH trip button	*R*	0.008	0.006	*R*	To RH trip button
Button	*G*	0.002	0.002	*G*	Button
Trip press	*M*	0.004	0.004	*M*	Trip press
Button	*RL*	0.002	0.002 (End of cycle)	*RL*	Button

9

Methods Design: Search for Alternatives

The Decision as to WHERE to Use Man in the Process

In addition to the aids to creativity described in Chapter 5, the methods designer has at his disposal a collection of generalities to guide him in the generation of alternatives. Before introducing these principles in detail, it is desirable to review some of the steps and decisions that take place in the course of designing a complete production facility. Recall that the manufacturing engineer decides what must be done to produce the product. Then, the manufacturing and methods engineers together decide which of the necessary operations will be performed by machine, by man, and by man and machine combined. During this decision phase, portrayed in the accompanying diagram, these specialists are designating the functions the human

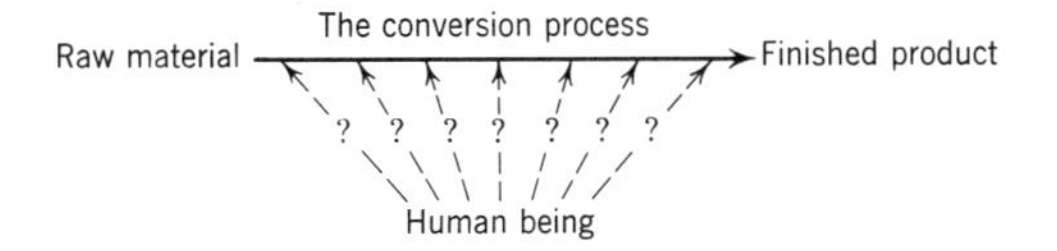

being will fulfill, the role he will serve directly as a transporter, measurer, manipulator, etc., and indirectly as a maintainer, trouble-shooter, monitor, and the like. Note that *man and machine compete for various tasks in the total manufacturing process.* There are some things that the man cannot do or does in a prohibitively inferior fashion, and likewise for the machine. To illustrate, relative to the machine, man cannot do the following.

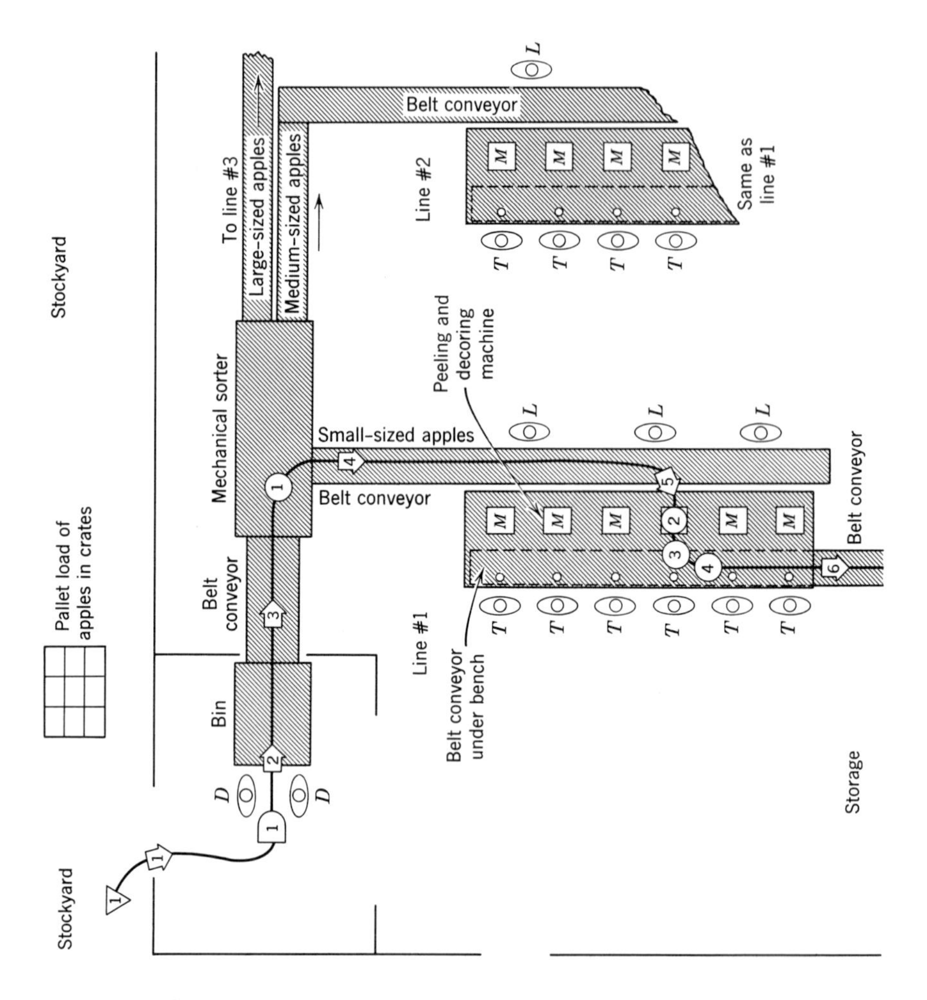

1. Exert large amounts of force, such as is needed in metal cutting and shaping.

2. Exert force smoothly or precisely, such as is needed in precision metal cutting.

3. Perform high-speed computation.

4. Perform repetitive, routine tasks without suffering certain side effects such as boredom, fatigue, and carelessness, which interfere with consistently effective performance.

5. Move at high speeds.

In contrast, the machine cannot, for example:

1. Learn, that is, profit from experience.

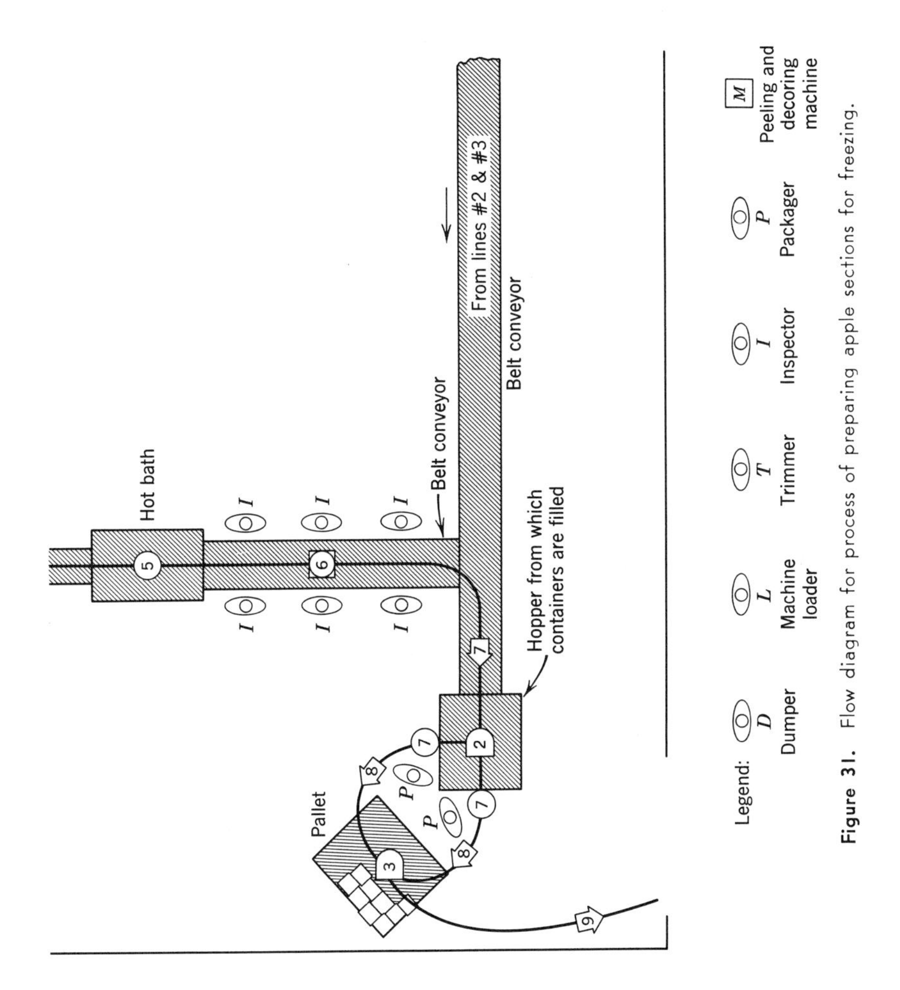

Figure 31. Flow diagram for process of preparing apple sections for freezing.

2. Generalize.
3. Cope with unexpected events.
4. Think creatively.
5. Learn of developments in its surroundings that it is not specifically constructed to sense.

Familiar illustrations of how man and machine have successfully bid for various tasks in competition with one another are to be found in the typical modern office. As a specific example, consider a plant engaged in the preparation of apple sections for freezing. Briefly, the apples are peeled, decored, sectioned, treated to retard browning, inspected, packaged, and then removed to the freezing plant. A flow diagram of this process is presented in Figure 31, a flow chart in

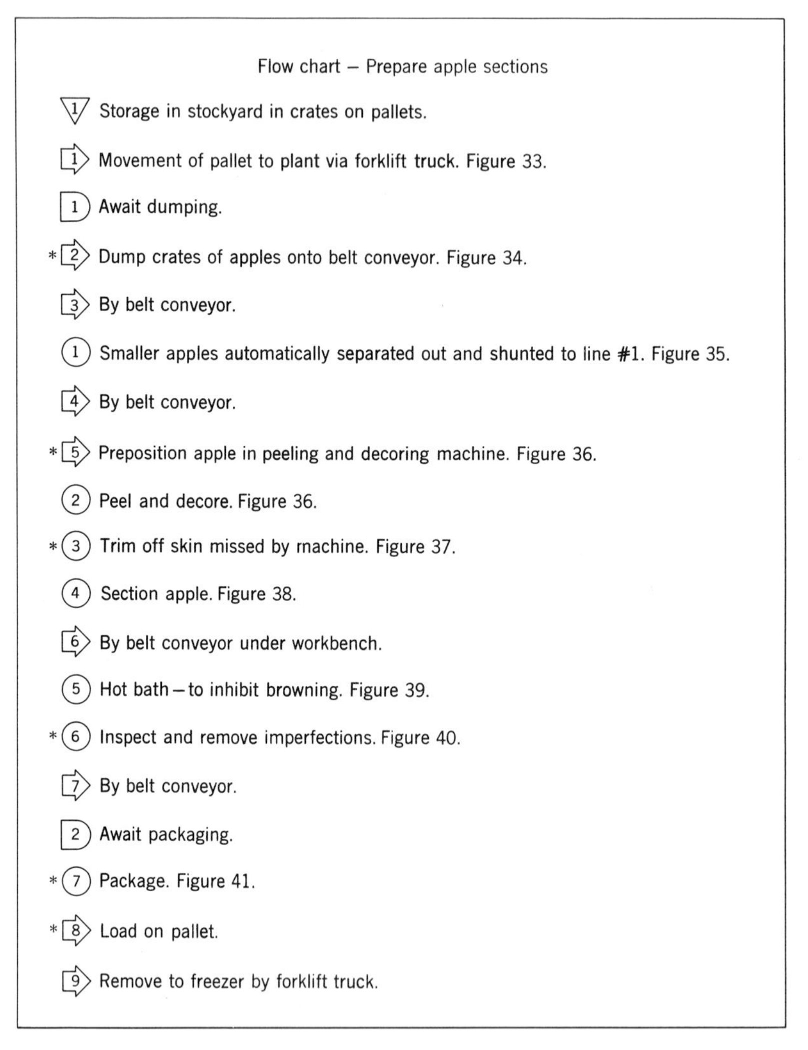

Figure 32. Flow chart for process of preparing apple sections for freezing.

Figure 32. The primary matter of interest here is the manner in which the human operator has successfully bid for some of the tasks, as indicated by the asterisks in Figure 32, and the machine for others. The transporting is primarily by machine, either by forklift truck or by belt conveyor. A machine grades the apples into three size classifications and discharges them on different belt conveyors which transport them to their respective processing lines. A rather ingenious

Figure 33. Forklift truck transporting apples from stockyard to dumping operation at start of process. (Courtesy of William E. McIntosh Company.)

Figure 34. Dumping apples onto start of belt conveyor system. (Courtesy of William E. McIntosh Company.)

Figure 35. Machine sorts apples into three size categories and distributes them on separate belt conveyors. (Courtesy of William E. McIntosh Company.)

Figure 36. Loader placing apples into "hand" of automatic peeling and decoring machine. (Courtesy of William E. McIntosh Company.)

Figure 37. Trimmer removing pieces of skin remaining after apple is discharged from peeling and decoring machine. Trimmed apple is dropped through hole in bench onto spindle of sectioning device pictured in Figure 38. (Courtesy of William E. McIntosh Company.)

Figure 38. Apparatus for mechanically sectioning apples. (Courtesy of William E. McIntosh Company.)

Figure 39. Machine that hot dips apple sections. (Courtesy of William E. McIntosh Company.)

Figure 40. Inspectors check for and remedy defects. (Courtesy of William E. McIntosh Company.)

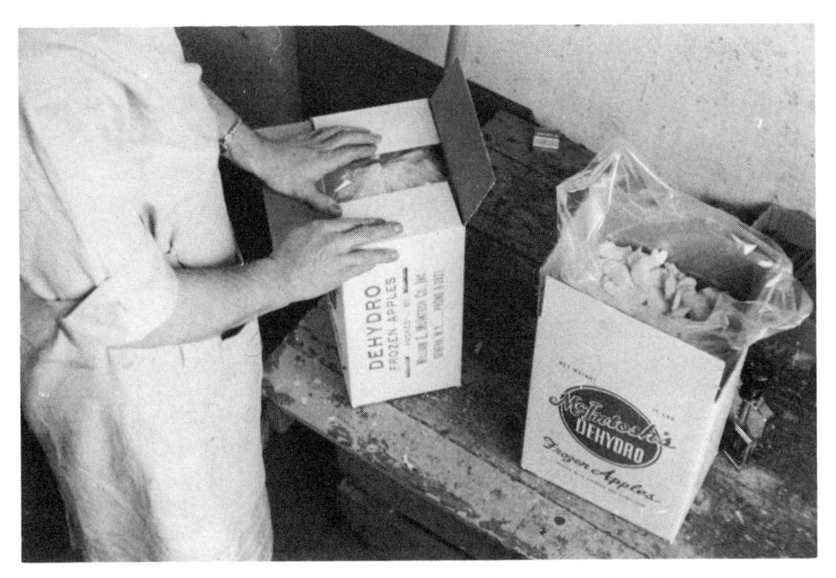

Figure 41. Packaging prepared apple sections. (Courtesy of William E. McIntosh Company.)

machine peels and decores the apples; another one automatically cuts the peeled apple into sections. A machine submits the apple sections to a hot bath and discharges them onto a belt conveyor.

Note that it is primarily variation or irregularity of one type or another that accounts for man's presence in this process. For example, the human is used to transfer apples from the conveyor and preposition them in the peeling and decoring machine, primarily because of the variation in size and especially shape of apples. Similarly, the human operator is used to detect and remove pieces of skin the machine misses as a result of variation in apple size and shape. Because of variation in frequency and location of imperfections, the human is used to detect and remove same. Because of variation in the types of containers used, man rather than machine does the packaging. In each case it is probably not impossible, only uneconomical, to develop a machine to perform the task that has been delegated to man.

In manufacturing in general, the relative roles of man and machine vary from the one extreme in which the entire process is manual, to the other extreme in which the process is completely automated and man is present only for monitoring, maintenance, troubleshooting, and the like. This spectrum of degrees of involvement of the human being is pictured in Figure 42 with several distinguishable levels of mechanization indicated.

Man is a readily available, extremely flexible tool, capable of an enormous number and variety of accomplishments with a period of training and practice that is often inexpensive compared to the cost of creating machines for the same purposes. It is primarily man's flexibility, adaptability, sensory and mental faculties, and relatively low procurement cost that establish him in a strong competitive position for the low, medium, and even many high production volume tasks. However, although the cost of developing, creating, and installing machines for many purposes is usually very high, the fact that they

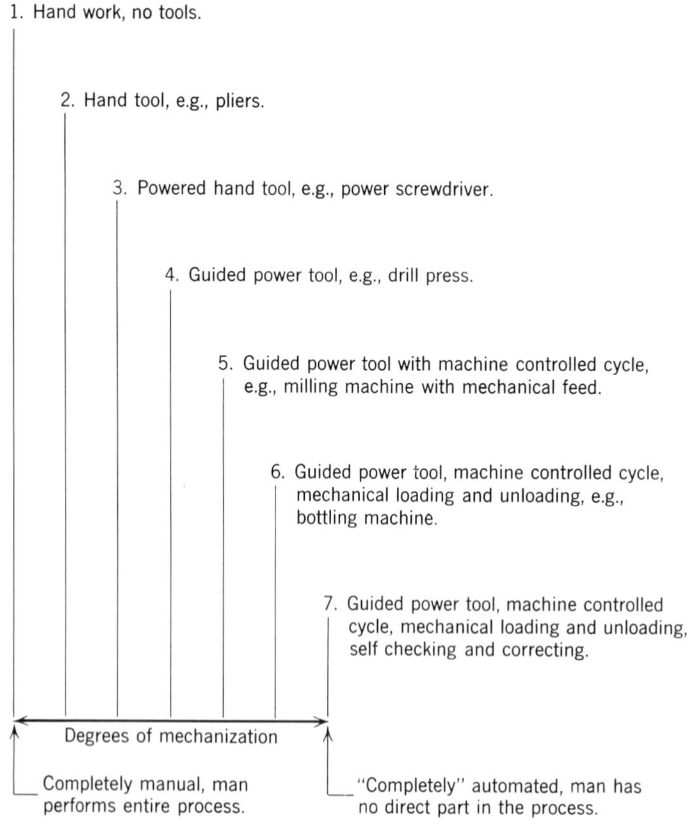

Figure 42. Range of degrees of mechanization along with examples found at various points along this scale.

can usually produce at higher efficiency, speed, consistency, and quality means that if volume of production is high enough, the higher initial cost of the machine is justified and machine replaces man.

Along with his remarkable capabilities, man also has some drastic limitations. Through history man has constantly strived to develop means of compensating for his various physical and mental limitations, and in fact a majority of the devices created by man are exactly that: means of compensating for one or more of man's physical or mental inadequacies. As a tool and weapon man's hands are extremely inadequate, which gave rise to prehistoric man's substitution of a stone for his bare hand for many tasks. Then to provide the leverage and force that man could not provide unaided, he added a handle to the stone to create the stone ax. Modern day examples of mechanisms and devices developed to compensate for various of man's physical and mental limitations are as follows:

1. Means of extending man's sensory faculties: radar, sonar, loudspeaker, telescope, thermometer.
2. Means of extending man's mental faculties: slide rule, desk calculator, electronic computer, magnetic tape memory.
3. Means of extending man's motor faculties: lever, hammer, vise, pneumatic and hydraulic cylinders, motor.

Each of the foregoing devices has been developed as a means of performing a function man cannot perform, or of performing a function man cannot perform adequately in view of the requirements involved.

Man's limitations with respect to various production tasks are quantitatively definable. With this information the designer can objectively select a manufacturing system that capitalizes on man's extraordinary capabilities and avoids or compensates for his limitations. Some of the basic tasks to be performed in the ordinary production process are presented in Table 4, along with some of the performance characteristics of these functions that are definable and measurable with respect to what can be expected of man. To be able to make an intelligent choice between man and machine for a given transporting task, positioning task, or any other type of task, the designer should have quantitative information available as to what he can reasonably expect from the human in these respects. Quantitative generalities of this type, defining the performance capacities of the human being, would considerably enhance the objectivity of the designer's choice between man and machine for various tasks. Although this need exists, quantitatively satisfactory generalities of

TABLE 4. Several Basic Production Tasks and Associated Performance Characteristics

Task	The Designer Should Have Quantitative Information Describing Man's Capabilities With Respect To:
Transporter	Speed—maximum and average Weight—maximum and average capacity Size—maximum and average capacity Effects of repetition
Positioner	Speed—maximum and average Weight—maximum and average capacity Size—maximum and average capacity Effects of repetition Flexibility Control or precision
Integrator (a medium for combining bits of information into a meaningful and useful whole)	Threshold Reaction time Variability Capacity Reliability
Memory	Capacity Span Reliability Access time
Communicator	Speed Variability Reliability Range

this type are scant, unpublicized, and fragmented. Those that are available, originating in recent years with psychologists and physiologists, are not as yet widely used in industrial practice. Instead, designers rely heavily on personal judgment and experience, a body of empirical knowledge that has developed over the years, and various special methods of synthesizing performance time and cost (for example, predetermined motion times, to be discussed later), in the process of deciding whether to use man or machine for a given task.

The military services have been giving very thorough consideration to the relative performance capabilities of man and machine. (And incidentally, in modern military systems, numerous situations are being encountered at an increasing rate, in which the human being cannot meet the more severe demands with respect to speed, reliability, and complexity. Man is rapidly disappearing as a direct component in the military system, so that our army of fighting men is gradually being replaced by an army of technicians and maintenance men.) As a result of this interest, a considerable volume of data relative to human performance capacities has been compiled for certain types of tasks, by psychologists and physiologists under sponsorship of the military services. This data is available to the public through a variety of sources,[1] but industry has been slow to assimilate it.

The Decision as to HOW to Use Man in the Process

The most effective means of using man in a task for which he is being considered should be known before an intelligent choice can be made between man and machine.[2] The need for this second type of decision, *how* best to use man for an operation, gives rise to a body of specialized knowledge consisting primarily of principles, which serves as a guide to the methods designer in his search for the best work methods. These generalities that guide the designer in integrating man into the production process pertain to preferred work *procedure,* preferred workstation *layout,* and preferred *equipment* design insofar as the human user is concerned. These three categories of principles are illustrated below.

Principles pertaining to work procedure. Principles pertaining to work procedure relate to such matters as distribution of work over hands and feet, preferred type and sequence of motions to use, preferred body member to use, and so forth. Following are examples of such principles.

1. The sequence of motions should facilitate learning and rhythm, and minimize the total number of motions required.
2. The work should be distributed as equally as possible over the two hands and two feet.

[1] See for example Ernest J. McCormick, *Human Engineering,* McGraw-Hill Book Company, New York, 1957; or *Handbook of Human Engineering Data,* 2nd ed., Tufts College, Medford, Massachusetts, 1952; or W. E. Woodson, *Human Engineering Guide for Equipment Designers,* University of California Press, Berkeley, California, 1954.

[2] In practice these two decisions: *what* tasks the man will perform, then *how* he should perform them, seem in general to be made successively. Optimally, a decision as to whether or not man will perform a given function should not be made until it is decided *how* he can best fulfill that function.

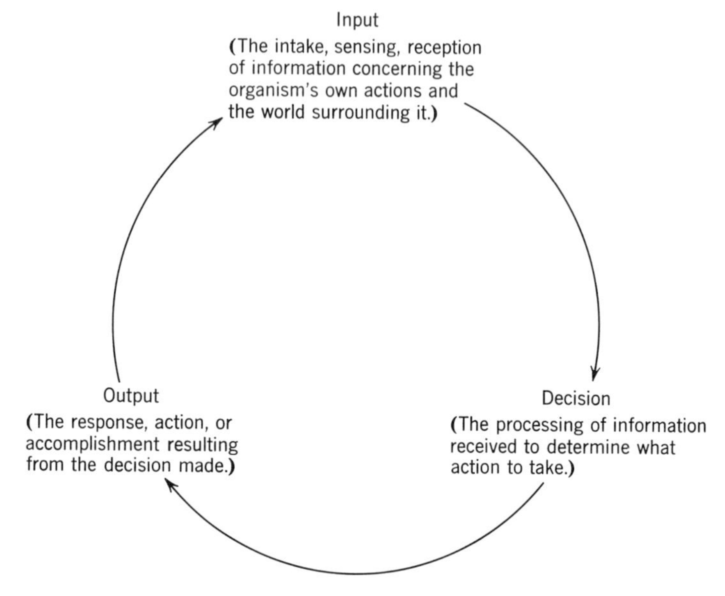

Figure 43. The basic control cycle as employed by the organism.

3. Relatively uncontrolled motions, such as those required for simple either/or functions, for example, on/off, open/close, should be accomplished by foot or leg action whenever possible.

4. Transports should be performed by natural forearm motion, with a minimum of upper arm movement.

5. "Drop disposal" should be used whenever possible.

6. Smooth, curved, ballistic motions should be used rather than stiff, constrained, sharp-angled ones.

The basic input-decision-output cycle. Before introducing principles concerning equipment ("hardware") and layout, it is desirable to recall the basic control cycle, diagrammed in Figure 43. The ability of a man or machine or combination of the two to satisfactorily and consistently achieve a goal is based on a continuance of the input-decision-output cycle. The application of this cycle to situations of particular concern to the methods engineer is illustrated in Figure 44, as well as by such familiar tasks as steering an automobile, reaching toward an object, walking to a particular destination, and so on.

Notice that in each of these illustrations, man comes in contact

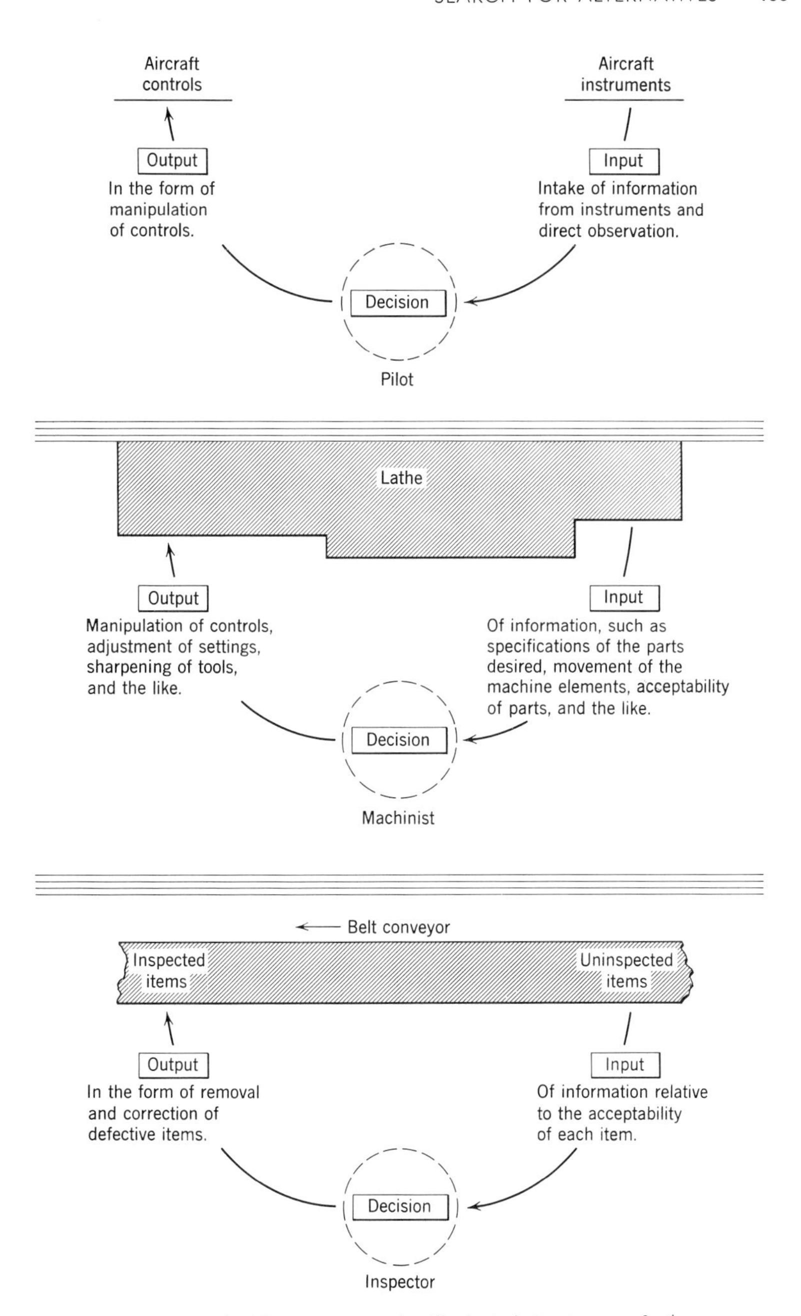

Figure 44. Basic input-decision-output cycle illustrated in terms of three common operations.

Figure 45. Standard knob shapes for aircraft controls.

with the situation he is dealing with at two points, namely:

1. The point at which he receives his information, for example, the instrument panel, the micrometer, the movement of machine elements; and
2. The point at which he affects the situation, as through controls.

The former is by means of sensors (vision, feel, etc.), the latter is by means of affectors (body limbs). The methods designer is primarily concerned with these two points at which the man comes in contact with the work situation. The designer is concerned mainly with the manner in which man receives information about the situation he is dealing with and with the manner in which he affects it. As a result of the fundamental difference between the "input" and "output" phases of behavior, the following principles pertaining to equipment and layout are classified accordingly.

Principles pertaining to equipment design. The following are principles relating to the *input* (observation, sensing, or reception) of information concerning the situation with which man is dealing.

1. *Use color, shape, or size coding to maximize speed and minimize errors in finding controls.* An application of this principle is illustrated in Figures 45 and 46. Figure 45 shows standard knob shapes developed by the Psychology Branch, Aero Medical Laboratory, United States Air Force, to assist in reducing blunders and location time in the operation of aircraft. Figure 46 pictures an application of these shape-coded controls in a commercial plane.

2. *Pre-position tools and materials whenever possible.* Figure 47 offers an illustration of a common method of pre-positioning portable tools at the workstation. A setup for pre-positioning materials is illustrated in Figure 48.

3. *Use simple on/off, either/or indicators whenever possible.* Examples of the on/off type of visual indicator are the high beam indicator on the automobile dashboard, the pilot light on the electric iron, radio,

Figure 46. Use of shape-coded controls in a commercial aircraft. (Courtesy of Douglas Aircraft Company.)

Figure 47. Pre-positioning device supports tool within convenient reach, renders tool almost weightless, and automatically lifts tool out of way when released by operator. (Courtesy of Gardner-Denver Company, Grand Haven Division.)

or electric coffee maker, the night alarm outside the bank, the alarm clock buzzer, and so on. In each such case the indicator offers information on two states of affairs: on/off, satisfactory/unsatisfactory, absent/present, emergency/no emergency. Whenever this type of information is adequate for the purposes at hand, this simplest type of indicator should be employed. Under these circumstances a more elaborate visual display, probably providing more information than is really necessary, is likely to require more time for interpretation and be more susceptible to misinterpretation than the on/off type. When constant or nearly constant visual alertness is not possible or desirable, an auditory on/off indicator should ordinarily be used.

4. *If a simple on/off indicator will not suffice, use a qualitative type indicator if adequate.* The qualitative indicator offers information on more than two but still a limited number of discrete conditions

or states, as illustrated by the inspection device indicator pictured below. Examples from the automobile dashboard instrument cluster

are the directional signal lights that indicate when the signal is on and whether it is the left or right signal that is flashing, or the temperature indicator which informs the operator of an abnormal engine temperature and whether the abnormality is in the hot or cool direction. When this type of information is necessary yet adequate for the purposes at hand, a qualitative rather than a quantitative indicator should ordinarily be employed for interpretation speed and error reasons.

5. *Present continuous quantitative information in lieu of the on/off*

Figure 48. Setup for pre-positioning materials for rapid and convenient assembly. (Courtesy of Alden Systems Company.)

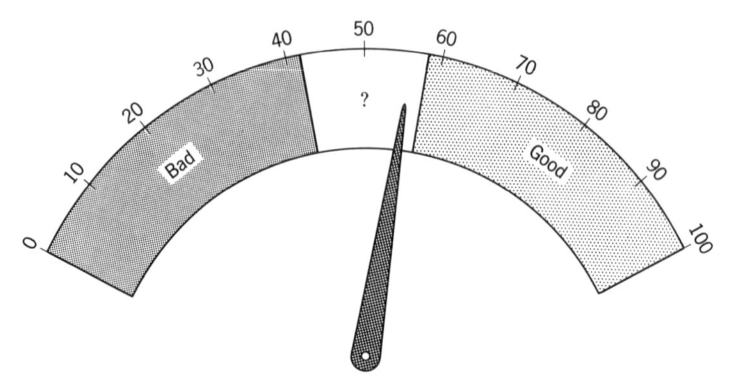

Figure 49. Combination qualitative and quantitative visual display, a promising possibility for many continuous quantitative information displays.

and qualitative types only when essential. The ordinary thermometer, speedometer, and radio dial, are examples of indicators presenting continuous quantitative information. Obtaining information from this type of indicator requires considerably more time and is much more susceptible to error than are the simpler types of information displays. If quantitative information really is required for the purposes at hand, of course it should be provided. Numerous experimentally based suggestions as to how best to present continuous quantitative information, such as the proper design of dials, are available in handbook and textbook form.[3] However, there is an inclination to call for a continuous quantitative information display when it is not required, thus making the task of interpretation more difficult and erratic.

Of course, it is possible to combine these types of indicators in such a way as to permit their being used according to the needs of the occasion. For example, the visual displays on electron tube testers are usually similar to that pictured in Figure 49. The continuous quantitative information is provided if needed, but ordinarily the user can rely on the rapidly and easily interpreted qualitative display, that is, bad—questionable—good, superimposed on the quantitative scale. This idea could be beneficially applied to many common types of quantitative indicators.

[3] See for example, *Handbook of Human Engineering Data,* 2nd ed., Tufts College, Medford, Massachusetts, 1952; or Ernest J. McCormick, *Human Engineering,* McGraw-Hill Book Company, New York, 1957; or W. E. Woodson, *Human Engineering Guide for Equipment Designers,* University of California Press, Berkeley, California, 1954.

6. *Provide adequate glare-free illumination.* Handbooks[4] provide suggestions on adequate amounts of illumination for different types of situations. Prevention of glare depends primarily on alertness of the designer in anticipating glare-producing situations.

The following principles pertaining to equipment design relate to the manner in which man affects (responds to) the situation with which he is dealing.

7. *Place materials in lipped bins or trays to facilitate locating and grasping of parts.* Figure 50 illustrates a common type of materials supply bin used to maintain a moderate stock of parts close to the point at which they are to be used, in a form quickly and easily grasped. A variety of designs and sizes of bins are commercially available.

8. *Utilize a quick-acting vise, clamp, collet, or fixture to support material upon which work is being performed.* There are a number of means of securing a part or assembly in the desired working position by a simple motion of the hand or foot. A common form of quick-acting work-holding device is the air operated vise exemplified in Figure 51. If possible these work-holding devices should

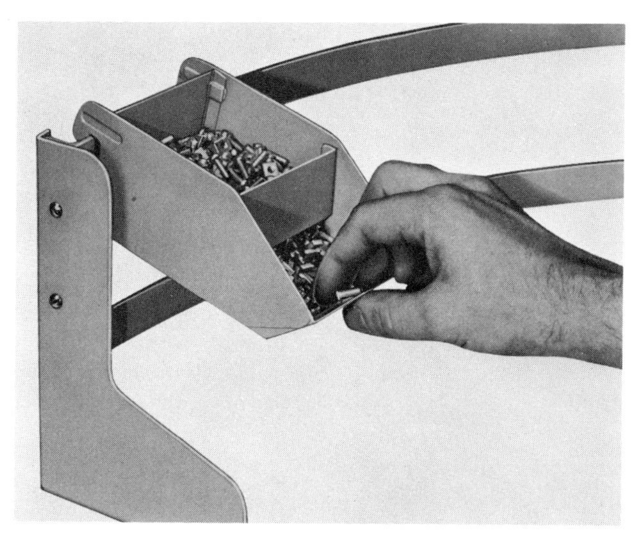

Figure 50. Lipped bin permits quick and easy sliding grasp of part. (Courtesy of the Bathey Manufacturing Company.)

[4] *Ibid.*

should be leg or foot controlled. Other quick-acting devices rely on the cam, the spring, hydraulic pressure, magnetism, or various special and sometimes rather ingenious mechanisms for rapidly securing and releasing the material and holding it in a convenient position to be worked on.

9. *Use stops and guides to reduce the control necessary in positioning motions.* An application of this principle is illustrated in Figure 52. The same principle is used in the typewriter, where there is an adjustable stop that greatly reduces the time necessary to return the carriage to the beginning of the next line.

10. *When adequate, controls should be of the simple on/off, either/or type.* Everyday illustrations of this type of control are not difficult to find. The two-way electric switch is the most common. If this type of control will serve the purpose at hand it should be used in preference to anything more elaborate. On occasion, several on/off type controls can be used to replace a more elaborate single control, with resulting increase in simplicity and decrease in operation time.

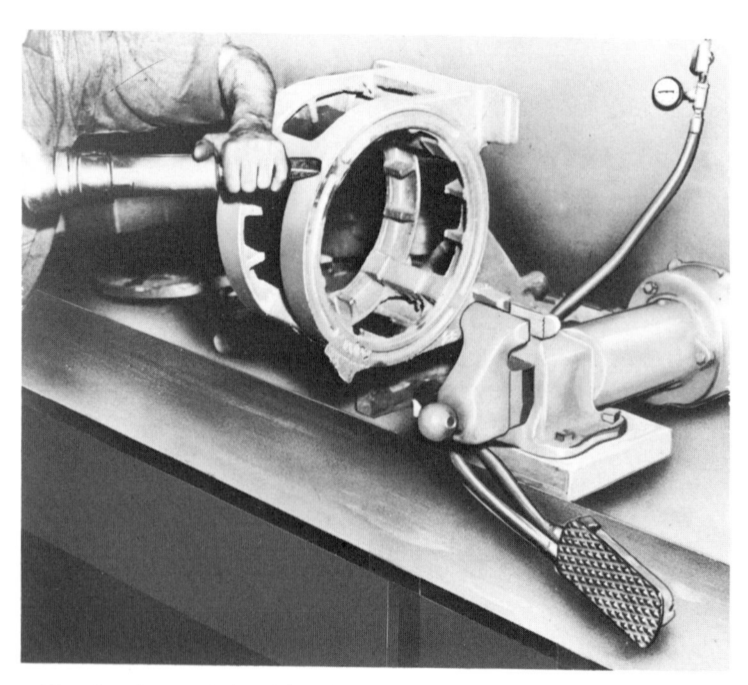

Figure 51. An air-operated quick-acting vise, set up for operation by the operator's knee. (Courtesy of the Van Products Company.)

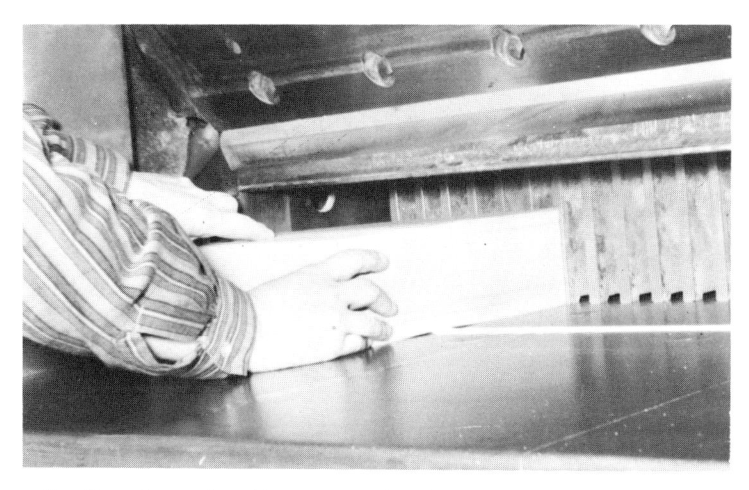

Figure 52. Use of an adjustable stop greatly reduces time required to position stack of paper in a paper-cutting machine. (Courtesy of F. Ellsworth Miller.)

11. *If an on/off type of control will not suffice, use a control with a limited number of discrete settings if adequate.* A familiar example is the channel selector switch on a television set. The device pictured in Figure 53 employs a selector-type control, (output selector, lower left).

12. *Use a control requiring continuous adjustment only when necessary.* The station selector knob on the ordinary radio offers continuous adjustment. The device in Figure 53 employs a continuous type of control (frequency, upper left). Making a setting with reasonable accuracy and precision with this type of control requires at least several-fold the time that would be required using a selector switch. If extreme care is required, the difference in time is considerably greater than this. Whenever the flexibility offered by a continuous control is really essential, it should of course be employed. However, the ability to make an infinite or even large number of different settings is often not essential, and a selector-type control would be adequate. The push-button radio station selector is an illustration of the use of a selector-type control to replace or supplement a continuous control. If the ability to make continuous settings is important but rarely needed, both types of controls can be provided, offering the major advantages of both: speed and flexibility.

Textbooks and handbooks[5] offer extensive amounts of information

[5] *Ibid.*

Figure 53. A device employing the three basic types of controls: on/off, selector, and continuous adjustment. (Courtesy of the Kay Electric Company.)

on the preferred design features of controls, especially of the continuous type. Suggestions deal with coding, size, shape, markings, control ratios, and other design features of controls.

13. *Direction of motion of switches, levers, handwheels, knobs, and other controls should conform to stereotyped reactions.* As a result of early acquired habits, the natural response for most people is to push a switch up if they want the phenomenon controlled by the switch to begin, to turn a knob or handle clockwise if they want the phenomenon to increase, as in the case of the volume control on a radio. Natural responses like these are referred to as stereotyped reactions. A control that calls for a response that is contrary to the stereotyped reaction for that situation is likely to result in considerable inconvenience, additional time, and perhaps danger. Often, through a little forethought, a designer can anticipate the manner in which most people will react to a given control situation. In other cases he will find it necessary to resort to handbooks or experimentation to determine the stereotyped reaction.

Principles pertaining to layout. Principles pertaining to arrangement of facilities and materials, as it affects the input of information, are exemplified by the following.

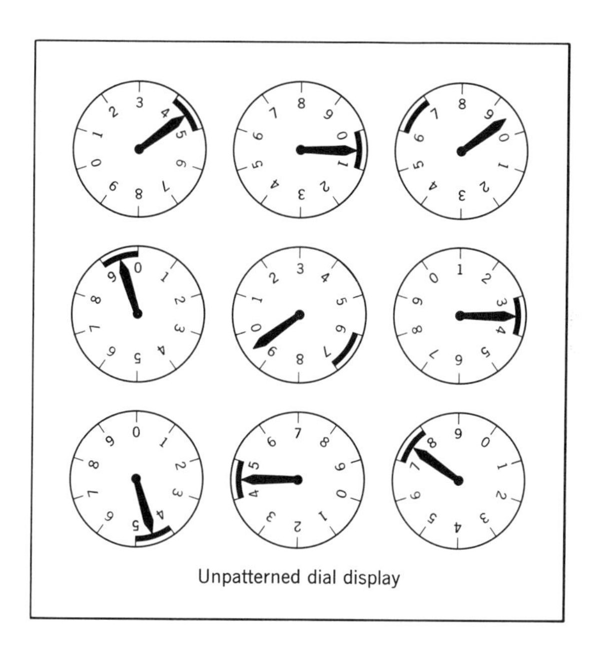

Figure 54. Comparison of an unpatterned and patterned arrangement of the same set of indicators. (Reproduced with permission from Chapanis, Garner, and Morgan, Applied Experimental Psychology, 1949, John Wiley and Sons.)

Figure 55. A case in which the operator has been provided with excellent visibility of the situation he is controlling. The Warner and Swasey Company's Gradall. (Courtesy of the Warner and Swasey Company.)

1. *Dials and other indicators should be patterned so that a maximum of information may be obtained with a minimum of time and error.* An example of the effect of proper patterning is demonstrated in Figure 54. The effects of patterning on interpretation time and error are significant.

2. *The physical relationship between operator and equipment should permit maximum visibility of the situation.* An example in which the operator has been provided with excellent visibility of the situation he is controlling is pictured in Figure 55. Two familiar situations in which designers have *not* yet been able to provide satisfactory operator

visibility is the common bulldozer and the forklift truck. In the latter it is often difficult for the operator to see the position of the fork tips in relation to the load he is attempting to pick up, and while in transit it is usually difficult for the operator to see what is before him because the load rather effectively blocks his view.

3. *Operating, setup (tuning), and emergency controls should be grouped according to function.* The operating controls of a television set are ordinarily grouped together and placed at a very convenient location on the set. The tuning (adjusting) controls are also grouped, but since they are very seldom used accessibility is not critical. In general, it is desirable to separate operating from emergency controls and to place both groups at locations very convenient to the operator. Setup controls are used to make settings and adjustments that hold for prolonged periods of operation and therefore have a much lower priority as to location. Whether or not setup controls are grouped is not always so important, but it is important, however, that they not be placed in and therefore dilute the operating and emergency control concentrations.

4. *Terminal points should be as close together as possible when visually controlled motions are to be performed simultaneously.* Assume that a workplace has been set up as shown in Figure 56*a* for assembly of four component parts A, B, C, and D. These parts are small and intricate so that the operator must look at each bin of parts before and while grasping a part from it. The intended procedure is indicated in the left-hand right-hand chart shown in Figure 56*b*. Notice that parts A and D are to be reached to and grasped simultaneously, and likewise for parts B and C. Locating these parts away from the "twelve o'clock" position increases eye travel time, head movement, and performance time. When two motions to be performed by left and right hands conflict for attention of the eyes, the distance between the location of these motions should be kept to a minimum. An improved layout is shown in Figure 56*c*. In this arrangement the locations of parts A and D have been moved to the "twelve o'clock" position as have parts B and C, but in a second tier of bins several inches above the B and C bins. In the original arrangement the least demanding motions insofar as visual direction is concerned, dropping the finished assemblies into the disposal chute, were located at the most valuable spot at the workplace as far as visual attention is concerned.

Principles pertaining to layout as it affects response, are exemplified by the following.

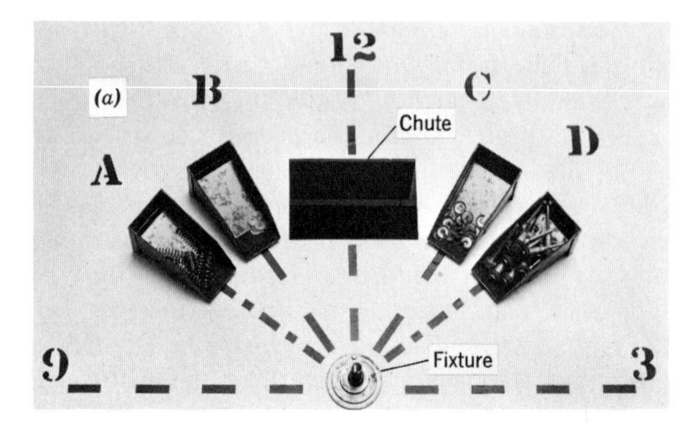

(b) Left-hand right-hand chart

LH			RH
Part A	*G**	*G**	Part D
Part A	*P*	*P*	Part D
Part A	*A*	*A*	Part D
Part B	*G**	*G**	Part C
Part B	*P*	*P*	Part C
Part B	*A*	*A*	Part C
Assembly down chute	*P*	*D*	Wait for LH

* Visual attention required simultaneously by left- and right-hand grasps.

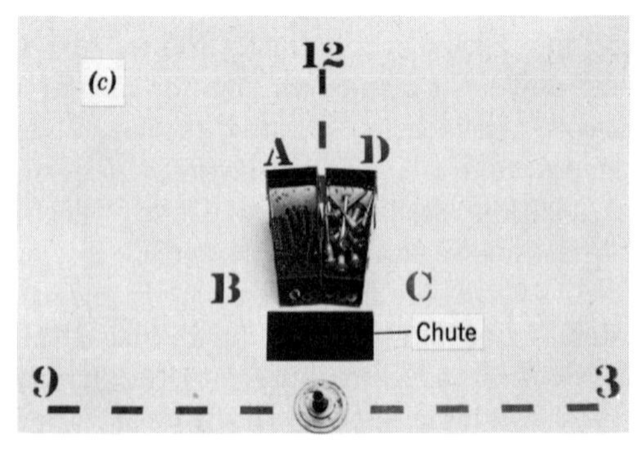

Figure 56. Undesirable (*a*) and preferred (*c*) workplace layouts for the operation described in (*b*), in which motions conflict for visual attention.

5. *Tools, materials, and controls should be located as close to the point of use as possible.* Two work zones traditionally referred to as the normal work area and the maximum work area are indicated in Figure 57. To reach an object beyond the limit of the maximum work area ordinarily requires a major change in posture, repetition of which would be very fatiguing. To reach an object in the maximum work area (lightly shaded area of Figure 57) requires considerable upper-arm movement and energy expenditure. Thus, when repetition is more than very infrequent, materials, tools, and controls should be located within the normal work area for the hand involved.

It is important to note that the outer boundary of the normal work area *is a limit.* It is the furthest that materials and the like should be located from the operator in order to be within normal and convenient reach. Oddly enough, many practitioners feel obliged to locate materials along the limit of this area. Why place materials out along this limit and arbitrarily and unnecessarily extend all transport motions?

Another unusual and undesirable practice is that of laying out bins of materials around one unnecessarily large semicircle as illustrated in Figure 58*a*, which in fact does not conform to the two arcs of the normal work area. An arrangement like this seems almost the rule rather than the exception. Note the relatively large wasted space. Tools, materials, and controls should be in as close to the point of use as is possible without causing congestion and interference, as demonstrated in Figure 58*b*.

6. *When feasible, motions of the left and right hands should be simultaneous and in symmetrically opposite directions.* This principle gives rise to the rather frequently encountered "symmetrical workplace" illustrated in Figure 59. Left- and right-hand motions are to begin and end as close to simultaneously as possible and are to be symmetrical about a vertical plane through the center of the workplace and perpendicular to the plane of the operator. Ordinarily this type of motion pattern involves a dual fixture and occasionally, duplicate tools and other accessories. However, a dual fixture does not necessarily follow if this type of motion pattern is employed. In the bimanual task illustrated in Figure 56, the motions are predominantly simultaneous and symmetrical about the vertical "twelve o'clock plane," but the parts handled by left and right hands are different. Thus a single fixture sufficed.

7. *Tools and controls should be placed so as to maximize speed and ease of location and minimize error.* A classical illustration of the need for and elaborate attempts to apply this principle is the cockpit

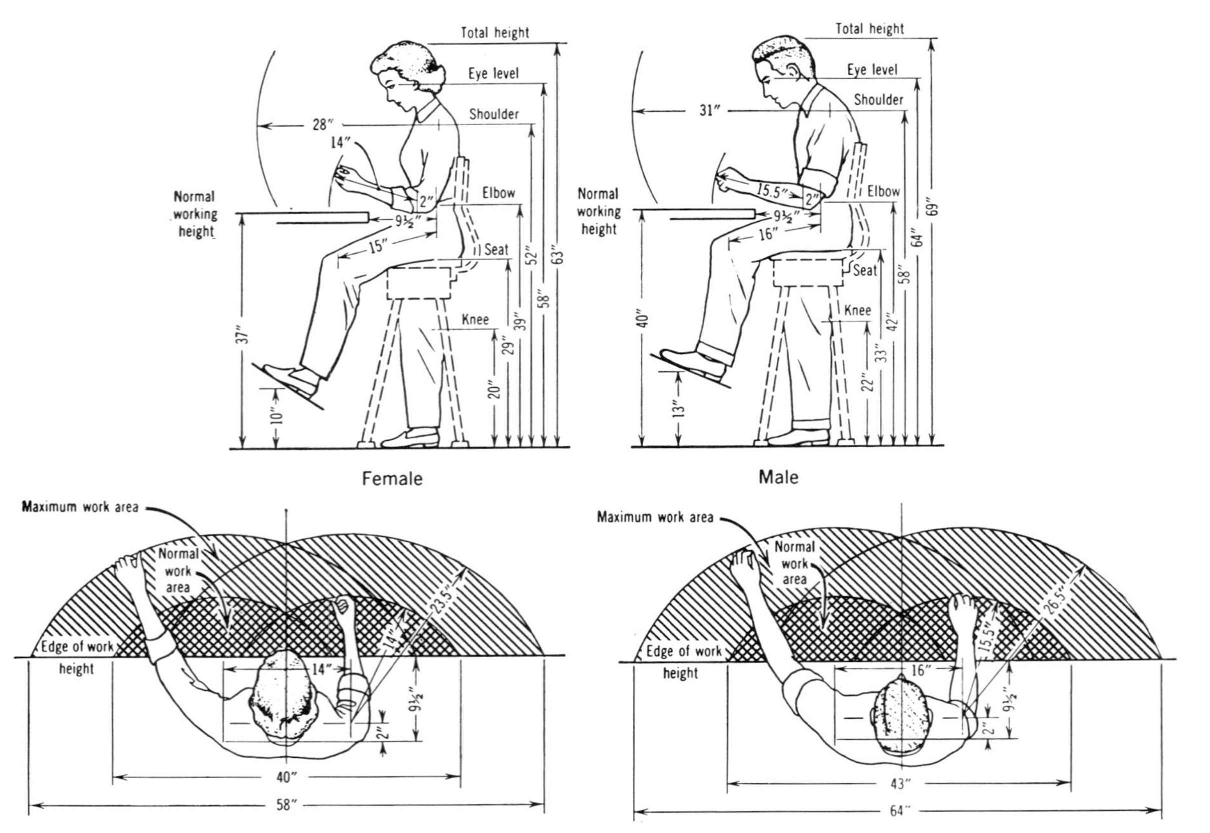

Figure 57. Normal (dark shade) and maximum (light shade) work areas in vertical and horizontal planes based on average male and female dimensions. (From R. M. Barnes, Motion and Time Study, John Wiley and Sons, New York, 1958.)

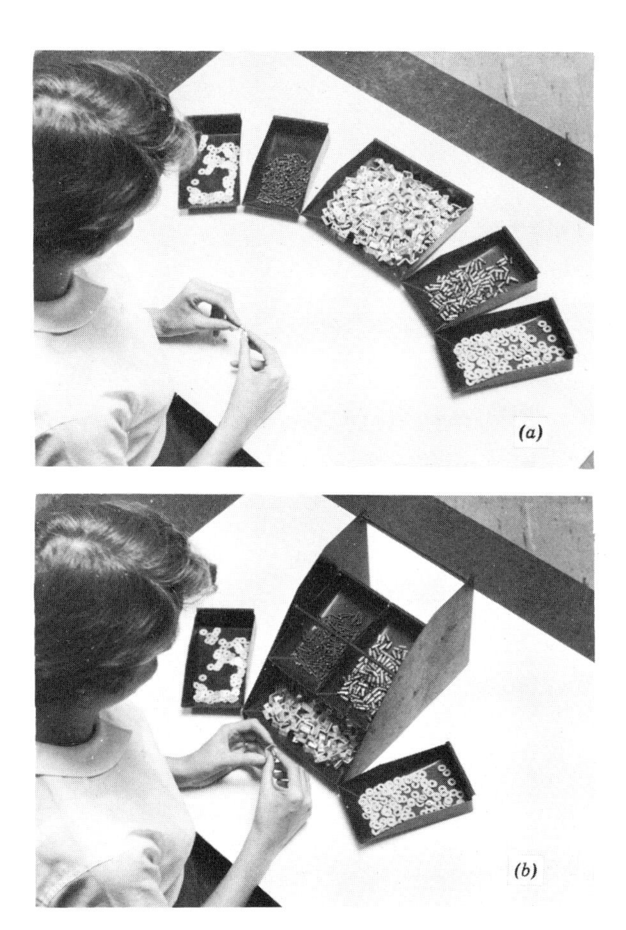

Figure 58. Common but unnecessarily spreadout arrangement of bins of materials at the workplace, (*a*), and the preferred locations, (*b*).

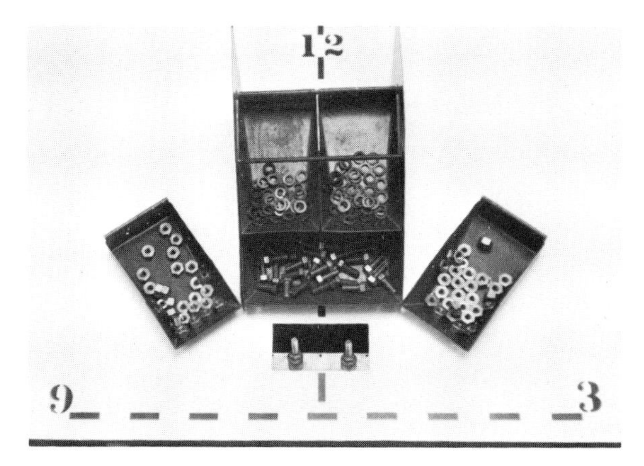

Figure 59. A workplace setup for a simultaneous and symmetrically opposite motion pattern in the assembly of nuts, bolts, and washers.

Figure 60a. "Tools and controls should be placed so as to maximize speed and ease of location and minimize error." This illustration shows the compact arrangement of controls at the headstock end of a Warner and Swasey turret lathe. (Courtesy of the Warner and Swasey Company.)

design of an aircraft. Here speed and a minimum of errors are critical. Another familiar situation in which design and location of controls are a critical problem given considerable attention is the console of a large organ.

Figure 60*a* pictures a machine tool well designed from the point of view of design and placement of controls. Operating controls are concentrated in a relatively compact area within easy reach of the operator, as indicated by the *cyclegraph* shown in Figure 60*b*. The ability to quickly and accurately locate a control without having to

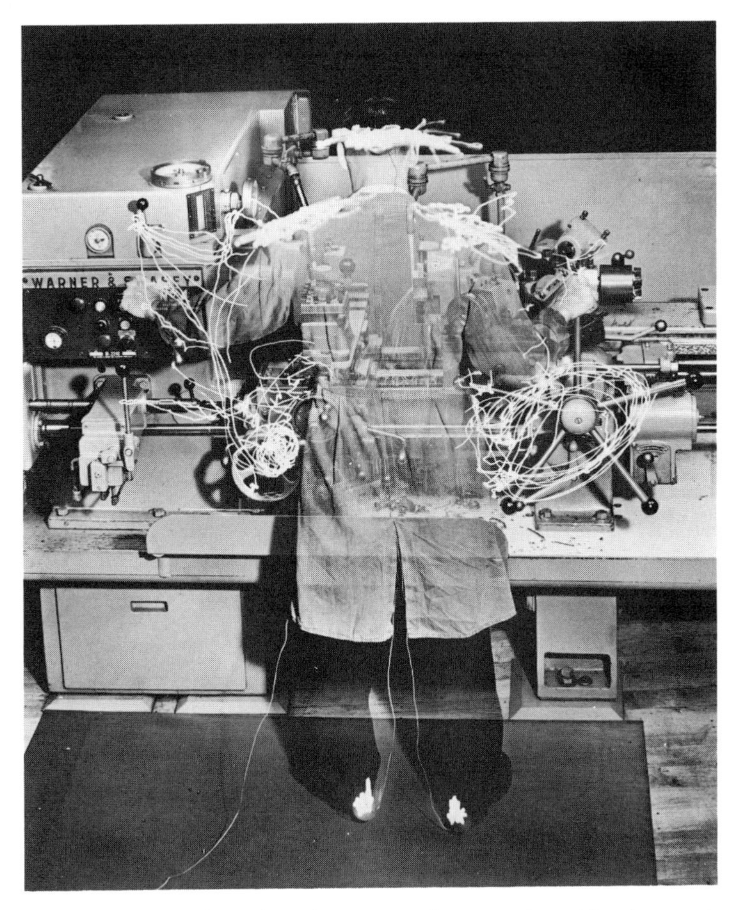

Figure 60*b*. A cyclegraph used to demonstrate the ease of reaching important operating controls on the turret lathe shown in Figure 60a. (Courtesy of the Warner and Swasey Company.)

search for it, since the operator should have his eyes on the point at which metal cutting is taking place, is vital in operation of machine tools of this nature.

Applicability of the Foregoing Principles

Few of these principles are appropriate in every situation they might logically be considered for. To illustrate, when a bimanual motion pattern is possible and desirable, it is not always best or even sensible that the motions begin and end simultaneously or that they be symmetrically opposite. There is a possibility of misapplication of most of the foregoing principles because they are not truly generalities. However, for practical purposes it is not desirable to withhold such "generalities" simply because they are not meaningful in *every* case to which they might be applied. On the other hand, such statements should not be published as generalities if in a majority of applications they will have zero or negative effect. A published list of such "principles" should represent a practical compromise between unrealistic restraint and unjustifiable generalization.

Another desirable feature of an aggregation of principles such as this is the omission of rarely useful principles to avoid dilution of the list. Although this may be achieved only after prolonged experience with a set of principles, a majority of currently published lists have apparently not benefited in this respect after decades of existence.

Sources of the Foregoing Principles

The collection of principles presented derives from two main sources. One is a body of knowledge traditional to the field of motion and time study, commonly referred to as the principles of motion economy. Many of these principles were originated by Frank and Lillian Gilbreth, later modified and supplemented by Barnes[6] and others. The principles of motion economy, commonly quoted or paraphrased in motion and time study textbooks, call for the use of pre-positioned tools, drop disposal, use of lipped bins, simultaneous and symmetrically opposite motion pattern, and the like. The principles of motion economy are predominantly empirical in nature, based on common sense, and supported by years of application in practice.

The second major source from which these principles derive is an extensive body of knowledge established and applied by psychologists and physiologists, pertaining to the human being and his integration into a "man-machine" system. This knowledge is derived and applied by a rather recent and rapidly growing specialty popularly referred to as human engineering. This specialty originated during the second world war, when the military realized that its equipment and systems were placing heavy demands on the human operator insofar as complexity, speed, and precision were concerned, and assigned psychologists to assist in the design of the human's role in the military system and of equipment with which he was associated. These experts assisted then, and still do, in the design of aircraft cockpits,

[6] R. M. Barnes, *Motion and Time Study,* 4th ed., John Wiley and Sons, New York, 1958.

radar and sonar devices, ship controls, and the like. Although the human engineer concentrated initially on the human being's role in the military system, his interest and activity since the war has expanded into design of industrial systems and design of producer and consumer goods. Note that the human engineer deals with the same basic problem as the methods engineer—fitting man into a system. However, it so happens that the human engineer has until recently concentrated on military problems whereas the methods engineer has been serving a similar role with respect to industrial problems.

A typical nonmilitary human engineering project is illustrated in Figure 61. Human engineers undertook redesign of the cab of the large dragline, Figure 61*a*. The cab (upper left of main unit) before it was relocated was in a position that restricted the operator's visibility of the field of operation. The new location shown offers excellent visibility of the situation the operator is controlling. The inside of the cab, Figure 61*b*, is well designed from the point of view of location, type, and compactness of controls. Most maneuvers of the shovel are controlled by two "joy sticks," one held in each hand. The layout of the control consoles is shown in Figure 61*c*.

There is a second major distinguishing feature between the fields of human engineering and methods engineering, and that is the quality of the body of knowledge applied by each. Whereas the generalities developed and applied by methods engineers are relatively scant, empirical, and in some cases of questionable validity, those derived and applied by human engineers are relatively extensive and have been experimentally established by scientific investigation. It behooves the field of methods engineering to assimilate the knowledge made available by human engineering and to begin to apply it in practice along with its traditional principles of motion economy. With such an integration in mind, the principles presented are an aggregation of some of the more useful principles from each of the two main sources. Heretofore these have been unjustifiably treated as separate bodies of knowledge. Only a relatively small sample of the wealth of available human engineering data has been included. Detailed recommendations are available in textbooks and handbooks.[7]

Importance of the Foregoing Principles

In addition to being directly useful to an engineer involved in methods design work, these principles are potentially useful in a

[7] Woodson; McCormick; Tufts Handbook; *op. cit.*

Figure 61*a*. Dragline used in strip mining, showing control cab (upper left of power unit) redesigned by human engineers. (Courtesy of International Minerals and Chemical Corporation and Dunlap and Associates.)

Figure 61*b*. Inside of dragline cab showing controls and their relationship to the operator. Operator's hands rest on the "joy stick" controls with which he controls all movements of the bucket and boom. (Jay B. Leviton for Fortune.)

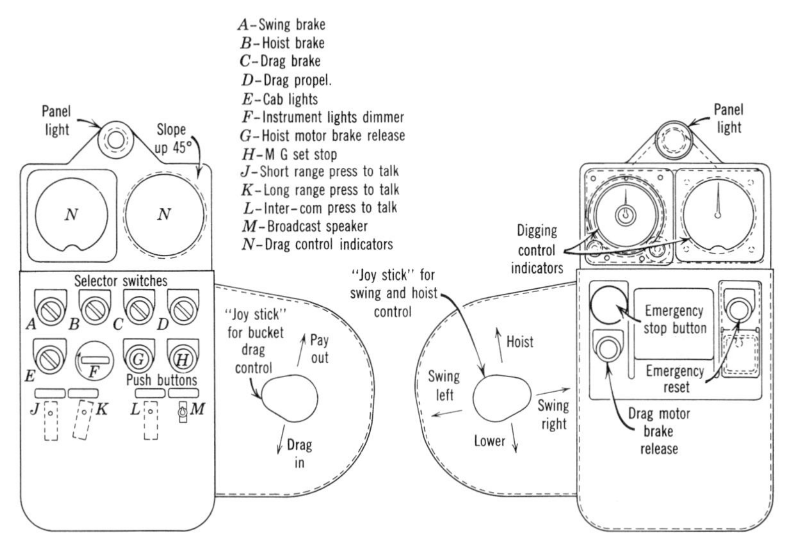

Figure 61c. Layout of left- and right-hand consoles from which activities of the dragline are controlled. Note compactness and simplicity of the controls. (From "Revolution in Control," Engineering and Mining Journal, April 1954. By permission of the publisher.)

variety of additional ways. For example, some readers will eventually, as supervisors or managers, be directly or indirectly responsible for overseeing the methods engineering function and will find it desirable to be able to intelligently evaluate the results being achieved by such a department. For this purpose it is necessary that you be able to discriminate between good and poor work methods. To do this some familiarity with the foregoing principles is necessary.

Many readers will become designers of producer, consumer, and military goods that in some manner directly involve the human being as a user and servicer. In any instance in which man must directly use or service a product, the designer of same should make extensive use of these principles and the philosophy underlying. To illustrate the need for careful consideration of the human as a user and servicer in product design, take the case of consumer goods design. Some typical "living testimonials" of what can result through neglect of this important consideration are shown in Figure 62. Numerous other examples can be found around the household. Apparently the designer often loses sight of the fact that ultimately a human being will attempt to use and to service the product. The designer must

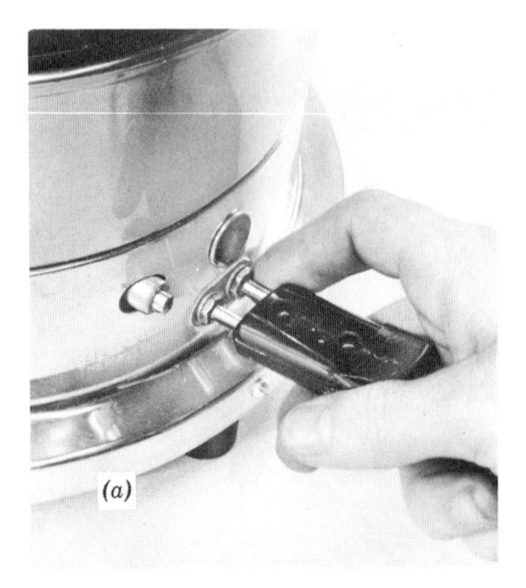

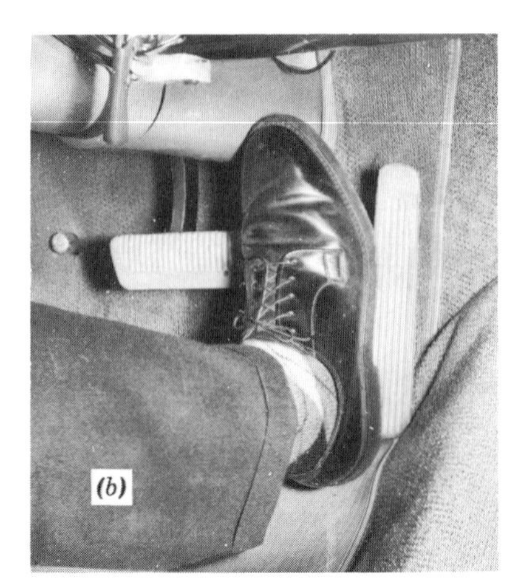

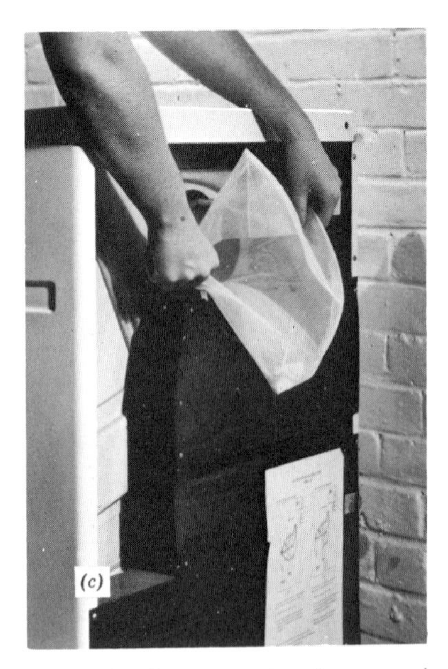

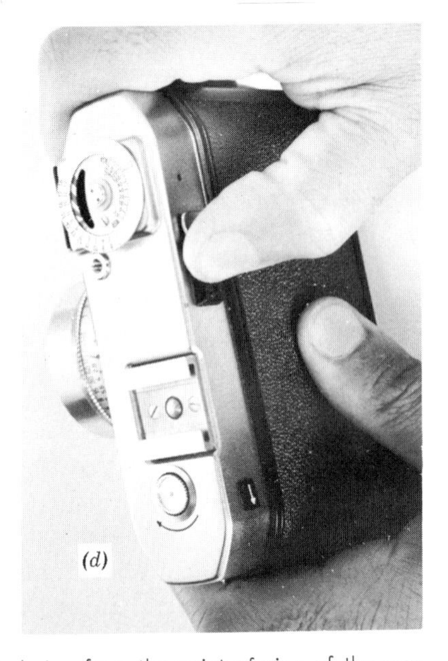

Figure 62. Some examples of poor product design from the point of view of the user. In (*a*) a shock hazard has resulted from failure to provide protective shielding around the terminal pins. In (*b*) the steering column offers a potential obstruction to movement of the driver's foot from accelerator to brake. In (*c*) the lint trap of an automatic drier is inconveniently located at the rear of the machine. Location of the film advance lever on the camera pictured in (*d*) requires the user to disturb his position and place thumb over the viewer in order to index the film. (From Consumer Reports, Consumers Union of United States, Mount Vernon, N. Y. By permission of the publisher.)

Figure 63. The main control room for a large wind tunnel. The designer of a complex control console such as this must give careful consideration to the means with which the operators will receive information about the system and the manner in which they can respond. (From the Westinghouse Engineer, March 1960. By permission of the publisher and the United States Air Force.)

devote attention to this matter if blunders are to be avoided and his product is to be capable of effective use by, and safe for, the consumer. Furthermore, as the complexity of modern equipment and systems increases, a trend that is especially noticeable in the producer and military equipment areas, the designer is forced to devote considerably more attention to the human's capabilities and to the human as a component, often a limiting one, in the system. In the course of designing a control console like the one shown in Figure 63, very thorough attention must be given to the type, patterning, grouping, coding, and other features of the information displays and operating controls, if the human operators of the system are to perform satisfactorily. In many of the high-speed manufacturing processes of today—for example the steel rolling mill, the plate glass forming machine, or the large printing press—a miscue, a few seconds delay, or a slight miscalculation on the part of the controlling operator can result in an astounding loss in damaged material and equipment. Design of man's task and the hardware he utilizes is very important to overall effectiveness of the system.

The preceding discussion has presumed that the human is a direct component in the system. In many modern-day manufacturing and military systems, man has only a minor direct part or no direct part at all. Certainly, though, he always has a vital indirect role as a servicer, troubleshooter, repairer, monitor, and the like. The company with a very costly automated production machine feels the pinch for every minute that this equipment is idle for repair and servicing. Rest assured, the management of this company is vitally interested in having that system designed so that man can troubleshoot, repair, and service it in a minimum of time. More and more the buyers of expensive production equipment demand that careful attention be given to design from the operating and servicing point of view.

Aside from designing the product for the ultimate consumer is the separate and crucial matter of designing the product for economy of manufacture. To be able to design products that are not unnecessarily expensive to assemble requires that the designer be familiar with economical work methods and thus the foregoing principles.

Other Aids to Generation of Alternatives

The preceding principles offer an important and useful source of alternatives. However, generalities can only assist up to a point, beyond which the designer must depend on his own ingenuity to generate alternatives. In fact, in many cases the major source of alternatives must be the designer's own ideas, so that the suggestions made earlier as to means of enhancing his creativity will be of considerable benefit here. One of these suggestions pertained to the use of check lists. Most textbooks of the field offer numerous check lists useful to the methods designer. Drawing on these published lists and especially on his own experience, a designer can prepare a useful check list of questions and ideas that will prove fruitful in leading to improvements in work method. Check lists for this purpose vary from very broad to very detailed in coverage, as illustrated by the following sample lists.

The broad and simple check list

Procedure
Equipment
Layout

might be used to systematically encourage the generation of a variety of possibilities under each of these categories.

A more specific check list is the following one included in most textbooks treating the subject of motion study.

1. Can this element (e.g. transportation, operation, grasp, hold) be eliminated?
2. Can this element be *combined* with another element?
3. Can the *sequence* of elements be changed to advantage?
4. Can this element be *simplified?*

The following series of questions illustrates application of the check list approach at the "microscopic" level.

Get

1. Can the distance be reduced?
2. Are tools and materials found at essentially the same location each cycle?
3. Are materials located in containers from which it is easy to isolate and grasp a part? Can lipped containers be used?
4. Can difficult-to-grasp parts (e.g. thin, tiny) be located on a sponge rubber mat to facilitate grasping?
5. Can shape or color coding be used?
6. Can a simpler type of control be utilized?

Place

1. Can the distance be reduced?
2. Can stops, guides, locating-pins, or the like be used to reduce positioning time?
3. Can sliding be used instead of carrying?
4. Can a funnel or taper be used to facilitate positioning?
5. Can the object be pre-positioned before or during transportation?

Use

1. Can two or more tools be combined?
2. Should power tools be used?
3. Are handles, handwheels, knobs, and levers properly designed for maximum speed and ease of operation?

Assemble

1. Can a device hold the material while the operator works on it? Can such a device be foot operated?
2. Can an electromagnetic holding device be used?

Hold

1. Can a clamp, vise, clip, vacuum, electromagnet, collet, or other device do the holding?

Dispose

1. Can a drop disposal be utilized?
2. Can the disposal be achieved while the hand is in transit?
3. Can the material be dropped directly from the holding device into a container?

Summary

Alternatives confronting the methods designer pertain primarily to (1) what role the human being will play in the productive process in terms of the functions he will perform, directly (as a measurer, manipulator, transporter, etc.) and indirectly (as a maintainer, trouble-shooter, monitor, etc.); and (2) how the human being can most effectively perform these functions. To assist him in the generation of alternatives the designer can resort to the general means of enhancing creativity discussed in Chapter 5 and a collection of principles describing preferred work methods.

Ordinarily, this seems to be the logical phase in which to analyze the existing solution, if such is justified. Frequently, the present method contains a number of inefficient features, such as poor layout or procedure, that can be corrected at little or no cost, thus making it more competitive in operating cost, still with negligible investment required. This being the case, an analysis of the present solution and appropriate challenging and scrutiny of its various features, especially the layout and procedure, is often worthwhile. Some of the conventional charts and diagrams may prove useful in such efforts.

EXERCISES

1. Engineers engaged in methods design work often exhibit a disturbing lack of imagination in the generation of improved methods throughout their plant. The methods they produce seem stereotyped. Yet, if for one reason or another they have an opportunity to display what they can achieve in improving work methods for types of operations they are relatively unfamiliar with, their solutions are usually gratifyingly original. How do you explain this? What do you suggest be done to improve this situation?

2. Provide a thorough written analysis and appraisal of the operating controls and information displays of one of the following pieces of equipment. Discuss types, strong as well as weak features, recommended alterations in design, etc.

a. Motion picture projector.
b. Motion picture camera.
c. Typewriter.
d. Bulldozer.
e. 35 mm. camera.
f. Automobile.
g. Electric kitchen range.

REFERENCES

Barnes, Ralph M., *Motion and Time Study,* 4th ed., John Wiley and Sons, New York, 1958.

Chapanis, Alphonse, Wendell R. Garner, and Clifford T. Morgan, *Applied Experimental Psychology,* John Wiley and Sons, New York, 1949.

McCormick, Ernest J., *Human Engineering,* McGraw-Hill Book Company, New York, 1957.

Schnorr, Charles G., "Human Engineering," *Journal of Industrial Engineering,* vol. 9, no. 6, November-December, 1958.

Woodson, Wesley E., *Human Engineering Guide for Equipment Designers,* University of California Press, Berkeley, California, 1954.

Handbook of Human Engineering Data, 2nd ed., Tufts College, Medford, Mass., 1952.

10

Methods Design: Evaluation of Alternatives and Specification of a Solution

In instances where evaluation is more than purely subjective, the process of evaluation proceeds generally as outlined earlier, namely:

1. Select specific criteria.
2. Predict performance of each alternative with respect to these criteria.
3. Convert these estimates to monetary terms.
4. Compare alternatives on the basis of quantifiable *and* unquantifiable criteria.
5. Select the preferred alternative.

Probably the most troublesome step in this procedure is prediction of performance. The designer is faced with such questions as these: How much production time will each alternative require? How much learning time? How much maintenance? How much effort? How fatiguing? And so on. Behavior of the human being is much more difficult to predict than physical phenomena. Performance differences between people, variation in performance of an individual with the passage of time (due to learning, fatigue, monotony, changes in motivation, chance causes, etc.), and intentional changes in behavior while being observed or measured, create special problems for the methods designer in his attempt to make measurements and predictions. To make matters worse, an unusually high percentage of criteria in this field are virtually impossible or prohibitively difficult to quantify. A number of frequently important criteria like fatigue, effort, monotony, and morale, must be treated as intangibles in making the final decision because there are no known means of measuring them or the only known ways are too expensive to use for practical purposes. To further complicate the matter, it is ordinarily necessary for the designer to make his predictions concerning performance while the alternatives are still in the idea stage, or at best merely on paper. Seldom does he have the opportunity to actually observe his alternative methods in operation in order to make measurements of time and other criteria. These three factors, the inherent difficulty of predicting human behavior, the extraordinary number of unquantifiable criteria involved, and the necessity of obtaining estimates while the method is still in the conceptual stage, make the whole evaluation process objectionably erratic.

Probably the most satisfactorily quantified performance characteristic is production time, and even this leaves much to be desired. The means currently available to the designer for quantitatively predicting performance time are to be described in detail.

Prediction of Manual Performance Time

Expected performance time, like any other criterion used to assist in selection of the final design, must ordinarily be estimated while the alternative methods are still in the blueprint stage. The problem is clearly one of predicting the performance that may be expected from alternative methods.

If the rate of performance of an operation is dictated by the machine or process, a prediction of performance time may usually be made with a relatively high degree of accuracy and precision. This is not true however, if rate of performance of the operation is dictated by the operator, for the reasons cited earlier. Some means currently available to the designer for obtaining predictions of manual performance time are outlined here.

1. By pure judgment, estimate the time necessary, on the basis of experience, comparison with similar cases, and common sense. The expected error associated with such a procedure is inherently a relatively large one, but the cost of making the evaluation is certainly low. There are at least two situations in which this type of estimating is appropriate. The first is that situation in which the consequences of making an erroneous decision are not costly, for example, an operation for which the total yearly labor cost is small. The second is the case in which an alternative is obviously inferior or superior and can be satisfactorily isolated on the basis of judgment alone.

2. Time a simulated performance of the proposal in question. Often the designer himself can "go through the motions" and time himself as he tries various alternative methods in question. Although relatively crude, this is a comparatively inexpensive procedure that is generally superior to pure judgment with respect to error.

3. Time an actual performance of the proposed method(s) using a trial setup or prototype of the layout and equipment. The limitations of this procedure are several. First there is the obvious cost of preparing the prototype, which drastically limits utilization of this type of evaluation to high-volume operations involving inexpensively obtained "props" (i.e., materials, tools, and other equipment necessary to reproduce the method in question). Second there is the cost of making the time trials. Third, almost of necessity, the person who demonstrates each alternative as the time studies are made is unskilled in the methods under study, thus increasing the error expected of this type of estimate. However, reliability of such a procedure should be superior to that of the first two described.

4. Synthesis of estimates using predetermined motion times. Different manual operations appear to be different combinations and permutations of only a limited number of unique body member movements, such as move hand to object, grasp object, move object, release object, and so forth. Because each of these small subdivisions is common to a large number of manual operations, it becomes technically and economically feasible to carefully derive an expected performance time for each. Using these basic subdivisions, usually referred to simply as motions, and their associated performance time values it becomes possible to:

a. designate the various motions required by a given method;
b. consult tables of time values to obtain the expected performance time for each of these motions;
c. sum these times to obtain a total expected performance time for that method.

A sample application of this technique to a simple motion sequence is illustrated in Table 5.

The error involved in estimating with the predetermined motion time technique is ordinarily less than that associated with the first two of the foregoing estimating methods, and this apparently compares quite favorably with the error of the "trial setup method." Furthermore, the cost of obtaining an estimate by the predetermined motion time method will often be considerably less than the cost incurred by the trial setup procedure. In fact, the expense associated with the trial setup method is ordinarily prohibitive. For these reasons the predetermined motion time technique is a very useful means of quantification in methods design. Consequently, this technique will be examined in considerable detail.

TABLE 5. Synthesis of Expected Performance Time for the Task Tighten Nut with Wrench

Motion	Description	Predetermined Time Value from Tables
Reach 10 inches	To wrench	0.42 second
Grasp	Wrench	0.10 second
Move 18 inches	Wrench to nut	0.74 second
Position	Wrench to nut	0.71 second
Move 1 inch	Wrench onto nut	0.06 second
Move 6 inches	To tighten	0.32 second
Apply pressure	Tighten	0.58 second
Move 10 inches	Wrench aside	0.44 second
Release	Wrench	0.06 second
		3.43 seconds

The predetermined motion time technique. A predetermined motion time is the expected performance time for a basic subdivision of manual activity, obtained by averaging the times required by many persons to perform the given motion. A predetermined motion time system is a set of these predetermined motion times from which it is possible to synthesize performance times for a large variety of manual operations. Unfortunately, there is no single predetermined motion time system that is accepted and used by all firms that have need for such a technique. Instead there are at least a dozen different predetermined motion time systems now being used. Some of these systems have been developed by large corporations for their exclusive use, others were originated by private individuals to provide the basis for or to augment their management consulting business. These systems differ both in structure and in results obtained. Furthermore, all appear inferior to a system that could be derived were sufficient quantities of time, money, and skill applied to its development.

One of the currently available predetermined motion time systems, Methods-Time Measurement, is described in Appendix A. The technique is further discussed and evaluated under the subject of work measurement.

Prediction of Fatigue

In evaluation of alternative work methods it is desirable that the amount of fatigue generated by each method be quantitatively predicted and compared. Unfortunately, quantitative prediction, even measurement, is impossible at present. Psychologists and physiologists, the experts on this matter of fatigue, have not been able to satisfactorily define let alone measure this phenomenon. Part of the difficulty is due to the fact that the term fatigue is commonly used rather loosely to describe at least four different cumulative, reversible effects of work. These effects are:

1. A diminished capacity to perform the task at hand, both in rate and in quality.
2. A diminished capacity to perform other tasks as well as the one being performed.
3. A number of physiological changes, for example changes in blood chemistry, the nervous system, and glandular secretions.
4. A feeling of tiredness, an experienced or perceived state of the organism. This is the meaning most commonly associated with the word fatigue in ordinary conversation. This feeling of tiredness is not closely correlated with the physical realities described in 1, 2, and 3. A person can feel very fatigued when his actual capacity to perform and the physical state of his body are relatively unaffected, and vice versa.

Thus, part of the difficulty in dealing with fatigue is attempting to measure in aggregate a number of different and only moderately correlated cumulative effects of work. If fatigue is ever to be satisfactorily defined and measured, it will probably be only by treating some or all of these different effects separately. It appears futile to seek a single fatigue measurement.

Each of the foregoing effects is difficult to measure. Changes in capacity to perform are not easily measured because of the simultaneous effects of changes in effort with the passage of time. Some of the physiological changes that take place are measurable, change in blood chemistry for example, but only by means of laboratory-type tests that are not practical for the ordinary practitioner to use.

Some authors report procedures that supposedly measure fatigue. However, as stated, no satisfactory measurement method has yet been found. One procedure that allegedly provides a measure of fatigue is referred to as the output decrement method. Here, the extent to which a worker reduces his rate of work after working for a prolonged period of time, say a day, is taken as a measure of the amount of fatigue generated by the task. The assumption made, and a fallacious one to be sure, is that the worker varies his *actual* production rate according to change in his work-rate *capacity,* to physiological changes in the body, and to feelings of tiredness. The worker is ordinarily unaware of the first two changes. Furthermore, these four phenomena are only moderately correlated among themselves so that production rate cannot be varied very closely with all three. Stated from another point of view, the primary assumption required by the output decrement method is that the worker does not significantly alter his effort level over the period measured nor in fact allow any of many other possible factors other than fatigue to affect his output rate over the day. This assumption will be specifically refuted in Chapter 15 under the topic of fatigue allowances in time study. Suffice it to state now, however, that the output decrement method of "measuring" fatigue is virtually useless for practical purposes.

Perhaps it is the fact that fatigue does not readily manifest itself in the worker's output rate, and the urge to evade the whole difficult problem, that give rise to the claim by some that fatigue is no longer a problem in ordinary industrial tasks and can therefore be ignored.

Even if satisfactory measures are developed, it will probably never be possible to readily predict the fatigue that will be generated by a method while that method is still in the conceptual stage. In all likelihood it will be necessary to actually experiment with the method under laboratory-type conditions and make direct measurements on subjects as they perform the task in question. For the bulk of practical methods design work, laboratory setups are not economically feasible. Note that unlike performance time the effects of fatigue cannot be readily synthesized.

In lieu of a satisfactory objective method of predicting or even measuring fatigue, the practitioner relies on judgment to estimate fatigue on a purely qualitative basis, or in many instances he unjustifiably ignores this criterion altogether.

Prediction of Effort

Effort likewise defies satisfactory definition and quantification. The most common interpretation of "effort" among psychologists and physiologists is that it is an experience, a feeling the person has as to how hard he is working, the degree to which he feels he is exerting himself. This feeling or sensation depends on the actual rate at which the person is working, his capacity for performing the task involved, the amount of physical work involved, the person's attitude toward the job, working conditions (e.g. temperature, humidity, noise level), how interesting the work is, and other factors. Using attitude as an example, two tasks requiring the same amount of physical work but one seeming to be a pleasant task to the individual (as ball playing to a boy) and the other one unpleasant (as a boy mowing lawn), can offer two very different feelings of exertion to the individual. Measurement of effort is complicated not only by its large number of determinants, but by the fact that under identical circumstances the degree of effort experienced differs radically among individuals and with the passage of time.

Another view of effort widespread among time study practitioners holds the term as synonomous with rate of work or speed. This is a careless and unjustified use of the term.

For evaluation purposes the methods designer should have an expression of the average effort required by a given method, relative to that required by other methods and tasks. This average reaction should reflect the amount of physical work required by the method, the pleasantness and intrinsic interest associated with it, conditions surrounding it, and other characteristics. Unfortunately, the degree of exertion a person feels he is putting forth cannot be directly measured or even satisfactorily communicated by him to another person. In lieu of direct measurement the best that can be expected is a ranking with respect to effort required, of alternative methods by persons who have actually tried them. However, since the methods designer must often make his evaluations while alternative methods are in the conceptual stage, no one has an opportunity to try the methods in order to judge the effort required. The designer's appraisal of effort then must usually be made on the basis of his own unaided judgment, yet this is certainly better than ignoring the criterion altogether.

Some hope for indirectly estimating effort lies in the use of various measurable physiological indices of level of exertion. Researchers have experimented with such indices as degree of tension in the muscles, resistance of the skin to an electrical current, blood pressure, and pulse rate.[1] The worker's pulse rate during and immediately after working has received serious consideration partly because of the ease of measurement under actual working conditions.[2] However, pulse rate, like most of these physiological indices, is a function of the effort exerted by the individual and of a whole host of other cumulative and noncumulative aspects of work, such as rate of energy expenditure, fatigue, monotony, and emotional state. Just what pulse rate specifically measures is not known.

[1] For a thorough discussion of most of the proposed work indices see T. A. Ryan, *Work and Effort,* Ronald Press Company, New York, 1947.

[2] See Dr. Lucien Brouha, "Physiological Evaluation of Human Effort in Industry," paper no. 57-A-55, ASME.

Prediction of Energy Expenditure

The rate at which the operator expends energy as he works is a purely physical phenomenon, involving none of the experiential characteristics that make quantification of fatigue, effort, monotony, and job satisfaction so difficult. Rate of energy expenditure is quantitatively measurable in familiar units; calories per minute, foot pounds per minute, and the like. A variety of methods of estimating rate of energy expenditure are described in the literature. The most direct one yet devised involves measurement of the carbon dioxide produced by work. The carbon dioxide expelled during and for a short while immediately following the work period is directly proportional to the quantity of fuel matter oxidized in the body and thus the energy expended to accomplish the work involved. Therefore under this procedure the carbon dioxide expelled by the worker is collected by means of special apparatus[3] while he is performing the task under appraisal. Then by means of standard conversion factors the number of calories expended is determined from the volume of carbon dioxide expelled. Of course this procedure requires that the job actually exist so that the measurements can be made on one or preferably several subjects as they perform it. This, plus the fact that the equipment and measurement process are moderately expensive, means that this procedure is rarely used in everyday methods design work. A disadvantage of lesser consequence is the fact that the attachments to the operator (mask over nose and mouth, or clip over nose and tube in mouth) might cause the person to behave unnaturally.

It has been proposed that the energy expenditure be indirectly estimated by measuring the mechanical work actually accomplished. The two are not the same. One is the input, what the worker actually expends in energy, what the carbon dioxide method measures. The other is the output, the useful work actually accomplished by the worker's energy expenditure. The work accomplished in lifting a fifty-pound weight through a vertical distance of three feet is calculated as Work = Weight × Distance, or 3 ft. × 50 lbs., or 150 ft.-lbs. On the other hand, to accomplish this work the operator expended 0.34 calorie (measured by the carbon dioxide method). Since one

[3] For a description of the most practical setup yet devised for measuring carbon dioxide output, see James H. Greene, W. H. M. Morris, and J. E. Wiebers, "A Method for Measuring Physiological Cost of Work," *Journal of Industrial Engineering,* vol. 10, no. 3, May-June, 1959.

calorie is equivalent to 3085 ft.-lbs. of work, the ratio of output to input is

$$\frac{150 \text{ ft.-lbs.}}{0.34 \text{ cal.} \times 3085 \text{ ft.-lbs./cal.}} = 0.14.$$

Thus the energy actually expended is considerably greater than the useful work achieved.

The attempt to estimate energy expended from the actual work accomplished as calculated by the standard work formula suffers from at least three shortcomings:

1. To estimate energy expenditure from work output requires the unrealistic assumption that the ratio between the two is reasonably constant for different methods under evaluation.
2. Most tasks involve many motions and manipulations for which the actual work being accomplished is extremely difficult to compute. For example, converting the movements of a worker in operating his lathe to foot-pounds of work would be a very laborious if not impossible task.
3. In static work, for example standing and holding a weight or exerting a force where no movement takes place, the distance in the work equation is zero so that according to this computation no work is involved. But tell this to the man standing there holding the weight.

A more practical means of measuring the work output or accomplishment is a specially constructed device called a Lauru platform (force platform).[4] The platform, shown in Figure 64, by means of piezoelectric quartz crystals detects forces exerted in the vertical, frontal, and transverse directions by a person standing on it. When a subject on the platform performs a task, the forces exerted in three dimensions by the person in executing that task are simultaneously measured. The results of an application of this device to the operation of two types of typewriters are shown in Figure 65.

This device does not measure energy expended nor does it measure effort. It measures the work *output,* not input, in terms of the forces exerted by the subject's muscular system. This method recognizes static work, which is a failing of the computational method discussed previously. Furthermore, techniques such as the carbon dioxide method are not sensitive to smaller differences in energy requirements of alternative methods that might arise because of different hand and arm motions. The force platform is very sensitive, supposedly

[4] Lucien Lauru and Lucien Brouha, "Physiological Study of Motions," *Advanced Management,* vol. 22, no. 3, March 1957.

Figure 64. When work is performed by a subject on the triangular platform (1), the forces exerted by the worker in three dimensions are measured by sensitive quartz crystals (one of which is indicated by the small arrow). The magnitudes of the resulting forces are recorded against a time scale, as pictured in Figure 65, by a multiple-channel recorder (4). (Courtesy of E. I. DuPont de Nemours and Company.)

capable of recording the forces exerted by a rat running across it. The main shortcomings of this method are the cost of the device and the fact that it measures the force component of work but not the distance component. The latter disadvantage can probably be satisfactorily overcome by manually making proper measurements of the distances objects are carried or pushed or turned in the course of performing the task, then correlating these with the force measurements recorded.

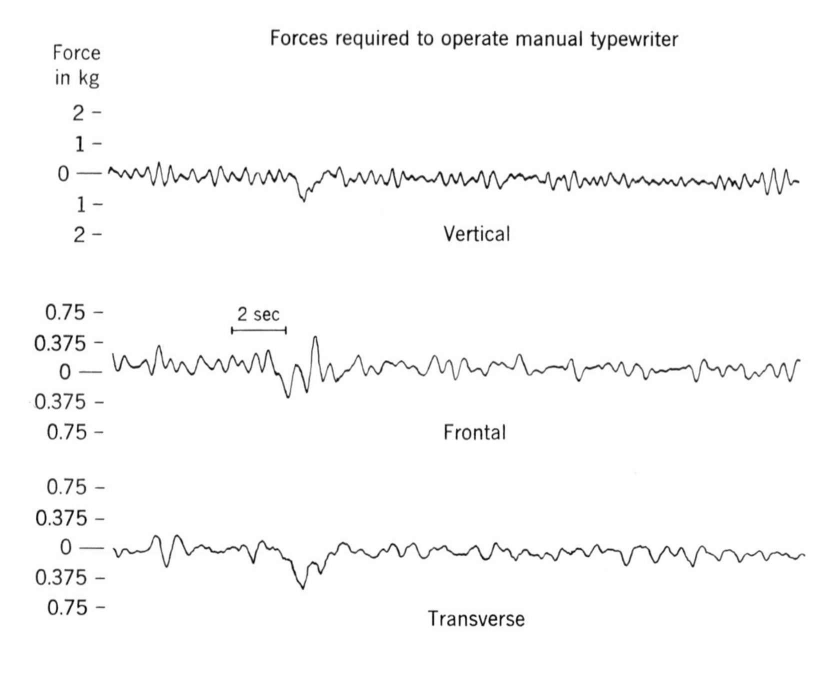

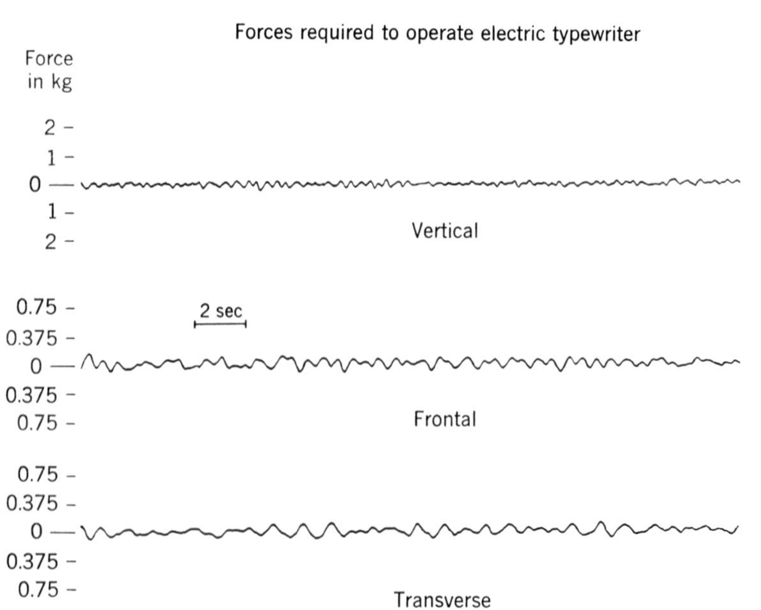

Figure 65. Recordings of forces exerted in operation of a manual typewriter and an electric typewriter as made by the Lauru force platform pictured in Figure 64. (Courtesy of E. I. DuPont de Nemours and Company.)

It appears that the energy expenditure required by a given method of work can be satisfactorily measured directly by the carbon dioxide method or indirectly by the force platform method. However, in spite of this measurability, the energy requirements of alternative methods are ordinarily estimated in a qualitative fashion by way of the designer's judgment, or ignored altogether. The reasons for this are the costs of making the measurements and the fact that evaluation must usually be made while the alternatives are still in the conceptual stage. To be sure, however, qualitative consideration of this criterion is superior to outright neglect.

Prediction of Monotony

Monotony (boredom) is generally viewed as the opposite of interest, a distaste for the task that accumulates as a result of lack of variety and mental challenge in the activity. The effect is in the increased effort required to continue working. Monotony is likely to be rather severe in highly specialized, highly repetitive manual tasks. Therefore, it is a criterion that should be given special consideration when making decisions concerning specialization (division of labor).

Like effort and one aspect of fatigue, monotony is an experience and therefore defies satisfactory quantification and prediction. In spite of the need to rely on judgment for prediction, monotony is not a criterion to be ignored in the comparison of alternative work methods.

Prediction of Job Satisfaction

Work method is an important although certainly not the only determinant of a worker's satisfaction with his job. Admittedly, job satisfaction is a very erratic matter to predict, yet it is a significant affector of the worker's ultimate productivity. It should, therefore, be given consideration in selection of the work method, even if only in a qualitative fashion.

It seems that management is ever attempting to motivate higher productivity on the part of its workers. Here is a means of building motivation into the job itself, by selecting a method that enhances job satisfaction. Conversely, here is a means of building automatic discouragement of higher productivity into the job, by adopting a method that deters job satisfaction.

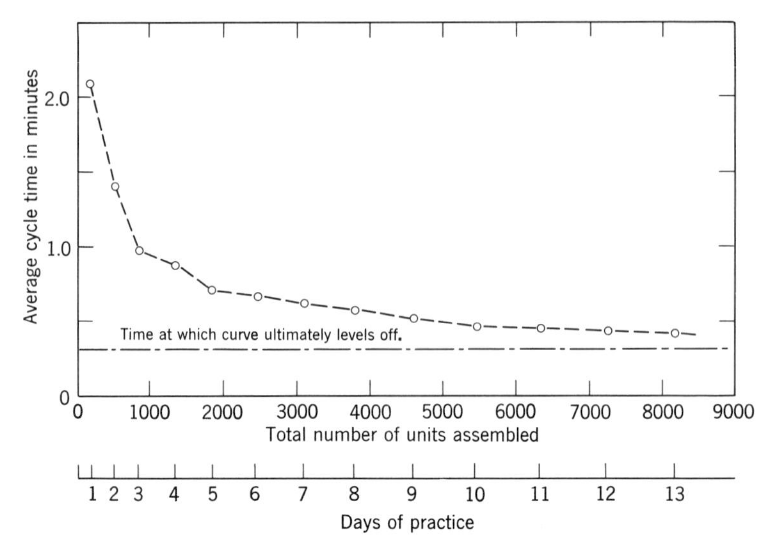

Figure 66. Learning curve for an assembly operation. The general form exhibited by this curve is representative of learning curves ordinarily encountered on manual operations.

Prediction of Learning Cost

Designers are often prone to forget, or deliberately choose to ignore, the fact that alternative methods of performing a task can require significantly different learning times. Furthermore, the period of learning and the associated learning cost for other than the very simple operations are often more than might be suspected. In fact, this cost might well be the largest single contributor to the initial cost of a proposed method.

Figure 66 shows a learning curve for an assembly operation. In this case the operator continues to display improvement up to 175 hours of practice. The total cost of learning—the value of the production lost over the learning period as a consequence of producing at something less than the production rate experienced under the replaced method—amounts to approximately $80.[5]

At present no satisfactory means of quantitatively predicting learning time is generally available. There is promise, however, that there

[5] This is the learning cost for one operator, calculated on the basis of 50 hours worth of lost production valued at $1.55 per hour, the operator's base wage.

will be such a technique in the near future.[6] In the meantime the methods designer must rely on judgment to predict the learning time requirements for different methods.

The Unquantifiable Criteria and Their Importance

Because of their experiential aspects, fatigue, monotony, effort, and job satisfaction must be treated as unquantifiables in the comparison of alternative work methods. Even if they were measurable there is the additional obstacle of very seldom having trial setups for alternative methods on which measurements can be taken. Current means of predicting learning time are crude but the potential exists for development of more objective procedures. Rate of energy expenditure is ordinarily treated as an unquantifiable because of impracticality not impossibility of measurement.

Why bother with these unquantifiable criteria? They apparently seem unimportant to many practitioners. So what if the worker is a little more tired at the end of the day, or has to put forth more effort, or the job is a little more monotonous, or results in less satisfaction? Is it not performance time, rate of production that management is really interested in? But it is in just this supposition that the practitioner so often makes his mistake. Such matters as fatigue, effort, monotony, and job satisfaction *do* affect productivity *in the long run.* On a job that generates greater fatigue or monotony, or requires greater effort or energy, or results in less job satisfaction, productivity will be less and *total labor* cost higher in the long run because of the likelihood of increased avoidable delays in production, increased employee turnover and transferring, increased supervisory costs, increased absenteeism, indifferent, apathetic, or even antagonistic employee attitudes, and other factors. Method *A* might require less time than method *B* when the performance times are based as usual on a short-run time trial, but since method *A* results in greater fatigue, in the long run *A* may not be superior at all. In general, methods designers make the mistake of putting too much emphasis on short-run performance time as a criterion at the expense of the unquantifiable criteria. Even though they cannot be put in terms of numbers, these

[6] On the basis of unpublished research conducted by the Department of Industrial and Engineering Administration, Sibley School of Mechanical Engineering, Cornell University, it appears feasible to predict learning time on the basis of certain characteristics of the method itself and to use a predetermined motion time system as the means of measuring these characteristics.

criteria can and should be given qualitative consideration in the evaluation of alternative work methods. As unsatisfactory as this may seem, it is certainly preferred to outright neglect.

Return on the Investment as a Function of Amount Invested in Improvement

For illustration, take the following case in which some alternative procedures have been proposed as improvements in method for a certain precision balancing operation. In Table 6 the expected investment, balancing time, net savings in operating costs, and return on the investment are given for the four methods. In Figure 67*a* the savings in operating costs for each alternative are plotted against the respective investments required, to indicate what is a common situation in methods design work. As this plot implies, savings do not ordinarily increase proportionately with increases in investment in method.

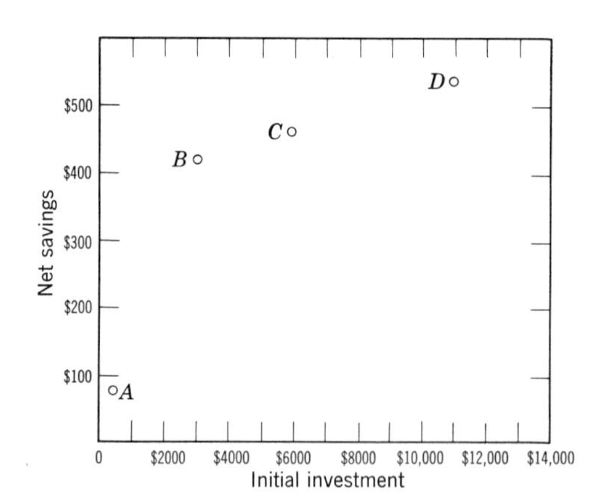

Figure 67a. Relationship between savings achieved and initial cost for the four balancing methods.

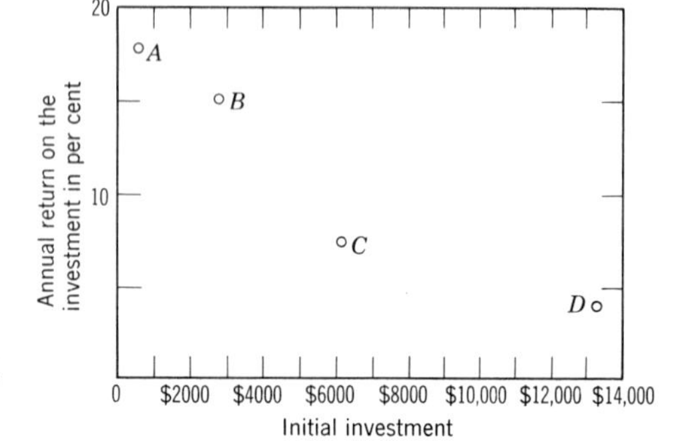

Figure 67b. Relationship between return on the investment and initial cost for the four methods of balancing.

TABLE 6. Comparison of Four Balancing Methods

	A Predominantly Manual	*B* Special Fixtures and Mechanical Aids	*C* Semiautomatic Machine	*D* Automatic Machine
Initial investment	\$450	\$2800	\$6100	\$13,500
Balancing time	0.73 minute	0.18 minute	0.14 minute	0.11 minute
Net saving in operating costs per year	\$80	\$420	\$470	\$540
Annual return on investment	$\frac{\$80}{\$450} \times 100 = 17.8\%$	$\frac{\$420}{\$2800} \times 100 = 15.0\%$	$\frac{\$470}{\$6100} \times 100 = 7.7\%$	$\frac{\$540}{\$13,500} \times 100 = 4.0\%$

Instead, as methods of greater initial cost are resorted to, the incremental increase in dollar savings achieved tend to decrease until additional expenditures accomplish virtually nothing in this respect. Thus, there seems to be some point in the course of investing in progressively more mechanized and thus more costly alternatives, at which it usually becomes more profitable to divert any additional funds available to other awaiting problems and their solutions. This is indicated by the plot of return on the investment for each balancing method against the respective investment required, appearing in Figure 67*b*. As the investment is increased and the cumulative return on the investment decreases, greater investment in solution of the problem involved should not be considered beyond the point where return on the investment becomes less than that of other investment opportunities of the enterprise. Surely the company can find avenues of investment offering better than a 4 per cent return, and the same is true of the 7.7 per cent return offered by the semiautomatic machine.

Recall the various levels of mechanization cited in Figure 42, page 130. Take the case of an operation which heretofore has been predominantly manual and which is in the process of being redesigned. Under such circumstances as these it is not uncommon to find that the greatest return on the investment will result from relatively simple and inexpensive improvements in workplace layout and procedure, and from the introduction of simple mechanical aids such as a work holding device. For the money put forth, such changes, especially those in layout and procedure, usually offer the greatest benefits. The addition of powered tools may well bring substantial and worthwhile savings but not with the same high return on the investment the simplest changes offer. The use of progressively more elaborate, more costly degrees of mechanization will probably continue to yield increased savings but not proportionately so, with the result that a point is reached at which further mechanization is not profitable in terms of return on investment. The level of mechanization at this point might be viewed as the optimum degree of mechanization for the problem at hand for the expected production volume. Because this type of situation is so common, this is a phenomenon that should be given careful consideration in the generation and evaluation of alternative solutions to methods problems. Attention should be given to generating a spread of alternatives covering a reasonable range of degrees of mechanization, reasonable in that in some problems certain degrees of mechanization are economically or technically out of the question. For example in some situations complete mechanization is obviously not feasible, in others, predominantly manual methods are out of the question.

Common Mistakes and Omissions

1. Claiming savings that actually do not exist. For example, because overhead is charged to the product on the basis of production time required, it is tempting to conclude that if this production time is reduced by a given alternative, overhead is therefore reduced proportionately. Actually, overhead of the plant is probably unaffected or negligibly affected in most such cases. Another example is that in which a particular method reduces the need for labor or space, but there is no way that "saving" can be used for any other purpose. Under these circumstances there is no cost reduction.

2. Ignoring labor fringe benefits that the company must pay for every labor dollar paid out. The amount of such benefits varies considerably from company to company, but averages about 20 per cent of direct earnings.

3. Comparing production time estimates that represent appreciably different rates of work. For example, production records often provide an estimate of the performance time that has been required by the present method. This time is then compared with times for alternatives that have been forecast by one of several methods described previously, predetermined motion times for example. These times might represent work speeds 20 or 30 per cent higher or lower than that represented by the time obtained for the present method. An erroneous conclusion might well result under the circumstances.

4. Ignoring or underweighting the unquantifiable criteria, such as fatigue, effort, monotony, and job satisfaction. This also includes what might be referred to as the hidden costs of change, the disruptive effects, the adjustment, the resistance, and the like, brought on by change in work pattern and habits. Certainly an easily overlooked or under-weighted criterion is worker acceptance of a proposed method. If the persons who must carry out the new procedure resent and resist it, the operating cost of that method might be thereby inflated even to the point where it becomes prohibitive. Worker acceptance is admittedly a difficult matter to predict. However, it can probably be anticipated more often than is commonly supposed and is therefore a matter worthy of considerable forethought.

Summary of the Evaluation Phase

It is difficult to generalize concerning evaluation procedure, for this varies considerably with the particular situation. Obviously judgment assumes a major role. In fact, in some instances it will be the sole method of evaluation because the economics involved justify nothing more elaborate. The criteria considered should not be limited only to those that can be quantitatively estimated. The unquantifiable criteria should be given adequate weight so that operating costs of the alternative methods under comparison reflect labor cost in the long run.

Notice, for example in the case of performance time, that some methods of prediction are quick and inexpensive but somewhat erratic, whereas others are more reliable but in turn more costly to use. Others fall between these extremes. A similar situation exists regardless of what characteristic is being evaluated. In general, the appropriate method of evaluation in a specific situation is that procedure for which the cost of making the appraisal is commensurate with the importance of the decision being made.

At the conclusion of the evaluation phase the alternatives conceived have ordinarily been reduced to a single preferred method, which remains to be specified.

SPECIFICATION OF THE SELECTED METHOD

Several purposes are accomplished by this specification process, namely:

1. Communication of the proposed method to those responsible for its approval.
2. Communication of the proposed method to those concerned with its implementation, especially for purposes of:
   a. actually setting up the operation insofar as the equipment and layout are concerned;
   b. instruction of operators.
3. Provision of an official record of the method specified.

For purposes of specification, an official document entitled *standard method description* is recommended. This document, illustrated in Figure 71, page 222, or whatever communication medium is used, should do the following.

1. Effectively communicate the procedure, layout, and equipment to be used.

2. Indicate the performance characteristics important to implementation of the method specified and to subsequent manufacturing programming and control. (For example, persons responsible for approval of the proposed design are interested in performance time, installation and operating costs, etc. Planners who schedule operations and plan labor and equipment needs are interested in the performance time, as are supervisors, instructors, cost accountants, estimators, and others.)

For effective communication of the work procedure the standard method description might well utilize a left-hand right-hand chart format to facilitate comprehension and instruction. A number of the other descriptive techniques previously introduced are occasionally useful for specification purposes. Considerable interest has been shown in the use of photographs, slides, motion pictures, and tape recordings as aids in specification and instruction.

At the completion of the more important projects, the specifications will probably be in the form of a formal report. The satisfactory preparation of reports that are effectively organized and illustrated, complete yet brief and to the point, and convincing in their argument, is a matter worthy of considerable attention.

Methods design is no different from other engineering specialties in that prediction of the actual performance of the designer's creation is imperfect. Ordinarily, as he oversees implementation and follows up on the method in the early stages of its use, the methods designer will find that minor modifications in equipment, layout, and procedure are desirable. The need for such changes arises as the result of oversight, poor judgment, attempting to fit the method to a particular operator, and the like. A certain amount of debugging is inevitable. After the operation is set up and running, he may also find it necessary to revise his original specifications of the method's performance characteristics, such as the performance time standard he originally established for the method. Some of these revisions are due to erratic original estimates, many others are due to changes that have taken place since the original specifications were made. All of this modification activity is part of the methods designer's follow-up responsibility.

SUMMARY OF THE METHODS DESIGN PROCESS

The highlights of the approach to a methods design problem as recommended herein are as follows.

1. Spend time initially in definition of the *problem,* adopting a broad, detail-free view. Avoid starting your attack on a problem by immediately trying to think of improvements in the present solution. Similarly, avoid starting by making a refined description of the present method.

2. Next, focus your attention on analysis of the problem. Ferret out and screen the restrictions so that insofar as possible you will be dealing only with real, valid restrictions. Attempt at this stage to determine all that you are free to alter.

3. Seek out numerous and varied alternatives, using volume and criteria as bases for directing this search. To maximize effectiveness of this phase rely on the general means of enhancing creativeness along with the body of principles pertaining to preferred methods of work.

4. In general, evaluate after and not during the search for alternatives. When feasible, use a predetermined motion time system to synthesize the time requirements for alternative methods and portions of same. Do not underweight the unquantifiable criteria; base total labor cost on productivity expected in the long run.

5. Adequately specify the selected method and its performance characteristics by way of the standard method description, augmented by various other communication media, especially those usable directly as instructional aids.

6. Follow-up on implementation and usage of the method to round out the design cycle as described in Chapter 6.

EXERCISES

1. A plumbing hardware manufacturer has hired a methods engineering consultant to improve work methods throughout the factory. One operation that has come to the consultant's attention is the inefficient manner in which pipe unions are assembled. These are standard ground-joint unions, appearing as shown in the accompanying sketch. Component parts are brought to the

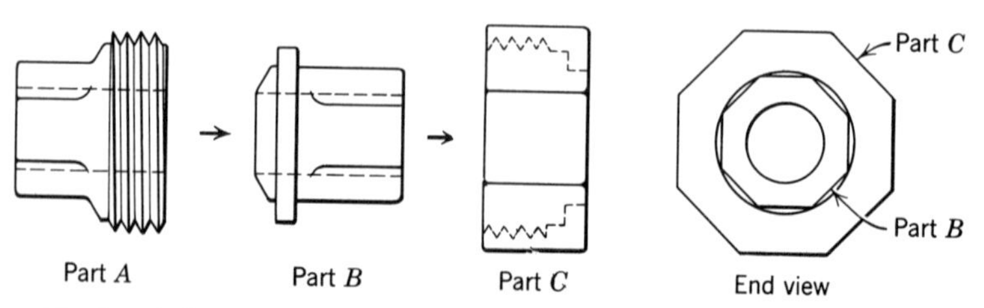

assembly bench in cardboard cartons. The operator places a carton of each of the component parts on the bench and removes parts as needed, as

Left-Hand Right-Hand Chart for Assembly of Pipe Unions

Left Hand		Right Hand	
Description	Symbol	Symbol	Description
Part *C* from carton	*G*	*G*	Part *B* from carton
	H	*P*	Part *B* in part *C*
	H	*G*	Part *A* from carton
	H	*P*	Part *A* to assembly
	H	*A*	Part *A* to assembly
	H	*P*	Assembly in carton

described in the preceding left-hand right-hand chart. The assemblies are placed in the same type of cartons the parts are supplied in, and periodically

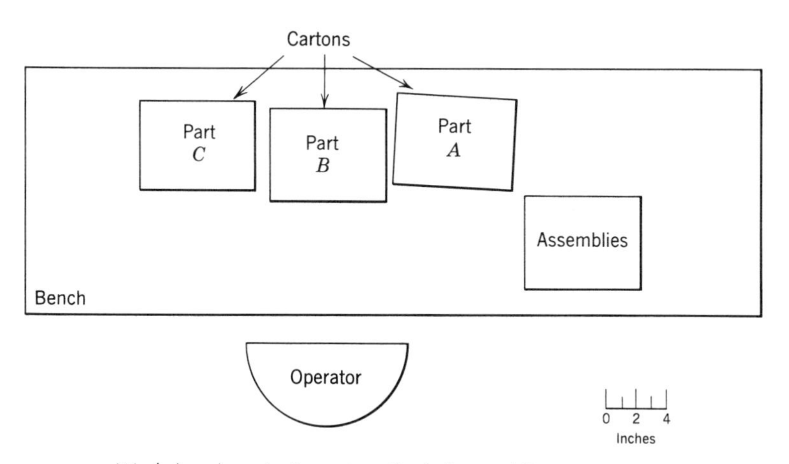

Workplace layout—Present method of assembling pipe unions.

are carried to the packaging operation where they are boxed by the dozen.

The usual assembly bench layout is as shown in the accompanying plan view. The assemblers stand while working. The labor rate is $1.62 per hour. The yearly volume for the eleven sizes of unions produced is 775,000 units. Design of the part is stable. Current production rate is about 500 unions per hour per assembler.

If you were the consultant what would you specify for this operation? (*Note:* It is highly desirable that the method specified be readily adaptable to all sizes of unions produced.)

2. A manufacturer of small electrical hardware items produces BX connectors, mechanisms used in household wiring to anchor BX cable to outlet and distribution boxes. The connector is a standard item available from most electrical supply houses and hardware stores. The present method of assembling these connectors is indicated by the accompanying plan view

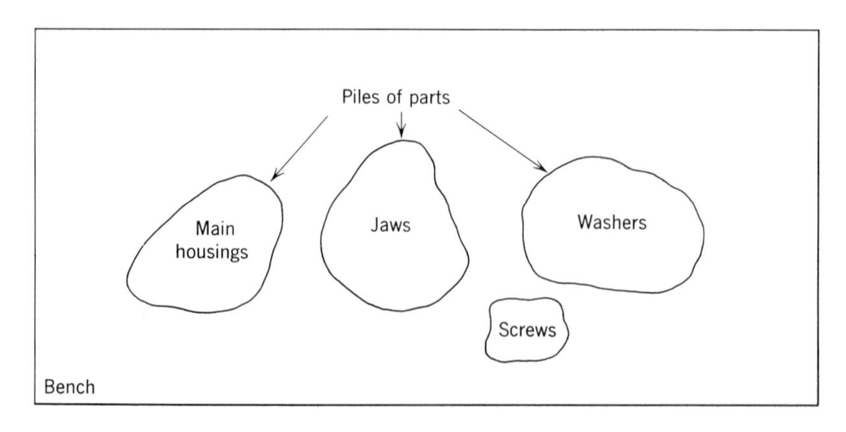

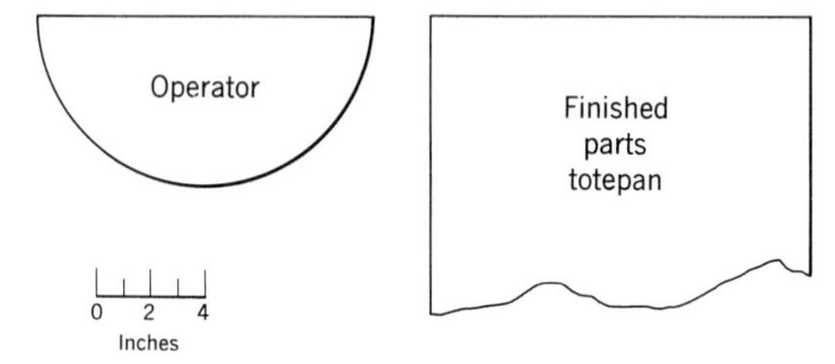

Workplace layout—Present method of assembling BX connectors.

Left-Hand Right-Hand Chart for BX Connector Assembly

Left Hand		Right Hand	
Description	Symbol	Symbol	Description
Housing	*G*	*G*	Jaw
Assembly	*H*	*P*	Jaw in housing
	H	*G*	Screw
	H	*P*	Screw
	H	*A*	Turn screw half-way down
	H	*G*	Washer
	H	*P*	Washer
	H	*A*	Hand-tighten
	H	*P*	In totepan

and left-hand right-hand chart. The present method of assembly requires approximately 0.21 minute. Yearly volume is 850,000. Wage rate of assemblers is $1.24 per hour. Parts come to the operator in 24″ × 14″ × 8″ totepans. Design of the product is stable.

Specify an improved method for this operation. Support your proposal with appropriate cost data. Cite some of the alternative solutions considered.

3. An electric razor manufacturer is in the process of designing production methods for a new product design. The overall production plan calls for a separate operation to subassemble the cutting head. Production requirements are expected to average 700 units per day over the 2-year life of this design. The labor rate for assembly operators is \$1.52 per hour. Parts are supplied to the operator in totepans 22″ × 12″ × 6″ and it is planned to transport the completed subassemblies to the final assembly operations on wooden trays 18″ × 18″, each holding 180 cutting heads. You are to design the assembly method for these subassemblies. (A sample assembly or specifications of same will be provided by the instructor.)

4. At a large fruit packing house, fancy apples are hand packed in bushel crates. The apples come to the packers on a belt conveyor. In the crate the apples are arranged in rows, each layer separated by a heavy paper partition. At present an objectionable amount of time is lost by the packers because the apples in a layer will not stay in place, unless held there by one hand, until that layer is practically complete. The filled crate is lidded and labeled at another workstation. The expected daily volume is 1100 crates a day over the 3-month packing period. Packers average \$1.95 per hour. The location of the belt conveyor and lidding and labeling workstations are fixed as shown in the accompanying figure. You are asked to specify the details

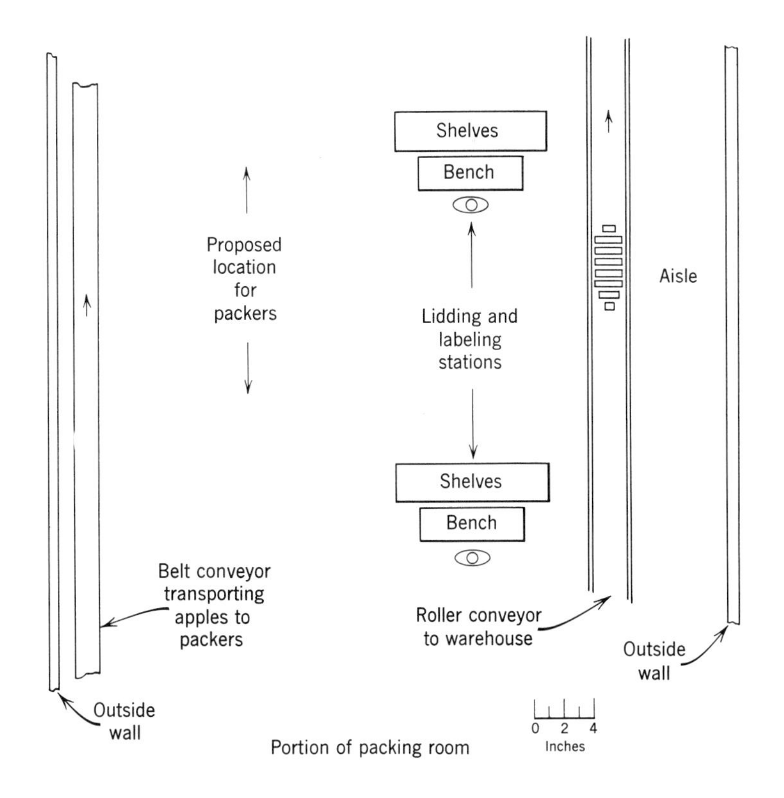

Portion of packing room

concerning what should take place between the belt conveyor and the lidding and labeling stations.

REFERENCES

Brouha, Lucien, "Physiological Evaluation of Human Effort in Industry," paper no. 57–A–55, ASME.

Brouha, Lucien, and Lucien Lauru, "Physiological Study of Motions," *Advanced Management,* vol. 22, no. 3, March 1957.

Greene, James H., and W. H. M. Morris, "The Design of a Force Platform for Work Measurement," *Journal of Industrial Engineering,* vol. 10, no. 4, July-August, 1959.

Greene, James H., and W. H. M. Morris, "The Force Platform," *Journal of Industrial Engineering,* vol. 9, no. 2, March-April, 1958.

Greene, James H., W. H. M. Morris, and J. E. Wiebers, "A Method for Measuring Physiological Cost of Work," *Journal of Industrial Engineering,* vol. 10, no. 3, May-June, 1959.

Ingenohl, Ingo, "Measuring Physical Effort," *Journal of Industrial Engineering,* vol. 10, no. 2, March-April, 1959.

Ryan, Thomas A., *Work and Effort,* Ronald Press Co., New York, 1947.

Ryan, Thomas A., and Patricia Cain Smith, *Principles of Industrial Psychology,* Ronald Press Co., New York, 1954.

Young, H. H., "The Relationship Between Heart Rate and the Intensity of Work for Selected Tasks," *Journal of Industrial Engineering,* vol. 7, no. 6, November-December, 1956.

11

Methods Design: Appraisal of Theory and Practice

The theories and practices of the field of methods design are the subject of numerous criticisms from a variety of sources. Some of these contentions are worthy of serious attention and are discussed in this chapter; others seem more of academic rather than practical significance.

Popular Approach to a Methods Design Problem

The "traditional," widely published and promoted procedure for attacking a methods design problem may appropriately be described as a rehash process. What might be referred to as the classics in motion and time study literature promote the following general procedure for solution of a methods design problem.

1. Document the present method in detail via one or more of the descriptive techniques traditional to the field, for example by a flow chart.
2. Question the present method in detail.
3. Devise a new method.
4. Specify the new method.

In most of the literature of the field, considerable emphasis is put on step 1 in this procedure—documentation of the present method—whereas definition and analysis of the problem are generally neglected. Not only does this procedure start with the present solution to the problem, it seems also to generally revolve about the present method, all to the detriment of the designer's long run performance. Another apparent shortcoming is the failure to give sufficient attention and emphasis to a thorough search for a number and variety of alternative solutions. What is lacking is a genuine engineering approach to what is truly a design problem.

Interestingly, however, the practitioner does not, in this author's opinion, generally conscientiously heed the recommendation of the classics to document the present method as the first step of his approach. However, the practitioner's approach does seem haphazard. Thus, writers and practitioners alike would do the field and themselves a service by putting more emphasis on methods *engineering*.

Principles Available to Guide the Choice between Man and Machine

Principles pertaining to man's capabilities in the performance of various production tasks, needed by the designer to guide his choice between man and machine, are relatively few. Those that are available are to the credit of other fields.

After the tasks to be performed by man in a given system have been decided on, these tasks must be aggregated into practical sized accumulations called jobs. A job is an aggregation of fundamental tasks comprising a given person's contribution to the system. The process of specifying the particular tasks that constitute a job is referred to by some as job design. Job design is a predominantly empirical process; there is little in the way of safe generalizations to guide the practitioner in objectively aggregating fundamental tasks into jobs. What is needed, and of course is not available, is a body of generalities to guide the practitioner in this respect.

One overshadowing "principle" serving to guide the aggregation of tasks into jobs is the so-called "principle of division of labor," which leads to the selection of small aggregations of tasks, that is, specialized jobs. The trend toward increased specialization has continued for centuries, to a point where in recent decades social scientists and others have raised doubts and apparently valid criticisms over the degree of specialization now found in many industrial jobs. It is true that many production processes have been subdivided into a large number of highly repetitive, unskilled jobs. In fact, jobs with a total work cycle of approximately 2 seconds are not unheard of. Specialization has a number of advantages among which are decreased learning time, less expensive labor, better utilization of equipment, and the like. However, whereas these costs tend to decrease as specialization is increased, a number of other costs simultaneously increase. For example, increased division of labor also results in increased unbalance in the production capacities of successive operations, increased handling, increased monotony, decreased flexibility, and decreased job satisfaction. If division of labor is carried beyond a certain point, these adverse effects will cancel out the benefits and total cost will begin to rise once again. Thus, there appears to be a degree of specialization for which total cost is a minimum, but what constitutes this optimum degree of specialization is currently a matter of considerable controversy.

Some writers and practitioners are of the opinion that in general the division of labor trend has progressed beyond the minimum cost point. The consequence of this belief is an interesting one: a countertrend to specialization commonly referred to as job enlargement. Job enlargement is an attempt to broaden a man's job by giving him a larger percentage of the total production task and by giving him increased responsibility for such matters as job setup, inspection, and equipment care. Attempts at job enlargement have been reported by a number of firms and investigators but so far the results are far from conclusive, in spite of some of the claims made in favor of such a policy. Although the results of experiments with job enlargement may not be significant, the rise of such a "de-specialization" trend certainly is. More questions are being asked, more doubts expressed, more inquiring articles written, on this subject of division of labor. So far, however, the uncertainty, confusion, and disagreement remain so that the designer must depend on judgment in deciding on the optimum degree of specialization for the process at hand.

Principles Pertaining to How to Use Man in the Process

The principles pertaining to preferred methods of work are inadequate with respect to number and quality. The only principles of this type that the field can lay claim to are the relatively few principles of motion economy. Of these, some are of questionable validity, many are rarely of use. This set of principles has remained virtually unaltered and unsupplemented for decades. Yet the quantity and quality of these principles, and the enrichment of knowledge in general that ordinarily takes place with the passage of time, leads to the expectation that these generalities would be considerably refined and expanded by this time. In contrast to the relatively large and objective body of comparable knowledge generated by the field of human engineering, the principles of motion economy look especially unimpressive.

For purposes of illustration, recall the principle that suggests the use of a simultaneous and symmetrical motion pattern. This is an idea that seems overpromoted and overapplied. It is not *the* solution to almost every assembly problem, nor is it as clearly superior to other alternatives as some textbooks indicate. In fact, some quotations of the time savings achievable by the use of a simultaneous symmetrical motion pattern are apparently based on inadequate experimentation. The type of procedure that does *not* prove the superiority of a simultaneous symmetrical motion pattern is illustrated by the following "experiment." A simple task is performed by a person using a method not considered to be simultaneous and symmetrical (henceforth referred to as the asymmetrical method), then the same person performs that task using the prescribed simultaneous symmetrical method. Performance times are recorded and then compared to indicate the superiority of the latter method.

A task often selected for such an experiment is that of placing wooden pegs in a board as shown in Figure 68. Typically, the investigation proceeds as follows. First the subject fills the board with pegs, say five times, using the asymmetrical method illustrated in Figure 68*a*. One hand places pegs "supplied" by the other hand. The time to fill the board is recorded each cycle so that an average time to fill the board by this method is obtained. Then the subject repeats this procedure while using the symmetrical method as illustrated in Figure 68*b*. Then the experimenter proceeds to compare the average times for the two methods and to draw conclusions therefrom. *Under this experimental procedure*, the symmetrical method will usually require approximately two-thirds of the time required by the asymmetrical method. However, the superiority of the symmetrical method is not expected to be this great in the long run for reasons explained below.

The experiment just described would appall the properly schooled experimentalist, and rightly so. His rejection of this investigation would be based primarily on the following objections.

1. Any confidence we might have in the results of this experiment can be readily dispelled simply by repeating the test and reversing the order in which the subject performs the two methods. If the subject fills 5 boards using the symmetrical method, then 5 boards employing the alternative method, the resulting average times for the two methods will ordinarily be about the same. The apparent substantial superiority of the symmetrical method found in the first experiment is attributable in part, if not in the main, to the fact that learning is taking place at a rapid rate (if the subject is inexperienced at this task). The symmetrical method benefits considerably from

practice the subject obtained in placing 150 pegs by the asymmetrical method. Some transfer of the effects of practice takes place between the two methods. When the order of the methods is reversed, the asymmetrical method assumes the advantageous position in this respect, and the superiority of the symmetrical procedure is partially or completely cancelled out.[1]

2. In this experiment, performance time is the only criterion of superiority used. However, especially for experiments that are to lead to far reaching generalizations, additional criteria such as fatigue and monotony generated, energy and effort required, and the like, should be considered. Suppose the subject were to perform these methods for prolonged periods of time, sufficiently so to permit the effects of fatigue to manifest themselves. Note that the asymmetrical method used involves considerably less arm movement and offers the possibility of alteration of method to provide some relief from fatigue. Thus the symmetrical method is not expected to be as superior in the long run as short-run time trials indicate. In fact, for the task employed in the experiment described, the long term difference in productivity of the two methods may well be negligible as a result of the effects of fatigue.

3. Because of marked differences between people, two methods of performing a task are expected to compare differently for different persons. For person *A*, the symmetrical method might be 25 per cent faster. For person *B* the two methods might be equally productive. For person *C* the asymmetrical method might be superior, and so on. Consequently, no valid generalizations can be made concerning relative merit of the two methods on the basis of a sample consisting of only one person. Before useful conclusions can be drawn, many persons must be included in the sample and generalizations based on the aggregate result.

4. A parallel situation exists for different types of tasks. Wide differences in results are expected as the nature of the task investigated is changed. Thus, before useful generalizations are justified, many different tasks must be experimented with.

5. The methods should be compared over a prolonged period for more than one reason. After a substantial period of practice, a person

[1] Under these circumstances, the experimenter could satisfactorily avoid the effects of learning by having the subject perform, say five cycles of the symmetrical method, then ten cycles of the alternative method, then another five cycles of the symmetrical procedure, then comparing average times for the ten trials of each. In this manner the confounding effect of practice is balanced out, or nearly enough so for practical purposes.

Figure 68a. The asymmetrical method of filling board with pegs. (From Ralph M. Barnes, Motion and Time Study, 4th ed., John Wiley and Sons, New York, 1958. By permission of the publisher.)

Figure 68*b*. The simultaneous symmetrical method of filling board with pegs. (From Ralph M. Barnes, Motion and Time Study, 4th ed., John Wiley and Sons, New York, 1958. By permission of the publisher.)

will probably find that the two methods compare differently than they did when that person was a novice at the task. The learning curves may well be significantly different for the two methods. For the task studied in the experiment described, a person is likely to experience a greater reduction in time to perform the symmetrical method through the effects of practice than he will for the asymmetrical method. This can only be determined by means of a comparison of the two methods over a long period with thousands of repetitions.

These objections to the manner of conducting the experiment described offer some indication of the type of investigations required to experimentally substantiate the "principles" of motion economy. Furthermore, these points should be borne in mind in interpreting the reported results of studies of this nature.

Other Criticisms

A survey of industrial practice indicates that methods engineering is still predominantly preoccupied with direct labor activity. For decades the field has concentrated on the relatively straightforward, repetitive, direct labor operations in the factory, whereas indirect labor, the handlers, the inspectors, the maintenance personnel, the office staff, etc., has exhibited a drastic increase in terms of numbers and of total cost. In manufacturing, for the nation as a whole, the number of indirect workers per direct laborer is rapidly approaching a one for one ratio. It behooves the methods engineering department, if it has not already done so, to shift a commensurate portion of its available resources to man's indirect role in the manufacturing system.

It has been said that this field shows a general ineptness for handling the human relations problems such work entails, especially in the manner of introducing changes to the supervisory staff and work force. In this author's opinion this is a valid criticism, although considerable progress has been made.

One criticism about which there is no doubt is that this field has been surprisingly stagnant since the time of the "pioneers" in the early part of the century. Substantial changes and contributions have been remarkably infrequent especially from within the field itself.

REFERENCES

Amrine, Harold T., and D. Edward Nichols, "A Physiological Appraisal of Selected Principles of Motion Economy," *Journal of Industrial Engineering*, vol. 10, no. 5, September-October, 1959.

Blair, Raymond N., "A Fresh Look at the Principles of Motion Economy," *Journal of Industrial Engineering,* vol. 9, no. 1, January-February, 1958.

Buffa, Elwood S., "Toward a Unified Concept of Job Design," *Journal of Industrial Engineering,* vol. 11, no. 4, July-August, 1960.

Carter, R. R., and L. E. Davis, "Job Design," *Journal of Industrial Engineering,* vol. 6, no. 1, January-February, 1955.

Carter, R. R., L. E. Davis, and J. H. Hoffman, "Current Job Design Criteria," *Journal of Industrial Engineering,* vol. 6, no. 2, March-April, 1955.

Davis, Louis E., "Toward a Theory of Job Design," *Journal of Industrial Engineering,* vol. 8, no. 5, September-October, 1957.

Gomberg, William, *A Trade Union Analysis of Time Study,* 2nd ed., Prentice-Hall, Englewood Cliffs, New Jersey, 1955.

part IV

WORK MEASUREMENT

12

Introduction to Work Measurement

Importance of Performance Time Standards for Manufacturing Operations

Visualize a medium-sized manufacturing firm producing a variety of heavy equipment items, some for stock and some to customer specification. Like most plants, this company requires time estimates for its manufacturing operations for numerous reasons.

When a potential customer submits the specifications of a piece of equipment to be manufactured, this company must quote a competitive price for that job. To arrive at this bid the company must estimate cost-to-manufacture which in turn requires a satisfactory estimate of the time this product will demand of the manufacturing process. A company that is poorly equipped with time estimates for various operations in its plant is handicapped when it comes to establishing bids on prospective jobs. The ordinary consequences of over- and underbidding are obvious. Thus, the availability of time estimates for individual operations from which total manufacturing time estimates can be derived are important in pricing a product. This is true for the product produced as a standard stock item as well as for the production-to-customer-specification situation.

This company has many pieces and types of equipment. Each of these is demanded by some products, not by others, and demanded for varying amounts of time in different sequences. Thus, there is a complex scheduling problem. If prohibitive amounts of unassigned machine time are to be avoided, if general confusion on the production floor is to be averted, if shipping dates to customers are to be quoted and adhered to, then the scheduling department must have reasonably accurate time estimates for activities in the plant. These times are needed so that schedulers can satisfactorily forecast job arrival and departure times at machines, plan the arrival of incoming materials, and otherwise obtain a semblance of order out of potential chaos.

Since production equipment and skilled employees are not ordinarily procured without considerable time lag, it is desirable that planners be able to predict equipment and manpower needs on the basis of long-range production forecasts. Projections of man-hour and machine-hour needs for future periods are derived from time estimates for operations and anticipated production volumes for those periods.

It is likely that this plant pays production employees according to an incentive wage plan. Under this system the employee is paid according to the extent to which he exceeds a certain standard output rate established for his job. Specifically, he is ordinarily paid on the basis of the following ratio:

$$\frac{\text{time officially allowed for work completed}}{\text{time actually required to complete that work}}$$

To operate such a plan it is necessary to have an "allowed time" for each operation against which to compare actual time consumed by the employee. This same basic ratio may be used as a basis for decisions concerning promotions, pay raises, and corrective action (e.g. further training, transfer) with respect to the employee. This same basic ratio may also be used to evaluate an entire manufacturing department and its supervisor, by accumulating the following ratio:

$$\frac{\text{total hours officially allowed for work completed over a given period}}{\text{total hours actually required by the department to complete that work}}$$

for a certain interval of time, say a month. If this ratio is significantly greater than one, the foreman of the department should be credited with doing an effective job. If it is significantly less than one, management will probably display considerable concern.

The company will be interested in these allowed times for its operations for other reasons. These allowed (standard) times are used as the bases for standard costs for operations and products throughout the plant. For example, if the standard time for an operation on a certain product is 0.10 hour and the worker on that job receives $1.50 per hour, the standard labor cost for that operation on the product is $1.50 per hour × 0.10 hour, or $0.15. (This is a very simplified explanation of the standard cost principle.) By comparing actual costs incurred with these standard costs for operations and products, management can pinpoint those cases in which operational and product costs are excessive and in need of corrective action.

Note that there are two basic types of uses of time estimates in a manufacturing enterprise: planning and evaluation. Uses falling in the planning category, such as scheduling, anticipation of labor needs, bidding, pricing, and deciding whether to "make or buy," are decisions concerning future courses of action of the enterprise—decisions as to *what* to do, *how* to do it, and *when.* Instances in which the time estimate is used as a basis for evaluation, as in the case of the incentive method of wage payment, the standard cost system, and control budgets, involve a decision as to *how effectively* the operator, foreman, machine, department, etc., are performing their assigned tasks. Uses in this category entail appraisal of operating performance.

The following example may clarify the distinction between these two types of uses of time estimates. Suppose that operator A customarily requires an average of 10 minutes to complete a unit of product on operation X. The estimated time for him to complete a scheduled batch of 100 units on this operation is

$$\frac{10 \text{ min.} \times 100 \text{ units}}{60 \text{ min. per hour}}$$

or 16.7 hours. The fact that operator A is expected to spend 16.7 hours on this job will be taken into account by the scheduling department as it schedules subsequent operations on this batch of 100 units, and as it schedules operator A and his machine for this and other jobs. Furthermore, suppose that in view of performance throughout the plant, 12 minutes per unit would be considered normal for this operation. When operator A subsequently performed operation X on the 100 units, he actually required only 15 hours in total or 9 minutes per piece on this occasion. The norm of 12 minutes indicates that operation X was performed at a relatively fast rate, for the ratio

$$\frac{\text{time officially allowed}}{\text{time actually taken}}$$

is $\frac{12}{9}$ or 1.33, and under a wage incentive plan, operator A would be paid accordingly. If the base wage for operation X were \$1.40 per hour, he would receive $1.33 \times \$1.40$ or \$1.86 per hour for his performance on these 100 pieces. Note that the 12-minute figure is being used to evaluate the time in which the operation has been performed, whereas the 10-minute estimate is a prediction of the latter.

Notice the fundamentally different ways in which these two time estimates are used, and that the type of time estimate most appropriate for planning is not equally appropriate for evaluation and vice versa. For planning, where an attempt is being made to anticipate how long the operation will take, a straightforward performance time forecast is logical. This performance time forecast is not appropriate for evaluation purposes, however; a different type of time is necessary. In this case the logical type of time estimate to use is a benchmark, a norm, what in practice is called a standard (the 12 minutes in the foregoing example). In the wage payment ratio cited earlier, it is the standard time we want for the numerator, while it is the denominator that we wish to predict.

Thus, a *forecast* is needed for planning whereas a *standard* is needed for evaluation. The actual-time forecast should represent what the particular worker assigned to the job will do. Regardless of whether this time is satisfactory or not, this should be known so that it can be taken into account in planning the flow of materials and utilization of equipment. On the other hand, the standard time is independent of the particular operator assigned to the job and the rate at which he chooses to produce. For job X in the preceding example, the standard time is 12 minutes, irrespective of who performs the job. This is a time that will be used to evaluate rather than predict the worker's rate of output.

Although two fundamentally different types of time estimates are logically needed, it is common practice to establish only the standard time for an operation, then use this same time for evaluation *and* planning purposes. Therefore, emphasis in the remaining discussion will be on methods of obtaining the standard time for an operation.

Time Study (Work Measurement)

In many instances the performance-time estimate originally made by the designer for a new or improved operation is found to be sufficiently accurate after the job has been set up and running, to be used for the planning and evaluation purposes described. However, this does not always happen; the designer's initial estimate of the standard time for his creation must frequently be revised after the operation is installed and running smoothly, to be satisfactory for the foregoing purposes. Much of the discrepancy between the standard time predicted and the one that subsequently becomes appropriate is due to changes in method that take place between the design stage and the time the operation is actually set up and running satisfactorily. Ordinarily there is an appreciable lapse of time between phases I and III pictured in the diagram on page 205. Frequently there is opportunity and reason for deviation from the designer's original specifications during this period.

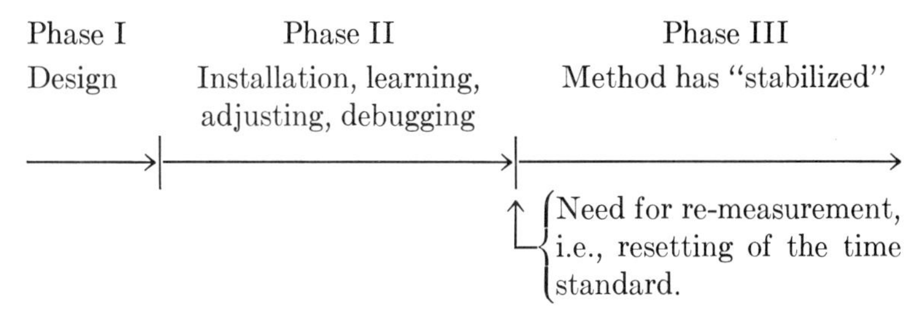

Some deviations from the specifications are made because the original design is unrealistic or because superior ideas are discovered in the interim. Other deviations are the result of attempts to adapt the specified method to a particular operator. Consequently, when the operation is in Phase III, installed and running smoothly, a special "time study" is often found necessary to establish a performance time standard that represents the method that has finally evolved.

Many such time studies are made entirely apart from the process of design, on those many operations about the plant for which the method has never been formally designed by an engineer. Management wants time standards for these operations as well. Therefore, because of the occasional need for remeasurement on formally designed operations, and because management wants time standards on all operations and attaches considerable importance to these standards, most firms have a time study department created for the sole purpose of establishing time standards for the company's manufacturing operations. The measurement process this department performs is referred to most often as time study. Although the alternate term work measurement is less common, its popularity is increasing.

Those relatively unfamiliar with this field are perhaps in for somewhat of a surprise, for what may seem at the moment to be a routine matter of watch reading and pencil pushing is in reality a very troublesome, controversial, explosive matter. It is the basis of an almost unbelievable amount of friction between labor and management.

One of the many factors that rule out any routineness in this process is the fact that the phenomenon being measured is affected appreciably by the very act of measurement. The process of making a time study is analogous to the police officer observing a group of mischievous boys. His very presence automatically brings about a change in behavior.

Another factor that complicates the matter is the fact that, in general, once this standard is set it cannot be arbitrarily reduced if the company eventually learns that the time is too liberal. There is a strict rule written into most company-union contracts that allows the company to reduce a standard time *only* if there is a significant change in the operation itself or if a clerical error was made in determining that standard.

An indication of its controversial and lively nature is the vital interest shown by unions in time study procedure and results. In fact, some unions have special departments created to advise and set union policy on matters pertaining to time study. Furthermore, there frequently are specialized union personnel called time study stewards, located at the plant to protect the rights of, to protest for, and bargain for workers on matters pertaining to time standards.

It is a field with a rather lively history, marked by a congressional investigation, by periods in which secret time studies were in vogue and outright cutting of standards at their discretion was practiced by managements, by scores of costly strikes and slowdowns, by myriad weird systems and schemes, scores of quacks, and so on.

Thus, there is considerably more to time study than first meets the eye. The problems arising from it are certainly numerous, far reaching, and often quite troublesome. In spite of its troublesome nature, the product of the time study process is a vital one to management. Do not underestimate management's interest in high rate of output, ordinarily a requisite to a favorable competitive position, or its need for advance estimates of what this productivity is going to be, so that activities and acquisitions may be planned accordingly.

INFLUENCE OF THE INCENTIVE METHOD OF WAGE PAYMENT

Because of the predominating influence the incentive method of wage payment exerts over time study practice, policies, and problems, it is desirable to digress momentarily to further discuss this matter. An understanding of the demands wage incentives make as to type and quality of time standards needed, and an appreciation of the magnitude and bases of the difficulties incurred if these demands are not met, are extremely helpful in understanding the ways and problems of time study.

The Principle

The incentive wage principle involves establishment of a standard rate of work for a job, the exceeding of which is rewarded by extra pay. For example, the standard rate of output for an operation has been established at 5 units per hour. If an operator produces 50 units in an 8-hour day, under the common type of plan his "basic wage" will be increased by a factor of

$$\frac{\text{actual production for the period: 50 units}}{\text{standard production for the period: } (8 \times 5) \text{ units}}$$

or 1.25 for that period. If we assume that his hourly base wage is $1.60, as a result of having bettered the standard by 25 per cent this worker will be paid $1.60 × 1.25, or $2.00 per hour.

Ordinarily this pay "bonus" varies in direct proportion to the degree to which the standard is exceeded. This is referred to as a one-for-one arrangement, and under it, for example, a rate of output 25 per cent above standard brings a 25 per cent increase in pay. Graphically, this plan appears as shown in Figure 69. Notice that the plan has a "guarantee." At or below standard the employee is guaranteed his base wage in the case illustrated.

Type of Standard Desired for Incentive Wage Purposes

The standard production rate or its reciprocal, standard time per unit, has some exacting requirements since it is a major determinant of take-home pay.

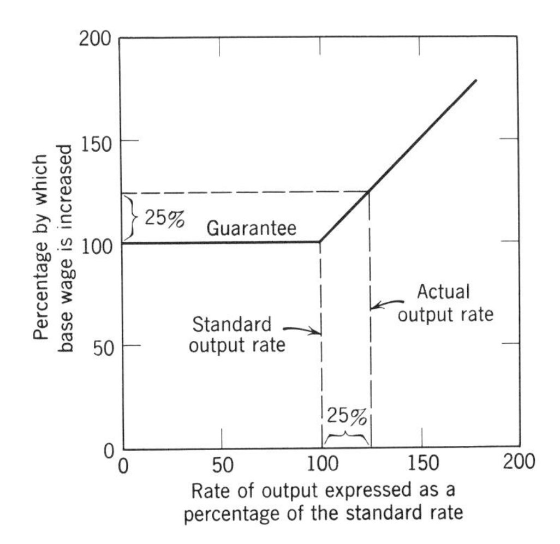

Figure 69. The incentive wage principle illustrated for the common "one-for-one" plan. Bettering the standard by 25 per cent gives the worker a 25 per cent increase in hourly earnings.

1. The standards on different operations should permit approximately the same opportunity for incentive earnings, that is, they should be *consistent* with one another with respect to incentive earning opportunity. If the standards on some jobs in the plant are such as to make it easy for persons to surpass and appreciably exceed the standard, and the standards on others are so tight that there is no incentive to even try to meet the standard, troublesome relations are going to develop among employees and between employees and company. Under such inequitable standards, an unbelievable amount of dissatisfaction, friction, transferring, juggling, and generally poor labor relations result, *at considerable cost* to the company.

2. The output rate the standards of a company represent should not be so low (loose) that very little effort is required by the operators to meet and beat those standards, and thus little or no increase in productivity achieved for the extra wages paid out. On the other hand, the standards should not be so high (tight) that workers feel that the extra effort required to meet those standards is excessive and consequently work slowly on the job, collect their guarantee, and thereby defeat the whole purpose of the wage incentive plan.

A survey of current practice indicates that standards are generally set so as to permit the average worker to excel the standard by 30 per cent. (But this 30 per cent is a figure that has "evolved" through the years, and is not the result of any formal decision nor is it necessarily optimum.)

Satisfaction of requirement 2 is not difficult for there appears to be a wide range of performance levels that are reasonable and seemingly satisfactory. But there is no equivalent leeway in requirement 1. The work force is very sensitive to inconsistencies. Operators can easily detect inequities between standards on different jobs. They are understandably quick to become dissatisfied with the situation and are not easily pacified—earnings are at stake. Therefore, if a method of setting time standards yields even moderately inconsistent results, costly difficulties will ensue. But not only is consistency a critical characteristic of a group of time standards, it is a difficult one to achieve in practice. Currently available methods of setting time standards unfortunately do not yield sufficiently consistent results.

In spite of the inability to establish time standards that are as consistent as they should be for this purpose, wage incentives are widely use in industry, especially among the large and medium-sized firms. The primary purpose of employing this method of payment is to achieve higher productivity and thus more efficient utilization of facilities. If the incentive wage plan can induce the worker, on the average, to produce 50 per cent more than he would under a straight hourly method of payment (e.g. produce 150 instead of 100 units per hour), presumably the company can obtain the output it wishes with only two-thirds the number of machines, amount of floor space, etc., it would have otherwise required.

Interaction Between the Incentive Wage Method and Time Study

The degree with which the incentive method of wage payment influences time study theory and practice is best illustrated by the fact that the type of time estimate ordinarily produced by the time study process, namely a time *standard,* and the level of performance this standard usually represents, are obviously attempts to satisfy the requirements imposed by the incentive wage scheme. In other words, every effort is made to produce a standard that will be the most suitable for the incentive wage plan as if the standard were to be used only for that purpose, which of course it is not.

It will soon become evident that wage incentives influence time study in numerous other ways. The fact that management cannot ordinarily tighten a standard once it is set, even if that standard is grossly in error, is traceable to the incentive use of standards. The same is true of the pronounced change in the behavior of an operator when a time study is taken of him, and of the numerous restrictive clauses in company-union contracts concerning the specifics of time study procedure; also of the severe opposing pressures exerted on the time study technician from management and labor for lower labor cost on the one hand and higher wages on the other, and of many other procedures and circumstances found in time study practice. In fact, general thinking in the time study area is obviously strongly influenced by wage incentives, as evidenced by much of the writing in the field.

It is also true that time study, primarily through its inadequacies, strongly affects the operation of an incentive wage plan. Some of the major difficulties arising from the use of wage incentives result because of inconsistencies in the time standards on which such schemes are based.

THE CONCEPT OF A STANDARD TIME FOR AN OPERATION

A measurement standard is a denominator (base) for expression of a characteristic or phenomenon in quantitative terms. Examples of common standards are the yard, the second, the degree, and the pound.

Like any standard, this unit of measure is *arbitrary,* the only requirements being that it be *agreed* on by the population using that standard and that it be *communicable.* We could satisfactorily express distance in terms of a certain elephant's tail, as long as the persons involved knew what was meant. Therefore, a standard need not be universal, it need only be a matter of agreement within the population that intends to utilize that standard. This population could consist of just two individuals, or as we will soon learn, of the personnel of a given plant or company.

The Standard Performance Time for an Operation

It may be helpful to express the concept of a standard performance time in "layman's terms" first, and to do so from several points of view. One such point of view utilizes the notion of a special and to be sure, hypothetical, worker who performs all operations in the plant at the same *rate* of work. For timing purposes he could always be relied on to perform with the same speed of movement on each operation he was asked to execute. The standard time for any given operation then is the time required by this special operator to perform a cycle of the job in question. It is, therefore, the time to complete that operation when performed at a *rate* of work common to all other operation standard times in the plant.

The second point of view employs a hypothetical device called a "standard-minute meter," with which the standard time for an operation supposedly can be measured directly. This hypothetical meter yields the time it would take to complete a given operation *as though* it were being performed at the standard *rate* of work, regardless of the rate at which the observed operator actually performs the operation. The operator may work very fast, very slow, or at any rate in between, but the meter ignores this and always reads the same for that operation as long *as the method remains the same.* Regardless of the speed of performance actually observed, this meter automatically reduces the actual time to standard minutes. The observed time is reduced to a time representing a certain level of performance uniform among all operation standard times throughout the plant.

Henceforth, the rate of work on which a company's time standards are based will be referred to as normal performance. There are a number of prevalent misconceptions concerning the nature of normal performance. Therefore, it should be stressed that:

1. The particular rate of work designated by a company as normal performance is purely an arbitrary matter. There is nothing scientific, nothing magic about it.

2. The rate of work selected is ordinarily one that may be exceeded by most workers without overexertion.
3. Normal performance is:
   a. Not the optimum rate of work;
   b. Not an average level of performance for a department, plant, industry, or nation. It does not derive from any specific population of workers.
   c. Not universal, it is a company-wide, "local," within-plant matter. There are substantial differences in the rates of work designated by different companies as normal performance. For example, at one company, walking at a rate of 88 yards a minute has been designated as normal performance. At another company, a rate of 94 yards a minute is called normal, at another 100 yards a minute, at another 112 yards a minute, and so on. These differences exist, they are here to stay, and contrary to the beliefs of some, the situation is not one to be concerned about.
   d. Not what some rather naively refer to as the "fair rate of work," a concept that apparently derives from the notion of a "fair day's work." This is a hopelessly impractical concept as long as rate of work and rate of pay are considered as separate matters. In fact, even when rate of work expected and rate of pay offered are considered simultaneously, the matter of fairness of the combination is rather elusive.

In addition to the concept of a certain arbitrary rate of work called normal performance, the commonly held definition of operation standard time recognizes that over the long run no one can repeatedly perform an operation without encountering some delays and retarding factors, and that some of these are unavoidable as far as the operator is concerned and should be provided for in a practical standard. Therefore, the standard time for an operation shall be defined as:

> The time to complete one cycle of an operation when performed with a *given method* at a *certain arbitrary* rate of work, with provision for delays and retardations beyond the operator's control.

This time is frequently expressed per unit completed, which will differ from the time per cycle when more than one unit is completed during that interval. Note that this is a within-plant standard, not expected to be transferable to another plant situation, far from universal.

General Methods of Measuring the Standard Time for An Operation

Unfortunately, there is no "standard-minute meter" or any satisfactory equivalent. Instead, the following methods are used in practice to estimate the time standard for an operation.

1. *Extraction from past experience.* Relying on records or memory of past production experience with the given operation or ones similar to it, to provide a performance-time estimate.
2. *Direct observation and measurement.* This requires direct observation of the operation as it is performed, and making appropriate measurements of that performance. These two basically different methods of direct observation are in common use:
   a. Stop-watch time study.
   b. Work sampling.
3. *Synthesis.* By the use of specially derived tables, graphs, and formulas, it is possible to "build," to synthesize the standard times for operations without the necessity of actually making direct measurements of or even observing the operations. There are two synthetic methods of setting time standards in common use. They are:
   a. Standard data.
   b. Predetermined motion times.

The remainder of Part IV is devoted to further description of these alternative methods of establishing time standards.

Establishing Time Standards on the Basis of Past Production Experience

This method of setting standards can take place in a number of forms, three of which will be described. The three are:

1. Direct extraction of past production rates from company records.
2. Use of production data as in 1 but adjusted on the basis of judgment to correct for nonnormal performance and unrepresentative methods and conditions underlying the data.
3. Direct estimate by way of the practitioner's judgment, relying on his general experience in such matters.

The first of these alternatives entails a direct reference to production records to determine the amount of time that *has* been required to accomplish the operation in question, or one resembling it. For example, job X has been run three times over its history, with the following results:

Date	Units Completed	Hours Required
January 5	50	9.2
February 16	100	16.6
April 7	100	15.7

$$\frac{41.5 \text{ hours}}{250 \text{ units}} = \text{Average of } 0.17 \text{ hour/unit.}$$

Under this proposal, 0.17 hour per unit would be the "standard" for the job. Note, however, that if straight, unadjusted, production data is used as the basis of a standard, that standard will be essentially whatever the operators on the job want it to be. The "standard" here is simply what "has been" and, therefore, permits people to create the basis on which they will be evaluated. Such a time might well be appropriate and useful for planning purposes but not so for evaluation, particularly if it is to be used as a basis for wage payment. A wage incentive system based on "standards" of this nature would soon bankrupt a company.

The second alternative is an extension of the first, in which the production figure extracted from records, the 0.17 hour per unit above, is adjusted on the basis of judgment. The intent of this correction process is to make the time representative of normal performance and of current methods and conditions. It is quite possible, and in fact very likely, that the level of performance embodied in the production data is not the desired normal, so that a correction is desirable. Similarly, since changes in methods and conditions may well have occurred over the period covered by the production data used, a correction for same should be attempted. This "corrected" production figure is preferred over the unadjusted figure produced by the first alternative.

Under the third alternative, the practitioner bases the estimate on his recollection of past production experience with the job in question or similar ones and on the standards already set on similar operations. With the aid of this background information, the estimator arrives at a standard for the job on the basis of judgment. After some years of experience in time study work, a practitioner usually becomes surprisingly expert in his ability to establish standards by pure judgment. Over the years, writers have unjustifiably condemned this method of setting standards to the point of virtual exclusion from consideration. As we will learn later, under most time study procedures it is not uncommon for the practitioner to decide beforehand, on the basis of his judgment, approximately what he thinks the standard should be. He then proceeds to carry out the time study procedure in such a manner as to substantiate his prior conclusion. This being the case, why not simply rely on the original judgment? Perhaps there are sufficient reasons for not doing the latter, but the point is that judgment *is* being used extensively and probably not with the large error the textbooks and some practitioners imply.

Obvious advantages of basing time standards on past experience are the speed and low cost of obtaining the estimates. There are some important disadvantages however. One is the possibility that operators can partially or completely control the standards they are to be judged by, which is especially likely under the first of the foregoing alternatives. Furthermore, since this estimate is based on data that represent a *prolonged* period of *past* experience, it is possible that in the interim there have been changes in method, personnel, business conditions, materials, quality, and so forth, that the estimator is unaware of and that make the estimate unrepresentative of current conditions. In fact, there are likely to be some unknowns with respect to conditions existing, level of employee application, method used, and the like, for the data period, so that just what the resulting standard represents in these respects is uncertain.

Although there are more accurate and precise methods of setting time standards, this procedure is probably not as inferior as often assumed. It has very strong advantages in the low cost and rapidity with which a standard can be set. This would certainly seem to be a competitive method of establishing standards in those instances in which the accuracy and precision demands are not severe, as in short run, low volume jobs, especially if the more demanding uses of standards like wage incentives are not involved. That this method can produce standards satisfactory for wage incentive purposes is highly doubtful. Under many circumstances, however, extraction from past experience will probably be the minimum-cost method of setting standards in the long run.

13

Stop-Watch Time Study: Introduction

To make a stop-watch time study it is necessary to go into the shop and time the operation in question with a watch or other timing instrument. An operator is timed as he performs the job, usually for a number of consecutive repetitions (cycles). Ordinarily the period of time studied is relatively short, often less than an hour. In contrast, work sampling involves intermittent observation over a prolonged period of perhaps weeks or months.

The Need for Rating

A practical problem arises when the observer goes into the shop with his watch to make a time study of an operation. It is fairly certain that the observer will not find the operator(s) working at the desired normal rate of performance. This is to be expected, for there is a wide range of individual differences in ability among workers, so that for natural reasons we would not expect to find very many operators working at normal rate. Perhaps more important, however, is the fact that most operators realize that it is profitable to intentionally alter their behavior during a time study. Most operators are inclined to use a method that differs from the customary one for the job and to work at a slower pace. The objective of this altered performance is to increase the prospects of obtaining a liberal time standard. How, then, can the observer get his standard time when there is this unlikelihood of being able to time someone performing the operation at the desired rate of work? There are a number of possibilities, all something less than completely satisfactory. For example:

1. The observer could time himself as he performed the operation

at the normal rate. The fact that the observer would lack the necessary skill in performing the many operations he must set standards on, plus inevitable union opposition to the idea, combine to eliminate this possibility.

2. The observer might time the operation as performed by a special "demonstrator" operator, a skilled worker who performs operations at the desired rate for such timing purposes. The appeal this alternative would have to workers and union is obvious.

3. The observer can and, as it turns out, must accept whatever level of performance the operator on the job chooses to present, then on the basis of judgment estimate the ratio of the actual *rate* of work observed to the normal *rate* of work, and then adjust the recorded *time* accordingly in order to obtain the desired time. This is what is usually done in practice. The process of arriving at this adjustment factor is ordinarily referred to as rating.

The Nature of Time Study Rating

Specifically, rating requires:

1. that the observer visualize in his own mind what the normal rate of work is for the operation in question;
2. that he then estimate the ratio of the observed rate to this mental image of normal rate.

Suppose, for example, that five cycles of an operation are observed and that the watch times for these cycles are 0.68, 0.73, 0.76, 0.69, and 0.64 minute, with an average of 0.70. During the same period, the operator's performance is rated. He is judged to be working at 90 per cent of the normal rate of performance, meaning that the rating factor is

$$\frac{\text{observed rate of work}}{\text{normal rate of work}} = \frac{0.90}{1} = 0.90$$

The time sought by the observer is $0.70 \times 0.90 = 0.63$ minute; that is, the watch time adjusted by the rating factor. Notice that this is a reduction process in which the time observed is reduced via rating to the desired time in normal minutes. In rating, normal rate of performance is ordinarily designated as 100 per cent or unity, and actual performance is judged with respect to this.

It is evident that there is no objective, easily communicated "normal rate of work." Unfortunately, it is necessary for the observer to visualize on the basis of his judgment what constitutes normal work pace for the operation in question. It is necessary also to rely on his judgment in converting the observed difference between normal performance and actual performance to a numerical ratio. This judgment is not unguided; elaborate efforts are taken to develop and maintain the observer's rating skill. Even with these efforts, however, the procedure leaves much to be desired. Rating and related matters are discussed at length in Chapter 14.

Allowances

The time derived in the preceding section applies to those periods in which the operator is able to work at the job. What about those periods in which he is prevented from working by such things as equipment failures, stoppage of material flow, defective parts, personal needs, the effects of fatigue, and the like? Since these are ordinarily beyond the operator's control, it is necessary to make some provision for such delaying and retarding factors in the time standard, at least if a wage incentive system is being used. Yet the actual observation period of the stop-watch time study is a relatively short span of time, such that the sample includes few or none of the numerous delays an operator usually encounters over a workday. Thus, the period of observation that is adequate for ordinary cycles does not suffice as a sample period for delays. As a consequence, a small amount of time is added to the rated time to provide for the unavoidable production delays, personal delays, and the effect of fatigue. This adjustment is called an allowance. The magnitude of this adjustment for a given class of operations is usually determined by a special observational study considerably longer in duration and less frequently made than the stop-watch time study.

The General Formula for Estimation of a Standard Time by Stop-Watch Time Study

Returning to the foregoing example and assuming that 0.10 minute per cycle is to be allowed for delays and fatigue, the standard time is

$$[0.70 \times 0.90] + 0.10 = \text{standard time per piece}$$
$$0.63 + 0.10 = 0.73 \text{ minute per piece}$$

In general terms:

$$\begin{aligned}\text{standard time} &= [\text{representative observed time} \times \text{rating factor}] \\ &\quad + \text{allowance for delays and fatigue} \\ &= \text{normal time} + \text{allowance}\end{aligned}$$

The rated time, the bracketed term in the general formula, the 0.63 previously calculated, will henceforth be designated as the normal time. The normal time for an operation is the time it would take to complete a cycle of that operation if performed at normal work pace.

Thus, two separate and basically different adjustments are made to the time recorded for an operation in the course of establishing a time standard for the job. The intent of the first is to correct the recorded time for nonnormality of the work rate observed. The magnitude of this adjustment is determined by the rating process. The intent of the second adjustment is to provide for delays and retarding factors that the ordinary stop-watch time study does not measure. The magnitude of this adjustment is determined by certain relatively long-term observational studies. Although the two adjustments are fundamentally different, both exist because the recorded time is ordinarily nonrepresentative, with respect to work pace and with respect to the consequences of delays and fatigue.

General Stop-Watch Time Study Procedure

The general procedure for a stop-watch time study is as follows:

1. Preliminary Steps
   a. Contact the persons involved, such as the foreman and the operator.
   b. Check the method, equipment, quality, and conditions against specifications for same. Seek out and remedy "inefficiencies."
   c. Record information pertaining to operation, operator, product, method, equipment, quality, and conditions.
   d. Break down the work cycle into elements.
2. Collect Data, by timing and rating the operator.
3. Process the Data
   a. Compute representative watch time.
   b. Apply rating factor.
   c. Apply allowance.
4. Present Results.

The "preliminaries" are discussed in the remainder of this chapter, "collection of data," "processing the data," and "presentation of results" are treated in succeeding chapters.

STOP-WATCH TIME STUDY: PRELIMINARY STEPS

Once the straightforward matter of obtaining the foreman's consent to make the time study, of obtaining his approval of the method, materials, etc., and of making necessary explanations to the operator(s) on the job, the time study observer prepares to time the operator in the manner described below.

Check the method, equipment, quality, and conditions. The procedure, layout, equipment, quality of incoming and outgoing materials, and conditions should be checked beforehand, to make sure they are as specified. In addition, the method should be scrutinized for "inefficiencies," with special vigor if it has heretofore been unspecified. Economically justified improvements should be made before setting the time standard. This matter of checking for and making improvements before the time standard is established is a very important one. *Setting time standards on operations without a reasonable attempt to make improvements beforehand virtually guarantees a collection of loose, inequitable, and troublesome standards,* especially under an incentive wage plan. The explanation is simple. If a time standard is set on an inferior method, which is susceptible to numerous improvements the worker can easily make, he *will* make them *after* the standard is put into effect. In this manner he is able to "loosen" the time standard and utilize the resulting savings in performance time to his financial advantage under the incentive plan. There is no objection whatsoever to worker-initiated improvements in method. The objection is to having such improvements used to "beat" the incentive wage system, eventually causing a grossly inequitable time standard and wage structure. The preferred practice is to *first* make the improvements that can be conceived of and economically justified, *then* set the time standard. (Whatever worthwhile improvements the worker should conceive of after the standard is established should be solicited and paid for under the employee suggestion system.) The practice of setting time standards on "inferior" work methods is a major cause of failure of incentive wage plans.

Record information. This is a routine matter of identifying the the operation, operator, and product studied, and of making a record of the method, equipment, quality, and conditions that exist at the time of the study and *on which the time standard is based.* Note that the time standard for an operation is for a given method only; the two go hand in hand. It often happens, however, that the same time standard remains on the books and in use for a prolonged period of time, whereas in the meantime the method has gone through a progression of changes, so that the standard is obsolete and hardly representative of the current method. Thus, there is the need for continuing efforts to keep time standards up-to-date. There is a corollary need, for evidence for the worker's and union's benefit, that a change in method *has* taken place, if and when the company wishes to tighten up on a time standard to bring it up-to-date with the method. These two factors, the desirability of being able to recognize when a change in method has taken place and of needing proof that a legitimate method change has occurred, require that there be a record of the method on which a standard was originally based.

Break down work cycle into elements. It is common practice to subdivide the work cycle into moderately short phases of activity called elements, and record the time for each of these instead of merely for the whole cycle. The reasons for this are primarily to facilitate comparison of similar elements between time studies of different jobs, to permit separate rating of different phases of the job if desirable, and to permit the subsequent development of standard data. There are occasions when out of tradition more than for utility, timing is done by elements.

When an elemental study *is* desirable, the elements should be as short as can be satisfactorily timed and recorded. The minimal feasible element time varies considerably with the skill of the observer, the method of timing and recording, the context of the element, and still other factors. Generally, the typical observer using the traditional stop-watch, clipboard, observation sheet, and "continuous" method of timing, should be able to time and record an element as short as 2.5 seconds without undue strain or error. Not so, however, if he should have to time a series of such relatively short elements in succession. Of course, special timing methods and equipment can be utilized to cope with the more demanding situations in this respect.

In selection of elements, manually paced (work rate determined by the operator) portions of the operation under study should be separated from those portions that are paced by the process (e.g. automatic machine feed or belt conveyor). In the selection of the endpoints (breakpoints) for elements, the points in the cycle at which the time is to be read and recorded, attention should be given to choosing as distinct points of demarcation as possible to facilitate reliable timing. Whenever possible, auditory rather than visual cues should be used to signify breakpoints, such as the sound made by a screwdriver as it is returned to the bench signifying termination of the element "tighten screws with screwdriver," or the noise made by the machine itself to signify the beginning and ending of the element "drill ¼-inch hole."

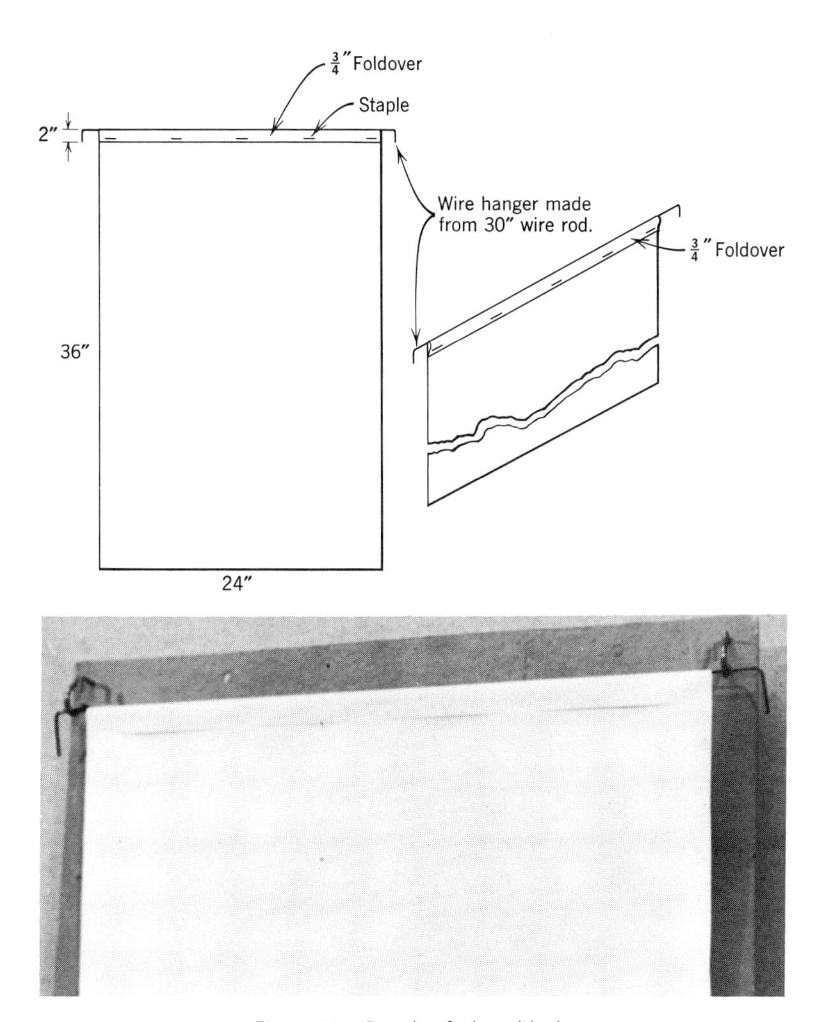

Figure 70. Details of chart blank.

Illustrative Case

At a certain office furniture and supplies plant a time study is to be made of the assembly of lecture chart blanks, 24- by 36-inch sheets of blank white paper with special wire attachments, to be sold for use with chart display easels distributed by the company. These chart blanks, made in a variety of standard sizes, or to special order, appear as illustrated in Figure 70 and are assembled as described by the standard method description, Figure 71.

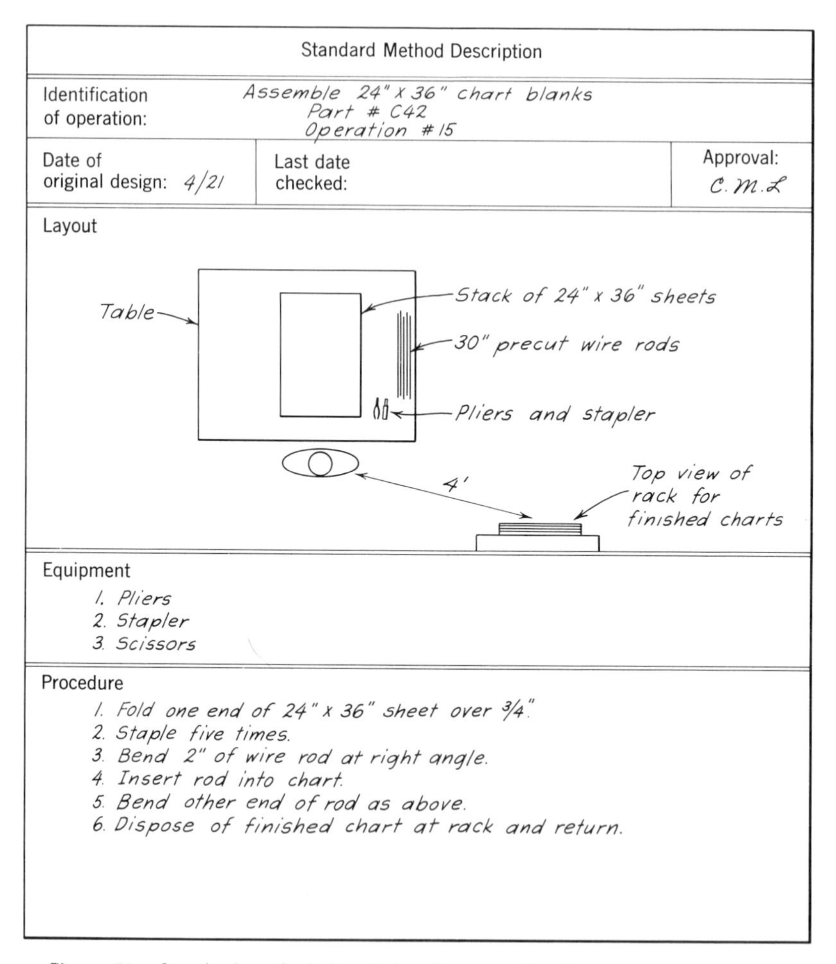

Standard Method Description

Identification of operation:	Assemble 24" x 36" chart blanks Part # C42 Operation # 15	
Date of original design: 4/21	Last date checked:	Approval: C. M. L.

Layout

Equipment

1. Pliers
2. Stapler
3. Scissors

Procedure

1. Fold one end of 24" x 36" sheet over 3/4".
2. Staple five times.
3. Bend 2" of wire rod at right angle.
4. Insert rod into chart.
5. Bend other end of rod as above.
6. Dispose of finished chart at rack and return.

Figure 71. Standard method description for assembly of 24" x 36" chart blanks.

To set a time standard on this operation the observer obtains a copy of the standard method description for this operation, proceeds to the shop, clears the matter with the foreman and operator, checks the method, etc., records what additional information he may need with reference to method, quality, etc., then breaks the operation cycle into elements in preparation for timing. In this case he has broken the operation cycle into the following elements with the breakpoints indicated:

1. Fold over end of 24- by 36-inch sheet.
   a. Beginning with "touch sheet."
   b. Ending with "grasp stapler."

2. Staple five times.
   a. Beginning with "grasp stapler."
   b. Ending with "drop stapler."
3. Bend and insert wire, bend other end.
   a. Beginning with "drop stapler."
   b. Ending with "drop pliers."
4. Dispose of finished chart.
   a. Beginning with "drop pliers."
   b. Ending with "touch next sheet at workplace," same as (1*a*).

Notice that dropping of a tool was used as a breakpoint wherever appropriate and possible. Element 3 would probably have been split up had there been a convenient breakpoint, but as it stands the motions involved in bending the wire, inserting it, and bending the other end blend together and overlap so that there is no distinct point for separation. After recording the foregoing elements on a standard observation sheet (illustrated in the following chapter), the observer is prepared to time and rate the operator.

EXERCISES

1. If the average observed time for a certain operation is 1.20 minutes and the operator is judged to be performing at 90 per cent of normal work pace, what is the estimated normal time for the job?

2. The normal time for a certain operation is 1.00 minute. *If* the rating process were errorless, what would the rating factor be in each of the following instances if the average observed time were as shown?

$$0.80 \text{ min.} \times (RF)_1 = 1.00 \text{ normal minute}$$

$$1.20 \text{ min.} \times (RF)_2 = 1.00 \text{ normal minute}$$

$$2.00 \text{ min.} \times (RF)_3 = 1.00 \text{ normal minute}$$

$$1.00 \text{ min.} \times (RF)_4 = 1.00 \text{ normal minute}$$

14

Stop-Watch Time Study: Timing and Rating the Operator

Timing Procedure

There are a number of accepted methods of using a watch to collect the necessary times, and numerous procedural details involved in the actual recording of data. A majority of the latter vary considerably from company to company. The preferred method of using the watch is referred to as the continuous method, where the watch is begun at the start of the study and remains running as the cumulative total time is read and recorded at each breakpoint.

This procedure is illustrated in Figures 72 and 73 for the operation "assemble chart blanks" introduced in the preceding chapter. Figure 72 shows a typical observation sheet used by the observer to record times during a stop-watch time study. The elements shown are those given earlier; the terminal points of these elements are given in parentheses. The times recorded on this sheet were taken by the continuous method of timing in the manner illustrated in Figure 73. Note that the watch is started at the beginning of element 1, at the instant the operator touches the sheet. At the end of element 1, signified by grasp of the stapler, the watch is read at 0.07 minute and recorded in the first "*R*" (reading) space under the first cycle. When the stapler is dropped, signifying the end of element 2, the watch is read at 0.23 minute and recorded in the "*R*" space opposite element 2. This process of recording cumulative time at the end of each element of each cycle is continued through a succession of cycles until sufficient data has been accumulated. During this period the watch is not stopped. In the present example the watch was started at 9:26 and remained running until 9:32. Since the observer must be able to account for all that

Time Study Observation Sheet

Identification of operation	*Assemble 24" x 36" chart blanks*		Date: *10/9*
Began timing: *9:26* Ended timing: *9:32*	Operator: *109*	Approval:	Observer: *M.W.S.*

	Element Description and Breakpoint		1 (0.00)	2	3	4	5	6	7	8	9	10	ΣT	$\overline{T}$	RF	NT
1	*Fold over end (grasp stapler)*	T	.07	.07	.05	.07	.09	.06	.05	.08	.08	.06				
		R	.07	.61	.14	.67	.24	.78	.33	.88	.47	.09				
2	*Staple five times (drop stapler)*	T	.16	.14	.14	.15	.16	.16	.14	.17	.14	.15				
		R	.23	.75	.28	.82	.40	.94	.47	4.05	.61	.24				
3	*Bend and insert wire (drop pliers)*	T	.22	.25	.22	.25	.23	.23	.21	.26	.25	.24				
		R	.45	1.00	.50	2.07	.63	3.17	.68	.31	.86	.48				
4	*Dispose of finished chart (touch next sheet)*	T	.09	.09	.10	.08	.09	.11	.12	.08	.17	.08				
		R	.54	.09	.60	.15	.72	.28	.80	.39	1.03	5.56				
5		T														
		R														
6		T														
		R														
7		T														
		R														
8		T														
		R														
9		T														
		R														
10		T														
		R														

Normal cycle time ________ + Allowance ___________ = Std. time ___________

Figure 72. Time study observation sheet for study of "assemble chart blanks."

transpires during this interval, this method of timing is preferred by labor unions. As indicated in Figure 72, the elapsed time "*T*" is subsequently determined by subtraction and entered in the appropriate block on the sheet.

There are other methods of timing, such as the snap-back method in which the hand of the watch is returned to zero after each reading. Similarly, there are numerous types of watches other than the one shown in Figure 73 in use, some offering various special features to eliminate the need for the observer to read the watch at the same instant the breakpoint cue is observed.

The recommended method of observing an operator is pictured in Figure 74. It is desirable that the observer keep the visual angle between operator and watch as small as possible so that while watching the performance, he may observe the position of the watch hand by means of peripheral vision.

How Long to Observe in a Stop-Watch Time Study

The element and cycle times observed in a stop-watch time study vary from repetition to repetition as demonstrated by the readings shown in Figure 72. Part of this variation is attributable to variability of the phenomenon being measured, due in this instance to natural variability that is characteristic of all human behavior, to variation in location of tools and materials from cycle to cycle, variation in material, and many other chance causes. The remainder of this variation in times is contributed by the measurement system itself, in this case the stop-watch and its reader. Figure 75 indicates the approximate nature of the frequency distribution of cycle times expected for a very large number of repetitions of a manual task by a given operator. The existence of this variation in times for repeated trials introduces a sampling error into the time-study process. In making a time study of an operator, the observer is in effect "drawing" a relatively small sample of N cycles from a distribution of cycle times, then on the basis of this sample, making an inference as to the magnitude of the mean ($\bar{\bar{T}}$) of this population of time measurements. Ordinarily it is certainly desirable to time more than one cycle. A representative number of cycles should be observed, but just what constitutes a representative number remains to be determined in each individual case. In practice, the decision as to an adequate number of cycles to observe is ordinarily based on the observer's judgment. In some instances a certain minimum observational period and/or number of cycles to be timed has been established as a matter of company policy or as a result of company-union agreement. Beyond this, however, the observer judges when he has accumulated sufficient data, taking into account such factors as the importance of the job

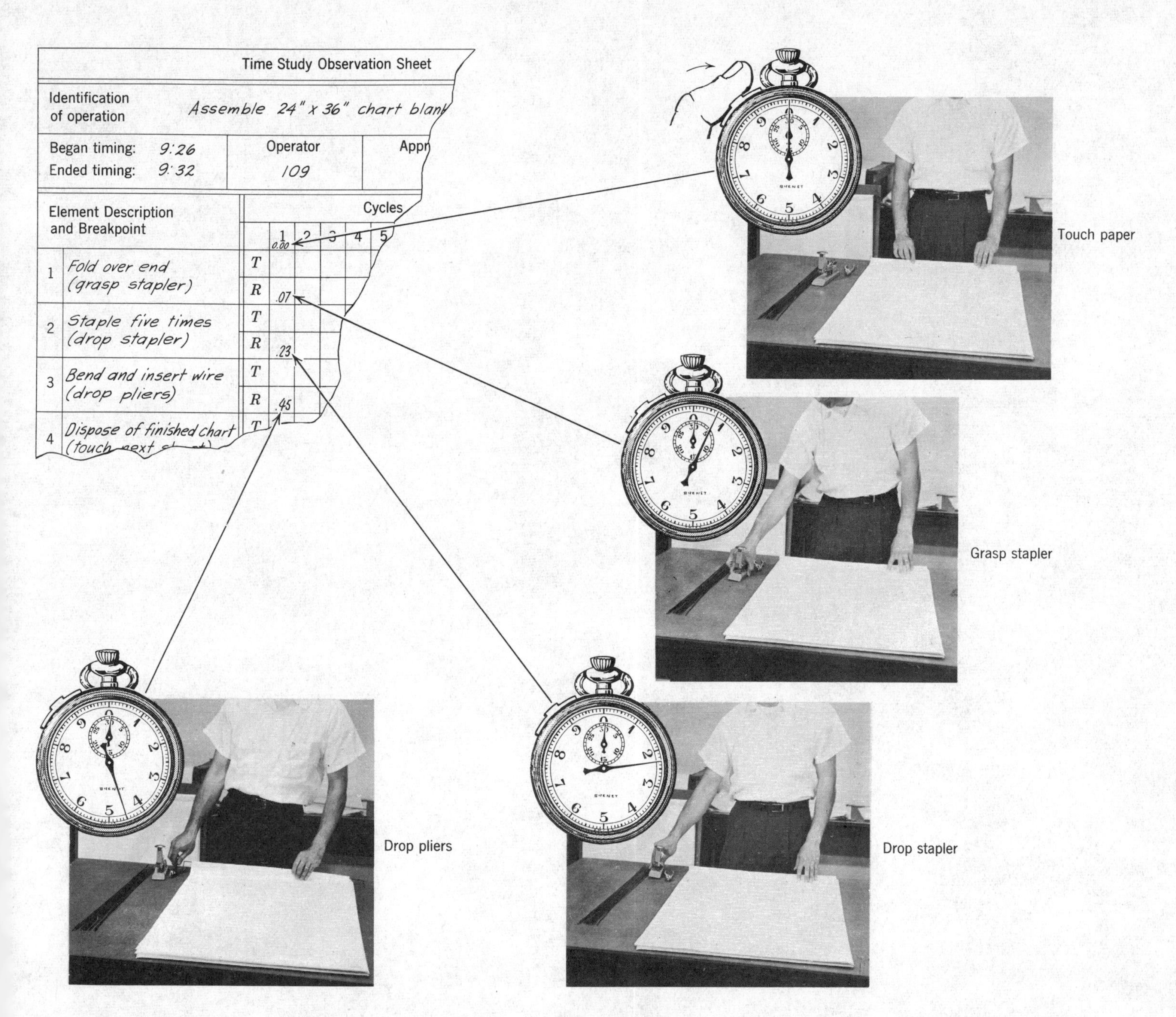

Figure 73. Illustration of the continuous method of timing the operation "assemble chart blanks." Watch is read in hundredths of a minute.

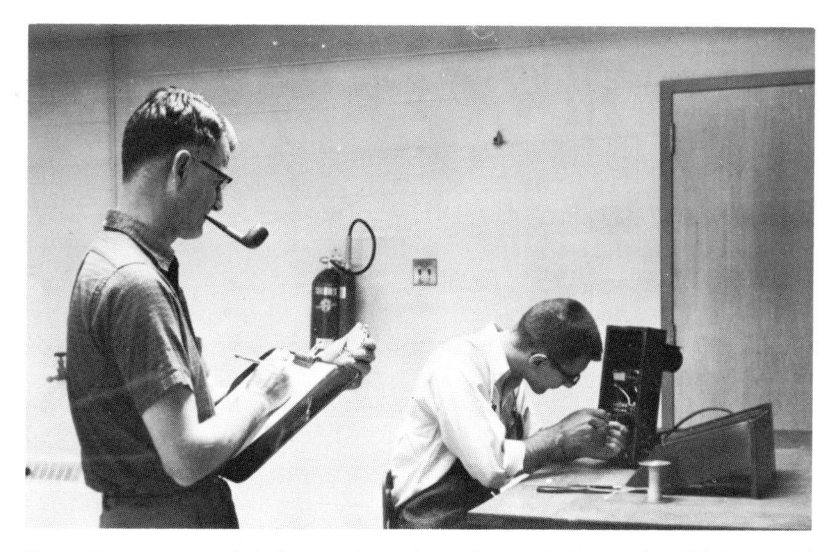

Figure 74. Recommended observer stance for a stop-watch time study. Clipboard with watch holder is conventional.

being time studied, smoothness of the operator's performance, typical cycle length, reputation of the operator timed, and other factors.

Statistical methods of estimating a satisfactory number of cycles to time. The choice of a satisfactory number of cycles to observe may be based on statistical methods of sample size estimation. Recall from elementary sampling theory that if many samples each consisting of N observations are drawn from a given population, such as the population of performance times shown in Figure 75, the means of these many samples will be approximately normally distributed as long as $N \geqq 4$, with a standard deviation of $\sigma/\sqrt{N}$, where σ is the

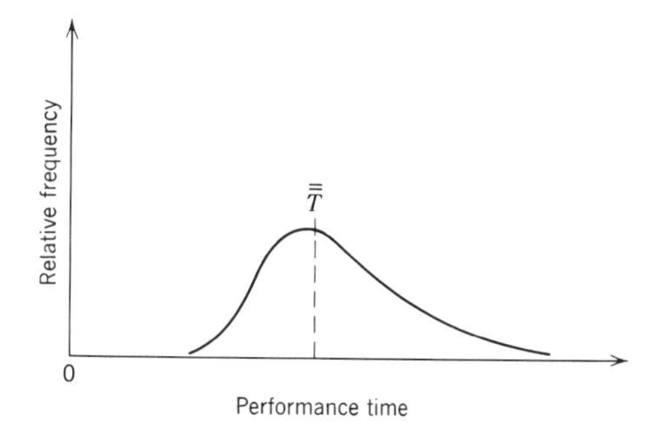

Figure 75. Generally expected frequency distribution of performance time measurements for a large number of repetitions of a given task by a given operator.

standard deviation of the population being drawn from. Recall also that if the standard deviation of the underlying population is not known and must be estimated from the sample itself, as is almost always true in practice, then Student's t-distribution should be used instead of the normal distribution to characterize the behavior of the sample means. Given the t-distribution as a model of the behavior of sample means, given a tolerable sampling error specified in terms of a confidence interval (I) and a confidence coefficient (C), and given an estimate of the standard deviation of the population of times being sampled, the engineer may utilize sampling theory to estimate the number of cycles required to satisfy that sampling error, by means of the following "two-stage" procedure.[1]

1. On the basis of the demands of the particular situation involved, specify a confidence interval I and a confidence coefficient C.
2. Time M cycles of the operation, say for $M = 10$.
3. Calculate the sample standard deviation (s) from the following formula:

$$s = \sqrt{\frac{\sum T^2 - \frac{(\sum T)^2}{M}}{M - 1}}$$

4. Calculate the confidence interval I_M provided by this sample of M observations from equation 1. The source and significance of this expression is indicated in Figure 76.

$$I_M = 2t_{0.90}\left(\frac{s}{\sqrt{M}}\right) \tag{1}$$

$t_{0.90}$ is obtained from a table of probabilities for the t-distribution for $C = 0.90$ and $M - 1$ degrees of freedom.[2] *If* $M = 10$ as suggested earlier, then $t_{0.90} = 1.83$ and equation 1 reduces to

$$I_{10} = 2(1.83)\left(\frac{s}{\sqrt{10}}\right) = 1.16s \tag{2}$$

If I_M is equal to or less than I, the specified confidence interval, the sample of M observations satisfies the sampling error requirement. That is, if $I_M \leqq I$, the M observations already taken are adequate. The sample mean ($\bar{T}$) may be satisfactorily based upon these observations.

5. If $I_M > I$, additional observations are required. The total number of observations required (N) may be estimated by

$$I = 2(t_{0.90})\frac{s}{\sqrt{N}}$$

$$N = \frac{4(t_{0.90})^2 s^2}{I^2} \tag{3}$$

[1] Based on Stein's two-sample test procedure as reported by Morton Klein, "Double Sample Estimation in Work Measurement," *Journal of Industrial Engineering*, vol. 10, no. 3, 1959.

[2] Values of $t_{0.90}$ for certain values of M are given in Appendix D.

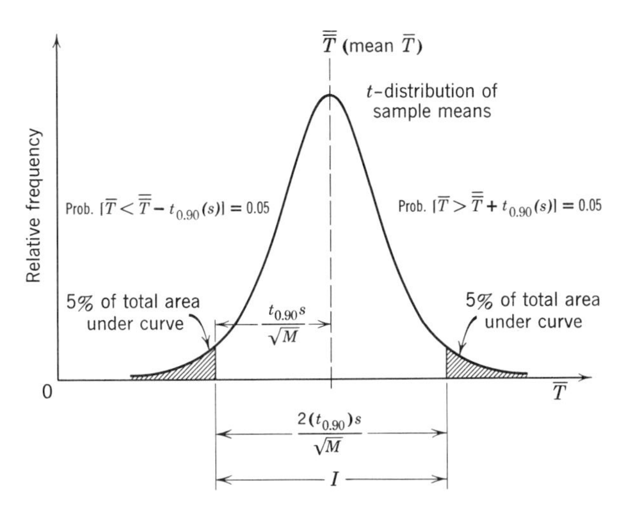

Figure 76. A distribution of means of samples of M observations each, drawn from a population of performance times exemplified in Figure 75. The significance and derivation of equation I, the sample size estimating equation, is indicated.

where $t_{0.90}$ is as before. If $M = 10$, this expression reduces to

$$N = \frac{4(1.83)^2 s^2}{I^2} = \frac{13.4s^2}{I^2} \tag{4}$$

The number of additional observations required is $N - M$. The final sample mean ($\bar{T}$) is based on the total number of observations (N) taken in the two samples.

The following case will serve to illustrate this procedure.

1. Considering the importance of the operation being studied, I has been set at 0.04 minute and C at 0.90. (Thus, the probability that $[\bar{T} - 0.02 \leqq \bar{\bar{T}}' \leqq \bar{T} + 0.02] = 0.90$, and the probability that the population mean will actually be outside the limits $\bar{T} \pm 0.02$ is 0.10.)
2. Ten cycles were timed with the following results:

T	T
0.35	0.32
0.33	0.39
0.40	0.30
0.37	0.39
0.34	0.41

$$\sum T = 3.60$$

$$\bar{T} = \frac{3.60}{10} = 0.360$$

3. The standard deviation of these ten times is

$$s = \sqrt{\frac{\sum T^2 - \left(\frac{\sum T}{M}\right)^2}{M - 1}} = \sqrt{\frac{1.3086 - \left(\frac{3.60}{10}\right)^2}{9}} = 0.037$$

4. From equation 2,

$$I_{10} = 1.16s = 1.16(0.037) = 0.043$$

which is greater than the 0.04 specified for I.

5. Since $I_M > I$, additional observations are required. From equation 4

$$N = \frac{13.4s^2}{I^2} = \frac{13.4(0.037)^2}{(0.04)^2} = 11.5$$

A total of twelve observations must be made. Since ten of these have already been made, only two additional cycles must be observed to satisfy the sampling error specified.

The foregoing procedure is time consuming, prohibitively so for many potential time-study applications. Fortunately, there are some shorter procedures that may be employed to reduce the estimation time and render the use of statistics more practical, in most instances, however, with a resulting increase in error of the sample size estimate. One that is slightly shorter and no less precise is the following procedure, generally recommended in time-study textbooks.

1. Proceed as before to select values for I and C that seem appropriate for the particular situation at hand.
2. Time M cycles of the operation, say for $M = 10$.
3. Calculate the sample standard deviation.
4. Compute the sample size required directly from equation 3, namely

$$N = \frac{4(t_{0.90})^2 s^2}{I^2} \tag{3}$$

where $t_{0.90}$ is obtained from a table of areas of the t-distribution for $C = 0.90$ and $M - 1$ degrees of freedom. As before, if $M = 10$, the foregoing expression reduces to

$$N = \frac{13.4s^2}{I^2} \tag{4}$$

If $N > M$, $N - M$ additional observations must be made to satisfy the sampling error requirement.

An advantage of the first procedure over the second is that computation of I_M after the M observations have been made may indicate that although $I_M > I$, the two are close enough (as the 0.043 and 0.040 in the example used) to make additional observations uneconomical. Both of these procedures are excessively time consuming for practical purposes, primarily because of the time required to compute s. The range ($T_{\max} - T_{\min}$) of the sample of M observations may be used to estimate s in a fraction of the time required to compute it from the expression

$$\sqrt{\frac{\sum T^2 - \left(\frac{\sum T}{M}\right)^2}{M - 1}}$$

The estimate of s is obtained from R/d_2, where R is the range of the observations and d_2 is a conversion factor, the value of which depends on M. For $M = 10$, $d_2 = 3.078$. Other values of d_2 for $M = 5$ through $M = 20$ are provided in Appendix E. Application of this method of estimating s to the example described on page 229, yields the following:

$$s = \frac{R}{d_2} = \frac{0.41 - 0.30}{3.078} = 0.036$$

Recall that by the long method, s was computed to be 0.037 for that sample of ten observations.

When this method of estimating s is used, the expression R/d_2 should be substituted for s in equations 2, 3, and 4. For example, equation 4 now reduces to

$$N = \frac{13.4\left(\frac{R}{3.078}\right)^2}{I^2} = \frac{1.42R^2}{I^2} \qquad (5)$$

Additional shortcut measures may be employed to further reduce the sample size estimation time, such as tables, curves, and alignment charts.[3] The usefulness, appropriateness, and theory of statistical procedures for sample size determination in stop-watch time study are appraised in Chapter 16.

RATING THE OPERATOR'S PERFORMANCE

At the conclusion of the timing period, the observer will have concurrently accumulated a certain number of actual performance times and an associated rating factor, now to be combined to produce an estimate of the normal time for the operation. Note that the rating factor is to be based on the same period of operator performance from

[3] For a description of the use of an alignment chart for this purpose see John M. Allderige, "Statistical Procedures in Stop Watch Work Measurement," *Journal of Industrial Engineering,* vol. 7, no. 4, July-August, 1956.

whence the times to be rated came. The observer should not time a series of cycles, then when the timing is complete, concentrate on and select a rating factor as if the two were separate and independent matters.

For the rating process to succeed and a useable standard to be established, the following three requirements must be reasonably satisfied:

1. The company must establish what it means by normal work rate, the level of performance it wishes its time standards to represent, and express this in some form that can be communicated.
2. A reasonable approximation of this concept of normal performance must be instilled in the mind of each rater.
3. The rater must develop the ability to apply this concept to various operations and produce reasonable numerical rating factors.

Unfortunately, the currently available means of accomplishing these three essential steps are relatively crude and leave much to be desired. The field does not possess objective means of deciding on, expressing, or applying a concept of normal performance. Some of the measures that are available are outlined below.

Establishing and Expressing the Company's Concept of Normal Performance

Just how a given company arrived at their particular concept of normal rate of work is a very mysterious matter and difficult to trace in most instances. The reason for this is that rarely has normal performance been established by an initial, carefully deliberated and specified decision. Rather, it is probably a matter of initially arriving at a vague and roughly specified notion of what the company wants normal performance to be, then letting the time study personnel determine through experience over a period of time the level at which normal will fall within these broad limits. In other words, if a general idea is given of what level is desired, after a period of time a certain concept of normal "evolves." To be sure, the level that does evolve is probably affected by the general level of worker performance that prevailed in the plant before standards were installed as well as by wage matters.

Although there is no one accepted or satisfactory method of selecting and expressing normal performance, these can be accomplished by the following means.

1. Arbitrary selection of a benchmark in terms of some common task, such as walking or card dealing. Thus a company might state that their concept of normal performance is walking at a rate of 3 miles per hour, or dealing a deck of cards at a rate of a deck per 30 seconds, and so on. The main limitation of such an expression is the error involved in extending the concept to a large variety of shop operations, most of which bear little resemblance to the relatively simple task selected.

2. Arbitrary choice of a rate of performing a representative shop operation or group of operations, as retained on motion picture film. By using actual, commonly encountered operations from the plant itself, and demonstrating normal performance for each of these, the limitation of the first alternative is mainly overcome. Of course, judgment is required in the selection of these representative normal performances, but this can be done with great care on the basis of pooled opinions.

3. Adoption of some packaged system such as a predetermined motion time system and the concept of normal performance embodied therein.

4. Employment of one or more experienced time-study technicians, letting their background be the primary determinant of normal, perhaps with some guidance by the management. This and many of the other means of settling on a rate of performance to be used as normal are predominantly "happen as it will" processes, that within limits leave the final result somewhat a matter of chance. Under these circumstances it is difficult or impossible for the practitioner to express what normal performance is.

Recall that the concept of normal differs considerably between companies. What company *A* calls 90 per cent performance, company *B* will call 125 per cent, company *C* will call 100 per cent, and so on. Although we might prefer it otherwise, this is not serious. But *within* the company every effort should be made to keep this concept uniform among raters and as constant with the passage of time as possible. Thus, the absolute level that a company selects for normal is not critical, but maintaining within company uniformity of interpretation and application of whatever level is chosen, *is* critical.

Instilling This Concept of Normal Rate of Work in the Mind of the Rater

Deciding on a particular performance level to use as "normal" is one matter. How to instill *and* maintain a clear and reliable image of this level in the mind of the rater is another problem. Words are of little assistance in this instance, the rater must be *shown* what is meant by normal performance. Apparently the most satisfactory means of achieving this is motion pictures of typical shop operations. By this medium normal performance can be demonstrated in a meaningful and convenient way. It is highly desirable that each rater study and practice-rate films of this type during his apprenticeship period, and in fact that he do so periodically even after years of experience.

Another method of instilling this concept in the rater's mind is by having him team up with an experienced rater and through his apprenticeship period rate with and learn from the "elder." This procedure seems acceptable but should be used in conjunction with and not as a substitute for thorough drilling with the filmed performances mentioned.

Developing a Reasonable Degree of Proficiency in Applying This Concept of Normal

A "definition" of normal performance satisfactorily imparted to the rater is necessary but not sufficient to satisfactory rating ability. The rater must develop the ability to satisfactorily compare the observed rate of work to his mental image of the normal rate for that operation, then convert this to a numerical rating factor. This is probably the most difficult and erratic phase of the rating process and, to be sure, is an ability that comes only with considerable practice. The most effective means of developing this ability seems to be drill with the previously mentioned motion pictures of shop operations. In addition to providing a means of demonstrating normal performance, these films may be used to provide practice in rating performances. The usual practice is to show the rater a variety of operations being performed at different rates, then to give him the known ratings for these so that from knowledge of his errors he can gradually improve the accuracy and precision of his estimates. Practice sessions of this nature should be repeated frequently at least until the observer's rating error has been reduced to a tolerable degree. Such drills should be supplemented by on-the-job practice under the guidance of an experienced rater.

Although these suggested methods of developing rating skill have limitations, they are certainly superior to merely pushing the trainee out into the shop and having him start to set standards, and forcing him to depend on some mysterious assimilation process to obtain a notion of what the company means by normal performance. Depending on chance instead of careful communication and thorough training in this matter of developing rating ability can be an expensive proposition to the company.

Specific Rating Systems

So far we have discussed rating as a general process. In practice one finds that there are a number of specific and rather different systems in existence to fulfill this need. These systems differ considerably in that a factor that one system recognizes and considers important, some or all of the remaining systems might ignore completely. These specific methods of rating can be classified according to the factors or variables they recognize, as follows:

1. *Pace rating,* which is a common term used to describe what has already been discussed in general terms. Thus, in pace rating the rater is judging the operator's work rate, his work pace, the *speed of his movements,* the rate at which he is applying himself. In contrast to some of the other rating systems, the rater is judging "how fast" the operator performs the motions involved, and *not* what movements the operator is using, *not* the operator's skill. Although this system is referred to as pace rating in this textbook and elsewhere, the same method of rating appears under such titles as effort rating, tempo rating, and speed rating (the latter being quite descriptive of the process that is taking place).

2. *Performance rating,* in which pace (see 1) *and* skill displayed by the operator are both judged. In one common form of performance rating these two factors are rated separately, whereas in another common version it is the net effect of the two that is judged. In the form requiring a separate judging of the pace and skill displayed, the ratings of these two factors are subsequently combined to yield a composite rating for the performance observed. The most common version of performance rating requiring a separate rating of pace and skill is the Westinghouse system of performance rating referred to also as leveling.[4] Under the Westinghouse system, a separate

[4] The confusion in terminology with respect to these various rating systems is extreme. For example, leveling is a term used by some as an alternate name for the Westinghouse system, by others as a general term for the process of rating. Pace rating is referred to by some as effort rating or performance rating. Performance rating is sometimes called effort rating. Leveling is sometimes referred to as skill and effort rating. As with a number of other topics in this field, a person must "clear the terminology air" before he can intelligently discuss the subject of rating with another.

adjustment factor for pace and a separate factor for skill are selected from Table 7. Assuming the observed performance is judged to be +0.08 with respect to skill and —0.04 with respect to pace, the resulting rating factor for that performance would be 1.00 + (0.08 — 0.04)

TABLE 7. The Westinghouse System of Performance Rating*

Skill			Pace†		
+0.15	*A*1		+0.13	*A*1	
+0.13	*A*2	Superskill	+0.12	*A*2	Excessive
+0.11	*B*1		+0.10	*B*1	
+0.08	*B*2	Excellent	+0.08	*B*2	Excellent
+0.06	*C*1		+0.05	*C*1	
+0.03	*C*2	Good	+0.02	*C*2	Good
0.00	*D*	Average	0.00	*D*	Average
−0.05	*E*1		−0.04	*E*1	
−0.10	*E*2	Fair	−0.08	*E*2	Fair
−0.16	*F*1		−0.12	*F*1	
−0.22	*F*2	Poor	−0.17	*F*2	Poor

* After Lowry, Maynard, and Stegemerten, *Time and Motion Study*, 3rd ed., McGraw-Hill Book Company, New York, 1940. By permission of the publisher.

† The authors of the Westinghouse system use the term effort instead of the more descriptive term pace.

or 1.04. Some users of this system rate conditions and consistency in addition to and in the same manner as skill and pace, and utilize tables of values similar to those in Table 7 for these additional factors. This author does not recommend attempting to judge the effect of conditions on performance or attempting to rate consistency independently of skill.

The other common form of performance rating involves rating the total performance—the net result of the skill and pace employed by the operator. Here it is the net effect of the operator's skill and pace that is being rated rather than the relative contributions of the two factors as in the Westinghouse system. The rater views the performance, compares what he sees to the performance he expects from an operator displaying normal pace *and* skill, then converts this to a numerical rating factor.

To compare the three methods discussed so far, suppose that a certain operator's performance is rated by the Westinghouse system at 0.06 for skill and 0.08 for pace, yielding a rating factor of 1.14. *Conceptually,* under the second performance rating method described, the total performance should be rated at 1.14. Theoretically, under the pace rating method this same performance should be rated at 1.08, recalling that pace only is rated.

3. *"Objective" rating,*[5] in which pace and job difficulty are rated. Under this procedure the operator's pace is rated just as in method 1, then a second adjustment factor for the difficulty of the operation is selected from Table 8. From these two separate estimates a composite rating factor for the observed performance is obtained. To illustrate this procedure, suppose that the performance of element *A* of a certain manual operation has been pace rated at 1.10, and that on the basis of the characteristics of this element the following selections were made from Table 8:

Category No.	Description	Reference Letter	Per Cent Adjustment
1	Amount of body used	*D*	5
2	Foot pedals	—	—
3	Bimanualness	*H*2	18
4	Eye-hand coordination	*K*	4
5	Handling requirements	*P*	2
6	Weight	*W*1	2
	Total adjustment for job difficulty 31%		

The rating factor for element *A* is the *product* of the pace rating of 1.10 and the difficulty rating of 1.31, so that the representative time for element *A* is multiplied by 1.44. Note that a rating for job difficulty must be obtained for each element of the operation. This is not necessary or likely for the rating of pace. Before attempting to apply objective rating the details of the procedure should be learned from a more complete exposition.[6]

Of the various rating systems just outlined, this author recommends pace rating but with certain additional provisions described in the remainder of this chapter, which concern means of treating the bother-

[5] M. E. Mundel, *Motion and Time Study,* 3rd ed., Prentice-Hall, Englewood Cliffs, N.J., 1960.

[6] *Ibid.*

TABLE 8. Table of Adjustments for Job Difficulty, Used in Objective Rating*

Category No.	Description	Reference Letter	Condition	Per Cent Adjustment
1	Amount of body used	*A*	Fingers used loosely.	0
		B	Wrist and fingers.	1
		C	Elbow, wrist, and fingers.	2
		D	Arm, etc.	5
		E	Trunk, etc.	8
		*E*2	Lift with legs from floor.	10
2	Foot pedals	*F*	No pedals or one pedal with fulcrum under foot.	0
		G	Pedal or pedals with fulcrum outside of foot.	5
3	Bimanualness	*H*	Hands help each other or alternate.	0
		*H*2	Hands work simultaneously doing the same work on duplicate parts.	18
4	Eye-hand coordination	*I*	Rough work, mainly feel.	0
		J	Moderate vision.	2
		K	Constant but not close.	4
		L	Watchful, fairly close.	7
		M	Within 1/64 inch.	10
5	Handling requirements	*N*	Can be handled roughly.	0
		O	Only gross control.	1
		P	Must be controlled, but may be squeezed.	2
		Q	Handle carefully.	3
		R	Fragile.	5
6	Weight	Identify by the letter *W* followed by actual weight or resistance.		Use Table below

Weight in Pounds	Per Cent Adj. Arm Lift	Per Cent Adj. Leg Lift	Weight in Pounds	Per Cent Adj. Arm Lift	Per Cent Adj. Leg Lift	Weight in Pounds	Per Cent Adj. Arm Lift	Per Cent Adj. Leg Lift
1	2	1	6	15	3	11	24	8
2	5	1	7	17	4	12	25	9
3	6	1	8	19	5	13	27	10
4	10	2	9	20	6	14	28	10
5	13	3	10	22	7	Etc.	Etc.	Etc.

* After M. E. Mundel, *Motion and Time Study*, 2nd ed., Prentice-Hall, Englewood Cliffs, N.J., 1955. By permission of the publisher.

some and inevitable variations in method. As for the other systems described, the Westinghouse system is subject to a strong and apparently valid criticism as the result of the seemingly tight adjustment values allowed for the higher levels of skill and pace. It does not seem likely, for example, that by increasing his pace from "average" to "excessive," Table 7, a worker could achieve only a 13 per cent increase in output. Furthermore, there is considerable doubt that worker skill can be judged satisfactorily for practical purposes. Although "objective" rating seems sound in principle, in substance it appears in need of further development with respect to the adjustments given in Table 8.

Deviations in Method

As long as the method observed is the method the time standard should be based upon, straight pace rating as described previously is ordinarily the procedure to employ. Unfortunately, however, to the never-ending dismay of the time-study technician, the method he is forced to time and rate often is not the one the standard should be based on or the one that will be subsequently used once the standard is set and put into effect. In time studying an operation, the observer is likely to encounter a variety of types of variation in work method. Four types of deviation from the standard method that are particularly problematical are described.

1. Random variations in method from cycle to cycle, such as variation in motion path, length and type of motion, number of motions, overlapping of motions, as well as miscues and irregularities. In reaching into a tote pan of parts in successive cycles, for example, the distance and motion path vary as location of the part grasped varies, and the type of grasp varies as a result of variation in the position in which the part rests, as do the number of regrasps, untangling motions, and fumbles for the same reason. The sources of random cycle to cycle variations in method are so numerous as to make the probability of any two cycles being exactly alike virtually nil.
2. An intentional and temporary change in method introduced by the operator only while the time study is being conducted, in order to inflate the cycle time with the hope of obtaining a more liberal standard on the job than would otherwise be obtained. This practice is common to the point where the observer should ordinarily expect it, at least whenever the wage payment method is an incentive scheme based on time standards. A classical example of this type of deviation

is found in an ordinary machining operation in which it is not uncommon practice to use a slower machine speed and feed during the time study than the operator expects to use after the standard is established. Or take a subassembly operation in which two hex nuts must be placed, turned down, and tightened. During a time study of this operation the operator uses a prolonged series of deliberate, individual finger motions to turn down the nuts. Ordinarily, he would skillfully spin the nuts down with several swipes of his index finger, in a fraction of the time the deliberate series of motions requires. In such cases the operator has no intention of using this special "temporary method" after the time standard is in effect.

3. A deviation from the standard method that inflates performance time as in 2, but in this case is unintentional. This may arise because the operator is improperly trained or because he is incapable of performing the job as it should be. Often this inability is due to insufficient opportunity to practice the operation under study, so that the operator cannot avoid the extra motions, fumbles, and the like.

4. A deviation in method that is an improvement over the standard method, an innovation, however, that is somewhat unique to the operator under observation and not expected of a majority of persons who might perform the operation under study. In this instance only the operator observed and possibly a small minority of other potential operators are capable of employing this particular method. This type of deviation may involve fewer motions, motions of a superior type, or a better integrated motion pattern, than should normally be expected for the task in question.

(This leads to a conclusion concerning the relationship between two commonly encountered terms: skill and method. *Skill is a matter of method.* Ordinarily, a manual worker is identified as exceptionally skilled because, perhaps in a rather subtle fashion, he uses fewer and/or different motions and/or a different arrangement of them than ordinarily expected or required. These are matters of method, of *how* the given task is performed. Herein the term skill will be used in reference to some aspect of work method that is expected only of a relatively small percentage of potential workers after fully trained and practiced.)

What is it that makes the foregoing types of deviation in method so troublesome? It is this: So often it is impossible or impractical to eliminate such deviations in method from the performance the observer is forced to observe. Obviously the random variations cannot be eliminated. In the temporary, intentional deviations, it is often impossible, many times infeasible, to attempt to force the worker to eliminate such a deviation *even if* the observer is certain that it is intentional and avoidable. Furthermore, in a majority of situations there is no opportunity for the observer to choose another operator if the one he has initially selected is not using the prescribed method, for in a majority of instances that operator is the only one performing that job. In addition, there frequently are restrictions on selection of the operator to be timed, imposed by the union. So in most cases the observer has little or no choice in the matter; it is that operator or none. The situation is aggravated by the fact the the operator that must be studied frequently deviates from the prescribed method as a result of inferior or superior skill. Yet if a time standard is to be set on the operation in question and based on the standard method, and if for any one of the foregoing reasons that standard method cannot be directly observed, what can the observer do? It is obvious that if the observer proceeds to time the nonstandard method and simply pace rates the performance, he is not going to get the desired result.

The Correction for Deviation in Method— A Third Basic Adjustment of the Observed Time

The performance an observer is confronted with as he is about to make his time study is likely not to represent normal work pace *or* the standard method for the task (and thus the method on which the standard time should be based). As a consequence, a third basic type of adjustment of the watch time is often necessary: one for deviation from the standard method for the task, this in addition to the pace rating and the allowance. The correction for deviation in method may be estimated by the following means.

1. If the deviation in method is extraneous and is sufficiently long enough to be satisfactorily timed with a watch, the time for this deviation can be subtracted out of the representative cycle time. For example, suppose that during the time study of a filing operation the operator uses approximately 20 strokes of the file when 15 would be adequate. Assuming the average time required by the operator for 20 strokes is 0.40 minute, the time for the necessary fifteen would be 0.40 minute × (15 strokes/20 strokes) or/0.30 minute. In this manner the time for the deviation, the 5 unnecessary strokes, is eliminated. Another illustration is provided by a drilling operation in which a small jig is used. Periodically, it is necessary for the operator to take the air hose and blow chips from this jig. Whereas it is ordinarily

necessary to do this about once every half dozen cycles, during a time study the operator does this at the end of each cycle. The normal time for the element "blow out chips" according to the time study is 0.05 minute. Since this element is necessary once every six cycles rather than every cycle, the time allowed per piece should in this case be 0.05 minute/6 or 0.008 minute.

2. If the deviation in method involves extra motions as in 1 but *too short* in duration to be timed satisfactorily by the stop watch, or if the deviation is a matter of using *less than* the expected number of motions, or if the deviation is a matter of using a *different* motion pattern than is prescribed, a predetermined motion time system may be used as the basis of the correction. A system like Methods-Time Measurement may be used to synthesize the normal time associated with the deviation in question and this added to or subtracted (as the case may be) from the normal time for the *cycle* obtained by timing and pace rating the work method actually presented. How predetermined motion times can be utilized in this manner is illustrated in the following examples.

In a certain short-cycle press operation the operator can and is supposed to pick up a part in preparation for the next cycle, while the part produced by the current cycle is being pressed. The operator observed during a time study of this operation cannot (or will not) do this. Instead, he waits until the stroke of the press is complete, and disposes of the finished part, *then* picks up the next part to place it in the die. The extra motions in terms of MTM are an $R6C$ and a $G4B$, both of which could be performed during rather than after the press time. If the normal cycle time determined by the stop-watch time study (again, in which *only pace* was rated) was 0.062 for the method the operator used, the normal time that should be allowed is

$$\begin{aligned} 0.062 - (R6C + G4B) &= 0.062 - (0.0061 + 0.0055) \\ &= 0.050 \text{ minutes.}^{7} \end{aligned}$$

For a second example, recall the assembly operation cited earlier in which two hex nuts were to be placed, turned down, and tightened. During a time study, the operator observed turned down the nuts by numerous deliberate motions rather than by spinning them with the index finger as could and should be expected under the circumstances. Assume that the normal time obtained from a stop-watch time study of the operation as performed by the inferior method is 0.26 minute.

[7] The times of the predetermined motion time system used for making such adjustments should be calibrated to represent the same level of performance used as normal in pace rating.

The use of MTM to synthesize the two methods of turning down the nuts yields the following:

"Deliberate" Motions

LH	Min.	RH	
Hold	0.0014	R1E	Repeat 6 times
↓	0.0010	G1A	
	0.0017	M1B	
	0.0010	RL1	
	0.0051		
	×6	(6 repetitions)	
	0.0306		
	×2	(nuts per assembly)	
	0.061	min.	

"Spinning" Motions

LH	Min.	RH	
Hold	0.0028	M2B spin	Repeat 3 times
↓	0.0028	M2B return	
	0.0056		
	×3	(3 repetitions)	
	0.0168		
	×2	(nuts per assembly)	
	0.034	min.	

To estimate from this data what the normal cycle time should be for the operation using the spinning method, the synthesized time for the deviation in method will be replaced by the synthesized time for the desired method, thus:

$$\text{Normal time} = 0.26 - 0.061 + 0.034 = 0.23 \text{ minute}$$

3. If the deviation in method cannot be avoided and cannot be satisfactorily compensated for by one of the foregoing means, the usual alternative is to attempt to adjust for the deviation as a part of the rating process. This is an undesirable practice at best, yet it may be the only alternative in some situations. For example, this procedure might be necessary if the observer encounters an operator that exhibits superior skill attributable to many small, subtle differences in motion pattern that cannot be satisfactorily synthesized and adjusted for by predetermined motion times. In a case like this, if that particular operator must be observed, performance rating must be resorted to. Recall that under performance rating, pace *and* skill are judged either separately or in aggregate. However, note what is being expected of the rater under this procedure, namely, that he "rate out" the effect of some deviation in motion pattern. Attempting to assign an appropriate number of rating percentage points to a deviation in method is an erratic procedure that probably greatly inflates the error of the rating process.

Thus, the only justification for using performance rating, which involves attempting to rate skill which is in turn a matter of method, is when the deviation in method cannot be eliminated or cannot be adjusted for by the more direct procedures recommended in 1 and 2.

Why attempt to estimate what effect a difference in number, type, or arrangement of motions will have on normal time for an operation by a gross judgment, when the effect can be measured directly in terms of time?

It appears then that if the observer can avoid the operators of inferior or superior skill and if the worker available makes no temporary change in method, or in other words, if the operator observed is employing the standard method (or approximately so), the observer can proceed to time the operation and employ simple pace rating. However, more often than not these conditions do not exist, the observed time must be adjusted for deviation in method by means of one or several of the alternatives discussed. Failure to make this adjustment introduces a sizeable expected error as the consequence of basing the standard time on a nonrepresentative method.

15

Stop-Watch Time Study: Processing the Data; Presentation of Results

Extracting a Representative Watch Time

The observation period yields a series of watch readings for each element. A single time value must be selected to represent each element and the whole cycle. Averaging is the preferred method of achieving this. To illustrate the recommended procedure, the observations taken during the time study of the chart assembly operation (see Figure 72) will be used. These times, reproduced in Figure 77, are processed as follows. The total of the elapsed times, designated as $\sum T$, is obtained for each element. For element 1 in Figure 77,

$$\sum T = 0.07+0.07+0.05+0.07+0.09+0.06+0.05+0.08+0.08+0.06 = 0.68 \text{ minute}$$

For each element, $\sum T$ is divided by the number of observations obtained for that element to yield the elemental average time ($\bar{T}$). $\bar{T}$ is often referred to as the representative time (observed time, watch time, recorded time). For element 1, $\bar{T} = \frac{\sum T}{N} = \frac{0.68}{10} = 0.07$.

A practical question frequently arises during this stage. One reading out of ten for element 4 is twice as long as some of the other times for that element. Nothing was observed to account for this; no timing or arithmetic error is apparent. Should this "abnormal" time be included in determination of the average time for that element? Practice differs widely on what to do with abnormal readings. In general, if the cause of the abnormal reading is a sufficiently rare event, or an event provided for through the allowance, or a timing error, it should be discarded. With the exception of ridiculously extreme cases,

Time Study Observation Sheet

Identification of operation	*Assemble 24" x 36" chart blanks*		Date *10/9*
Began timing: *9:26* Ended timing: *9:32*	Operator *109*	Approval *P.M.L.*	Observer *M.W.S.*

Element Description and Breakpoint			Cycles 1 (*0.00*)	2	3	4	5	6	7	8	9	10	Summary ΣT	$\bar{T}$	RF	NT
1	*Fold over end (grasp stapler)*	T	.07	.07	.05	.07	.09	.06	.05	.08	.08	.06	.68	.07	.90	.06
		R	.07	.61	.14	.67	.24	.78	.33	.88	.47	.09				
2	*Staple five times (drop stapler)*	T	.16	.14	.14	.15	.16	.16	.14	.17	.14	.15	1.51	.15	1.05	.16
		R	.23	.75	.28	.82	.40	.94	.47	4 .05	.61	.24				
3	*Bend and insert wire (drop pliers)*	T	.22	.25	.22	.25	.23	.23	.21	.26	.25	.24	2.36	.24	1.00	.24
		R	.45	1 .00	.50	2 .07	.63	3 .17	.68	.31	.86	.48				
4	*Dispose of finished chart (touch next sheet)*	T	.09	.09	.10	.08	.09	.11	.12	.08	.17	.08	1.01	.10	.90	.09
		R	.54	.09	.60	.15	.72	.28	.80	.39	1 .03	5 .56				
5		T														*0.55*
		R													*normal minute*	
6		T													*for cycle*	
		R														
7		T														
		R														
8		T														
		R														
9		T														
		R														
10		T														
		R														

Normal cycle time *0.55* + Allowance *(0.55 X 0.143) or 0.08* = Std. time *0.63* *min./pc.*

Figure 77. Completed time study observation sheet for study of "assemble chart blanks."

readings should not be discarded on the basis of their magnitude alone. Unrestrained arbitrary rejection of readings merely adds another major subjective element to the time-study process.

Calculation of the Normal Time

$\bar{T}$ is adjusted by the rating factor (RF) for the element involved to obtain the estimated normal time (NT). For element 1 in the foregoing study,

$$NT = RF \times \bar{T} = 0.90 \times 0.07 = 0.06 \text{ minute}$$

The normal time for the cycle is the sum of the elemental normal times, 0.55 for this operation.

In this particular study, the observer rated each element and in the course of doing so obtained some different elemental rating factors. In many cases no separate rating of elements is attempted by the observer. One aggregate rating factor for the performance is estimated, and this applied to the representative cycle time.

Application of the Allowance

After a representative time has been determined and the rating factor applied to this to obtain the normal time, the allowance for unavoidable production delays, personal delays, and fatigue must be added to yield the standard time. For the operation "assemble chart blanks" and similar operations at the company involved, specially conducted long-term surveys indicate that approximately 1 hour of the 8-hour workday is unavailable to the operator as a result of delays and the effects of fatigue, and that allowance should be made for same in the standard time. It remains then to determine how much time to add to the normal cycle time as a consequence of the 60 unavailable minutes in the day. Shown below are three acceptable and equivalent ways of prorating this 60 minutes.

1. $$NT + \frac{\text{Minutes in workday unavailable to operator}}{\text{Number of pieces operator can produce in workday, working at normal pace}} = \text{standard time}$$

$$0.55 \text{ min./pc.} + \frac{60 \text{ min./day}}{\left(\frac{480 - 60}{0.55}\right) \text{pcs./day}} = 0.55 \text{ min./pc.} + \frac{60 \text{ min.}}{764 \text{ pcs.}}$$

$$= 0.55 \text{ min./pc.} + 0.08 \text{ min./pc.}$$
$$= 0.63 \text{ min./pc.}$$

2. $$\frac{\text{Total minutes in workday}}{\text{Number of pieces operator can produce in workday, working at normal pace}} = \text{standard time}$$

$$\frac{480 \text{ min./day}}{\left(\frac{480 - 60}{0.55}\right) \text{pcs./day}} = \frac{480 \text{ min./day}}{764 \text{ pcs.}} = 0.63 \text{ min./pc.}$$

$$3.\ NT + NT\begin{pmatrix}\text{The ratio of time unavailable to}\\ \text{the time available for production}\end{pmatrix} = \text{standard time}$$

$$0.55 + 0.55\left(\frac{60}{480 - 60}\right) = 0.55 + 0.55(0.143)$$
$$= 0.55 + 0.08 = 0.63 \text{ min./pc.}$$

It appears that the most common method of expressing the allowance is in the form of a percentage, which indicates the amount by which the normal time should be increased to allow for the retarding and interfering factors beyond the operator's control. Thus, for a given job it is typical to say, "the allowance for fatigue is W per cent, for unavoidable delays is X per cent, for personal delays is Y per cent, and the total allowance for the job is Z per cent." Therefore, the third of the foregoing methods of applying the allowance is probably used the most frequently. Note that in method 3 the allowance was 14.3 per cent, that the normal time of 0.55 minutes was increased by an amount 0.55×0.143, and most important, that the 0.143 was computed from $60/(480 - 60)$ and *not* from $60/480$. (If, as is customary, this percentage is to be applied to the normal time, which includes no time for delays and fatigue, then the base for determining that percentage should be equivalent.) Application of the allowance for the time study of "assemble chart blanks" is illustrated at the bottom of Figure 77 where computation of the standard time is summarized.

In some instances the allowance will differ for different elements, as might be the case if one element in a cycle is clearly more fatiguing than the remaining ones. Under these circumstances the allowance must be applied separately to the element normal times rather than to the total normal time for the cycle.

Estimation of the Delay Allowance

Estimation of the allowance is not ordinarily an integral part of the stop-watch time study. Allowances are usually determined simultaneously for whole groups of operations which for this purpose may be considered homogeneous, for example, all operations of a given type, all operations in a certain department. In practice these allowances are estimated by a variety of ways, a number of which are not generally satisfactory.

Allowances for unavoidable production delays are sometimes established by overall company policy, by collective bargaining, or by similar procedures that frequently result in a flat delay allowance for all company time standards. For example, at the XYZ Company it is a matter of policy to allow 10 per cent for unavoidable delays in all time standards set. Under such a practice, delay allowances can hardly be expected to be representative of more than a small portion of the operations throughout the plant, for the expected length and frequency of delays vary considerably with the type of operation and the conditions surrounding it. In one department of the XYZ Company where the equipment is fairly new, operations are reasonably independent of one another, the supervision is excellent, and so forth, a production delay allowance of 5 per cent might be realistic. In another department of the plant where the circumstances are quite the opposite from this, a production delay allowance of 20 per cent might be appropriate. The 10 per cent allowance that is applied company-wide is not representative in either case.

It is desirable that an attempt be made to apply a production delay allowance that is realistic for each operation on which a time standard is set. To accomplish this it is necessary to estimate the appropriate delay allowance for each different set of circumstances. Although a uniform production delay allowance for all operations throughout the company probably will not suffice, it is not necessary to resort to the opposite extreme and make a separate determination for each job. A separate estimation for each group of operations that are essentially homogeneous with respect to delay conditions seems adequate. There are two widely used methods of estimating delay allowances: the production study (interruption study) and the ratio-delay (work-sampling study). Both are prolonged studies of production activity ordinarily extending over at least one workday.

The production study is an extended period of continuous observation in which all activities of the operator are timed and recorded. To illustrate, at the conclusion of a one-day production study, a tabulation of the data collected reveals that 440 of the 480 minutes observed were found to be available to the operator for production. Forty minutes were consumed by unavoidable production delays. On the basis of this one-day production study, the unavoidable delay allowance for this type of operation is estimated to be

$$100 \times \frac{\text{40 minutes of unavoidable delay}}{\text{440 minutes of production time available}}$$

or 9 per cent.

The ratio-delay study, to be elaborated on in a later chapter under its alternate title, work-sampling study, involves intermittent observations usually over a week or more. Such a study will yield an estimate of the unavoidable delay time and the production time available so that the unavoidable delay allowance can be computed as above. Since this type of study ordinarily extends over a much longer period than is feasible in a production study, the ratio-delay study is a superior sampling procedure and is generally conceded to yield a superior estimate. The proportion of time consumed by delays is known to vary considerably from day to day purely on the basis of chance. The production study, by encompassing only one day as seems to be the usual case, therefore becomes vulnerable to a rather large sampling error as a result of the between-day differences. The ratio-delay study is advantageous in this respect because more days are sampled but each to a lesser extent. For this and other reasons, the ratio-delay study is receiving increasing preference over the production study as a method of estimating delay allowances.

The discussion of delay allowances thus far has related to the unavoidable production delay. Estimation of the personal delay allowance is a different matter. This type of allowance is a provision for legitimate personal needs. The time required for this purpose is not likely to vary significantly with different operations and departments, and therefore is more amenable to determination on a plant-wide basis by such methods as company policy or collective bargaining. Therefore it does not seem unreasonable that the XYZ Company would have a flat policy of allowing, say, 4 per cent for personal delays on all operations employing male workers and, say, 6 per cent for same on all operations employing female workers.

Estimation of the Fatigue Allowance

The currently available means of estimating fatigue allowances leave much to be desired. Judgment must be relied on heavily. Apparently the most satisfactory method of estimating fatigue allowances for a company's operations is a scale of representative allowance values, as illustrated in Figure 78. Certain commonly occurring jobs in the plant are assigned positions on the allowance scale as demonstrated in Figure 78, on the basis of the most objective information available from the fields of medicine, physiology, and psychology, and from industrial fatigue researchers.[1] These key jobs, then, serve to guide the assignment of allowances to other operations throughout the company for which the fatigue allowance must be estimated. To

[1] As for example the unique investigations of fatigue, work, working conditions, and the like, conducted by Dr. Lucian Brouha and his staff at the E. I. DuPont Company's Haskel Laboratory.

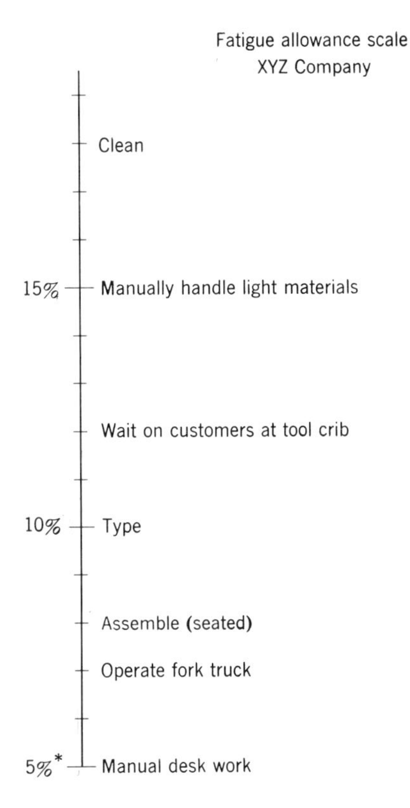

Figure 78. A simplified fatigue allowance scale, the suggested means of establishing fatigue allowances.

estimate the fatigue allowance for a job, the location of that activity on this scale must be judged. The best that can be said of this procedure is that it is a guided judgment process and the most satisfactory means of estimating fatigue allowances economical for practical day-to-day allowance setting. To be sure, there are many less desirable procedures in common use, some of which are to be described.

Some companies have a company-wide fatigue allowance, established arbitrarily or by collective bargaining, which is applied uniformly to all standards set. The same criticism that was raised concerning this practice in the determination of production delay allowances applies here. Since different jobs throughout the plant vary considerably in physical demands, repetitiveness, and the like, the fatigue allowances should vary accordingly. The absolute level of a company's fatigue allowances is not critical. The important matter is that the allowances applied to different operations be consistent with respect to one another, for example that a job that is twice as fatiguing as another job be assigned twice as much fatigue allowance.

Some authors claim that the fatigue allowance for a job can be estimated by means of a prolonged study of rate of output on that job, by a production study for example. The purpose of the prolonged study (of at least one workday) is to determine to what extent total output and rate of output are diminished over a prolonged work period presumably as the result of fatigue. Such studies of "output decrement" do not provide satisfactory estimates of fatigue allowances. A worker's rate of output *might* exhibit a general decline through the day, presumably but certainly not necessarily a result of fatigue. More likely, however, there will be *no* significant downward trend in production rate through the day, which does not mean that the worker has not suffered from fatigue. It is likely that as his fatigue increased, he increased his effort to compensate for the effects of same, and kept his output rate at or close to the same general level through the day. Feelings of tiredness are not necessarily reflected in the worker's output rate. Apparently many workers learn through experience approximately how much they want to produce in a day, then spread this fairly evenly over the day. Furthermore, the worker takes a certain number of breaks in a workday, some of which are primarily for social purposes, others primarily for resting (recovering from fatigue). Actually, the worker might keep his production rate at an even keel through the day by taking sufficient strategically placed rest pauses, which the ordinary time-study practitioner discounts as avoidable delays. Thus, the effects of fatigue may be absorbed by increased effort, rest pauses, or slowdown in rate of work, in varying proportions. *Only if* fatigue were primarily reflected in the production rate, could drop-off in output rate through the workday be considered a valid index of fatigue.

The general inability to estimate the fatigue allowance satisfactorily leads to some extraordinary practices in this respect. Some rather weird and fancy formulas, graphs, charts, etc., can be found. They give an illusion of objectivity that does not exist, since these devices themselves were originally based primarily on judgment.

Special Allowances

On occasion, allowances are added for other than delays and fatigue. For example, in a case in which the volume of production is very low, there may be a special "small-lot allowance" to provide for the fact that the length of run is so short that the operator never gets out of the initial learning period. Or in the case in which the equipment paces the operator during part or all of the work cycle, a special allowance is sometimes added to permit the operator to "earn" incentive pay even though the output rate of the machine is limited. (Adjusting earnings by altering the time standard is poor practice.) Still other special allowances are encountered in practice, some sensible and necessary, others not.

FINAL PHASE OF A STOP–WATCH TIME STUDY: PRESENTATION OF RESULTS

Frequently, a time standard does not go into effect until it is approved by one or even several different persons in managerial and supervisory positions. The results, data, and computations associated with a time study should be clearly presented, explained, and justified to facilitate securing this approval, for purposes of explanation to workers and union personnel, and for future reference. Furthermore, as mentioned earlier, a specific record of the method and other circumstances underlying the standard given should accompany that time value.

EXERCISE

1. A stop-watch time study has been made of a light machining operation. The results are given in the following figure. For this type of operation it is estimated that 405 minutes of the 480 minute day are available to the operator for production purposes. Complete the computations to determine the standard time and the number of pieces per standard hour.

Time Study Observation Sheet			
Identification of operation: *Operation 10, Part # 10725A1*			Date: *7/29*
Began timing: *3:22* Ended timing: *3:34*	Operator: *1065*	Approval:	Observer: *P. Maguire*

	Element Description and Breakpoint		1 (*0.00*)	2	3	4	5	6	7	8	9	10	ΣT	$\overline{T}$	*RF*	*NT*
1	*Chuck part and engage feed*	*T*														
		R	*.14*	*.32*	*.49*	*.62*	*.79*	*.96*	*7 .11*	*.28*	*.45*	*.59*			*0.85*	
2	*Cut 7/8 inch diameter*	*T*														
		R	*1 .06*	*2 .24*	*3 .40*	*4 .54*	*5 .71*	*6 .87*	*8 .03*	*9 .19*	*10 .36*	*11 .52*			*1.00*	
3	*Remove part and dispose to rack*	*T*														
		R	*.15*	*.33*	*.48*	*.62*	*.81*	*.96*	*.11*	*.29*	*.45*	*.62*			*1.05*	
4		*T*														
		R														
5		*T*														
		R														
6		*T*														
		R														
7		*T*														
		R														
8		*T*														
		R														
9		*T*														
		R														
10		*T*														
		R														

Normal cycle time ________ + Allowance ____________ = Std. time _____ *min/pc.*

16

Appraisal and Improvement of Stop-Watch Time Study

ERRORS IN MEASUREMENT SYSTEMS

Before engaging in a detailed discussion of stop-watch time-study errors and their reduction, it is desirable that some of the general characteristics of errors and measurement systems be introduced.

Two Types of Error

A measurement system, a stop-watch and its user, for example, contributes two fundamental types of error to the measurements it makes: a constant error (bias) and a variable (chance) error. To facilitate introduction and explanation of these two types, consider a situation in which test samples are being weighed on a sensitive balance as a step in the analysis of these samples. The balance and the analyst who uses it constitute a measurement system, and like all measurement systems, this one contributes error to the measurements made.

The variable error. This type of error may be isolated in the sample weighing process by running the *same* test sample through the process a number of times, preferably without the analyst being aware of the re-entry. The results of such an experiment are presented in Figure 79. In this instance, the same test sample has been weighed 150 times by the same measurement system, involving the same balance and same analyst. Since it is the same quantity that is being measured, the source of variation in the readings is the measurement system itself.

Thus, this weight measurement process and in fact all measurement systems contribute variation to the measurements they yield. To each measurement the system contributes an error that varies *by chance*

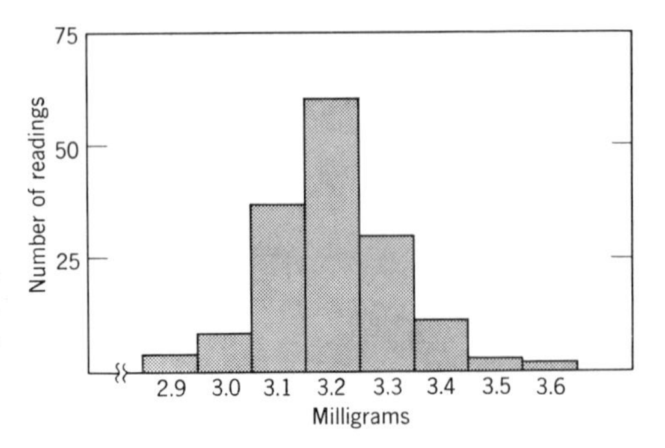

Figure 79. One hundred and fifty measurements of the same quantity. Same sample repeatedly weighed by the same analyst using the same balance.

in magnitude and direction (positive or negative) from one measurement to the next. The degree of variable error is generally referred to as the precision of the measurement system. Precision may be illustrated by target shooting. Suppose two riflemen fire a series of rounds and produce the targets pictured in Figure 80. The spread of man B's shots indicates that he is more precise in his firing than is man A.

Accuracy, bias, constant error. Notice in Figure 80 that although man B is more consistent than man A, both have an inclination away from the bull's-eye, both tend to shoot to the right. Such a tendency away from the bull's-eye is referred to as a bias or a constant error. Notice that if each man appropriately adjusts the sight on his gun, he is able to correct for this bias and bring the "average" of his shots

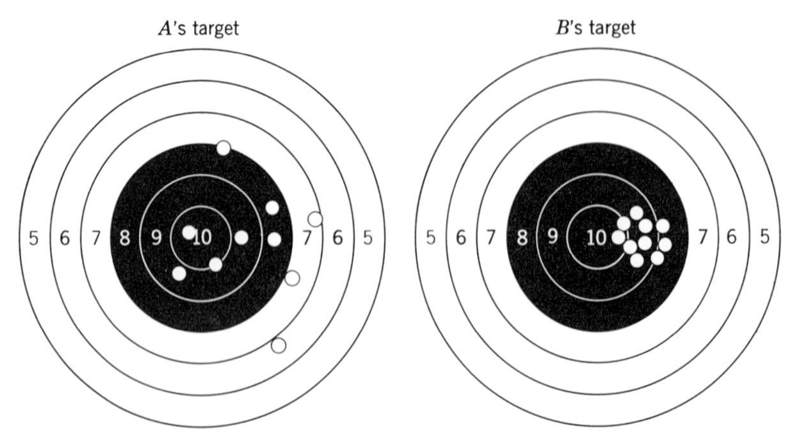

Figure 80. Man B appears to be a more precise rifleman than man A. (From Chapanis, Garner, and Morgan, Applied Experimental Psychology, John Wiley and Sons, New York, 1949. By permission of the publisher.)

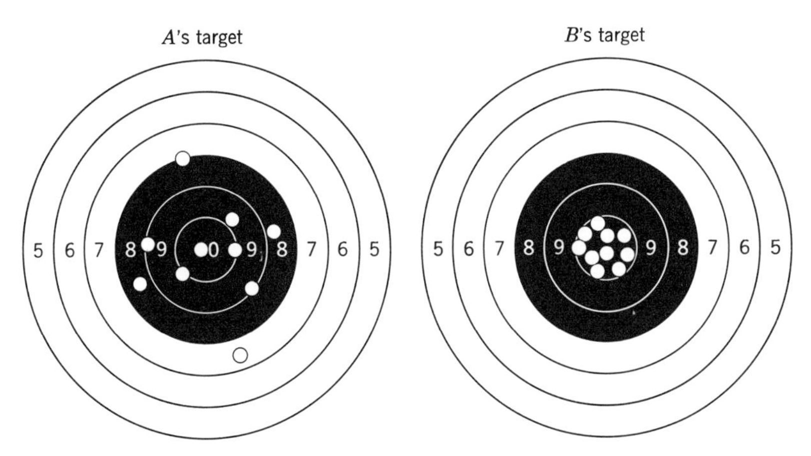

Figure 81. The same series of shots shown in Figure 80 after elimination of the constant error. (From Chapanis, Garner, and Morgan, Applied Experimental Psychology, John Wiley and Sons, New York, 1949. By permission of the publisher.)

to the center of the bull's-eye, as pictured in Figure 81. By adjusting the sights, he alters each shot by a constant amount, that amount being the distance between the center of the bull's-eye and the original location of his "average shot" as pictured in Figure 80. This introduces the concept of accuracy. Just as the rifleman and his rifle have done, a measurement system almost always contributes a certain uniform amount of error to each measurement, as well as a variable error. Phrased in another manner, the system is seldom "centered" on the correct or true value. This type of error is most frequently referred to as a bias, and the degree to which a system is biased is commonly referred to as its accuracy.

The two types of error then are:

c, which is a constant error, a bias, a uniform error present in every measurement made by the system, and

v, a variable error, an error that varies by chance from measurement to measurement.

Thus, the measured quantity (X_m) of an actual quantity X is

$$X_m = X + c + v$$

The Determination and Estimation of Accuracy

Accuracy is not estimated as easily as precision, but like precision, it is never *known,* never *determined,* only *estimated.* Let us refer to

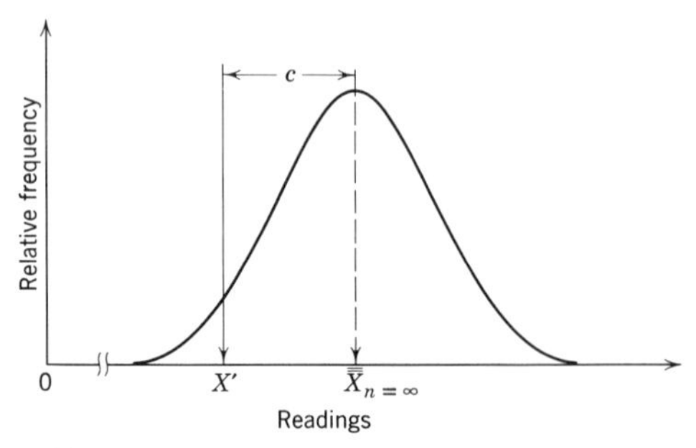

Figure 82. Determination of accuracy (constant error or bias) for a hypothetical case.

the initial example of weighing test samples. *If* the precision of that system were perfect and therefore the system contributed no variation to its measurements, then to determine accuracy it would only be necessary to weigh the original standard pound (or a reasonable facsimile) once and note the deviation of the reading from one pound. This would give us the accuracy of our system. But all measurement systems contribute some chance variation to their measurements; chance variation is always superimposed on the constant error. Therefore a sampling error is introduced, and to obtain a satisfactory estimate of the accuracy of our weighing system, it will be necessary to weigh the standard pound a number of times. Because of this sampling error, accuracy could be *determined* (as opposed to estimated) only by making an infinite number of measurements of a given quantity, the true value of which is known. Then the bias of the system would be the difference between the true value (X') and the mean of the infinite number of measurements ($\bar{\bar{X}}_{n=\infty}$), thus

$$c = \text{constant error} = \bar{\bar{X}}_{n=\infty} - X'$$

The manner of determining accuracy of a system is illustrated for the hypothetical case in Figure 82.

Inability to obtain an infinite number of readings is not the only obstacle to the determination of accuracy. Ordinarily we would need a second measurement system with zero bias with which we could determine the true value X' of the quantity being measured.[1] How-

[1] This zero-bias system is not necessary if the object being measured is an original arbitrary standard, such as exists for the foot and the pound, but for practical purposes this alternative is seldom feasible.

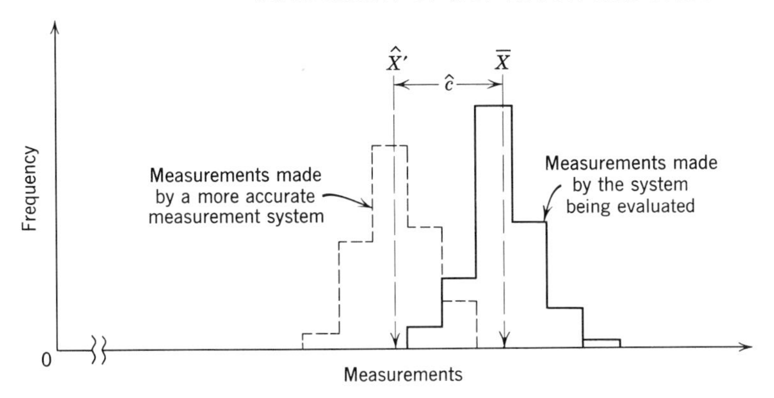

Figure 83. Estimation of accuracy for a hypothetical case.

ever, there is no measurement system that is perfectly free of bias. In practice, accuracy is estimated as follows.

1. Obtaining a "known" quantity, which is either an arbitrary standard such as the pound, or a quantity that has been estimated by what is believed to be a more accurate measurement system than the one being evaluated. Since the latter is an estimate of the true value X' it will be denoted by $\hat{X}'$, based on a finite but substantial number of measurements as pictured in Figure 83.
2. Next making a finite number of measurements of this "known" quantity $\hat{X}'$ with the system being evaluated. The mean of these measurements is represented as $\bar{X}$ and is an estimate of $\bar{\bar{X}}_{n=\infty}$.
3. Then an estimate ($\hat{c}$) of the accuracy of the system is $\hat{c} = \bar{X} - \hat{X}'$.

The method of estimating accuracy of a measurement system for a hypothetical case is illustrated in Figure 83.

Accuracy and Precision in Summary

The precision of a measurement system may be described as the degree of chance variation (chance error) a measurement system contributes to the measurements it yields; or as the degree to which a measurement system can reproduce the measurement of a given quantity. Precision is commonly measured in terms of the variance (σ^2) of repeated measurements of the same quantity. Accuracy of a measurement system may be described as the degree to which the system is biased; the constant error of the system; or the degree to which the mean of an infinite number of measurements of a quantity differs from the true value. Any measurement made by a system is influenced by the accuracy and precision of that system, so that the measured quantity differs from the actual or true quantity by a certain amount, a part of which is the same for every measurement made by the system and the remainder of which is variable, so that $X_m = X + c + v$. Ordinarily, when quoting quantities it is the X_m, the measured quantity, that is being dealt with and not the actual quantity.

Accumulation of Errors in a Measurement System

In general, measurement systems have three major components:

1. A *sensing mechanism* (sensor), the means by which the system determines the limits of the quantity to be measured. For example, in measuring time, the beginning and ending of the event must be detected. In ordinary stop-watch time study the observer is the sensor, and in an automatic timing system a photoelectric circuit, or micro-switch, or some such mechanism is the sensor. In measuring the size of an object with a micrometer, the jaws of the micrometer act as the sensor.
2. A *translation mechanism,* the means by which the sensed quantity is converted into equivalent units of measurement. In timing, the interval being measured must be translated into seconds, minutes, etc., and this is accomplished by a watch, or electronic timer or similar mechanism. Size measurement by ruler or micrometer is accomplished by a scale indicating inches.
3. A *readout mechanism.* The number of measurement units to which the sensed quantity is equivalent must be extracted from the translation mechanism. This is the function of the readout mechanism. In ordinary stop-watch time study the observer accomplishes this visually as he reads the watch. In an automatic timing system this might be accomplished by a device which ultimately prints the result on paper tape. In measurement by ruler this is accomplished visually.

Each of these components is capable of contributing some bias and chance variation to the measurements, resulting in a total error for the system. Denoting the constant error of the components as c_s, c_t, and c_r respectively, the net bias of the system is the algebraic sum of these individual biases, thus

$$c = c_s + c_t + c_r$$

where any of the c's may be positive or negative. This statement applies to the accumulation of bias in any system. Bias of the components of a system, then, *may* partially or completely compensate for one another, so that the net accuracy of the system is better than the accuracy of any of its component parts. And, unfortunately, the reverse situation can be true.

The accumulation of variable errors of the system is different. Using the variance as an index of precision, the total variance contributed by the system is the sum of the component variances, provided that the variable errors are independent of one another. Thus, the precision of a system is

$$\sigma^2_{\text{system}} = \sigma_s^2 + \sigma_t^2 + \sigma_r^2$$

under the assumption that the variable errors of the individual components are independent of one another. The variance of the *observations* is:

$$\sigma_o^2 = \sigma^2 + \sigma_s^2 + \sigma_t^2 + \sigma_r^2$$

where σ^2 is the variance of the variable itself. If the variable errors of some or all of the components are correlated with one another, the variance contributed by the system is

$$\sigma^2_{\text{system}} = \sigma_s^2 + \sigma_t^2 + \sigma_r^2 + 2\rho_{st}\sigma_s\sigma_t + 2\rho_{sr}\sigma_s\sigma_r + 2\rho_{tr}\sigma_t\sigma_r$$

where ρ is the coefficient of correlation between the respective variable errors of the system's components. However, a more than negligible degree of correlation between any two component variable errors is unlikely.

Reduction of Error

The main difficulty in improving the accuracy of a system is in obtaining a satisfactory estimate of the degree of bias; but once this is available, correction is usually relatively simple. This correction may be achieved by a simple adjustment of the system, as would be done to your watch to correct for slowness or fastness and to the rifle sights to compensate for a bias, or by making a compensating adjustment to the measurements after they are made. For example, the more expensive thermometers have marked on them a certain correction, determined by the manufacturer's calibration, to be made to each temperature reading. Thus, the estimated bias of a system may be compensated for by adjusting (calibrating) the system or by adjusting the readings, with a relatively small expenditure.

Reduction of bias may be illustrated by an example from time study. Suppose the plot of estimated errors a time-study man has made in rating a number of operations appears as shown in Figure 84. Apparently he has a tendency to rate high. This being so, we may improve the situation by informing him of this and requesting him to attempt to correct for this bias as he rates, or by allowing him to continue on as he has been and always subtract an amount $\hat{c}$ from his ratings. The former alternative seems more practical.

An important and frequent misconception concerning correction of error is that a bias may be compensated for by taking additional readings. This is not so. This error is the same regardless of what the sample size is. And note, reducing the bias of one component of a system may actually *increase* the net bias of the system. For example, if the biases of three components of a system are $+2$, -3, $+2$, yielding a net bias of $+1$, and we reduce the bias of the second component from -3 to -1, we have increased the net bias to $+3$. Therefore, due consideration should be given to the biases of all components of a system before tampering with any one of them.

The main difficulty in improving the precision of a system is in reducing the error, not in estimating its magnitude. The variable error of a mechanical component of a measurement system is commonly inherent in the device, in its very physical characteristics, such as the quality of bearings, slack, susceptibility to effects of atmospheric changes, quality of workmanship represented, and so on. More often than not the situation can be improved only through replacement by a more expensive mechanism. All or a major portion of the variable error of a human component is ordinarily inherent. In many instances

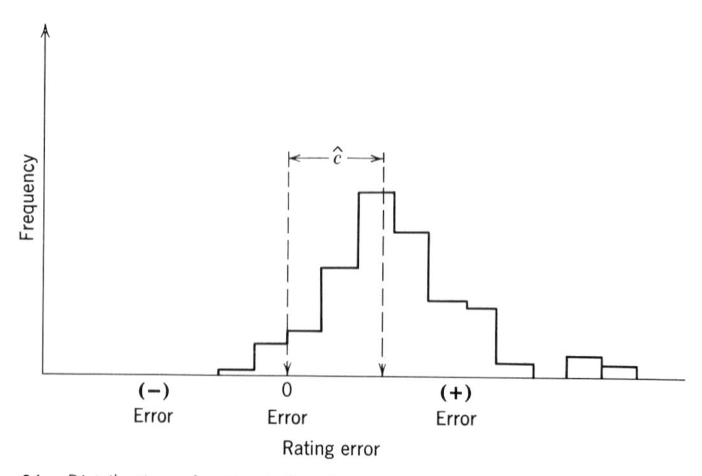

Figure 84. Distribution of estimated rating errors made by a certain time-study man.

substantial improvement may be achieved only by replacement of the person.

Frequently it is more economical to take additional readings to reduce the effect of the variable error, rather than to attempt to alter the source of the error. Both of these consequences of an objectionable variable error ordinarily involve a large expenditure relative to the cost of improving accuracy.

The problem involved in attempting to improve precision may be illustrated by the time-study case cited in Figure 84. If this rater's variable error is objectionable as it appears it might be, there is little that we can do to improve the situation except replace him or hope that additional experience will reduce his variability. Asking him to make repeated time studies in the course of setting a time standard is not only prohibitively expensive, but futile, for once an observer has rated a performance, any succeeding attempts he makes at rating that performance will produce results closely correlated with the first value he obtained.

Error in Summary

Accuracy and precision are matters of degree; they are not bilateral. Thus it is meaningless to speak of a system as "accurate" or "inaccurate," "precise" or "not precise." In fact, what a person might actually mean by the expression "inaccurate" is any one of a wide range of degrees of bias, depending on his ability to judge accuracy and on the stringency of the standard he uses for judging it. In different situations, different demands are made on a given type of measurement system, so that the particular system that suffices in one situation (and thus is described as "accurate") will not be satisfactory in others (and in such cases would be described as "inaccurate"). Without qualification, "accurate" and similarly vague descriptions of the error of a system are uninformative.

The distinction between accuracy and precision apparently is not generally understood in practice or by writers. The two terms are commonly and carelessly used interchangeably. Most authors in the time-study field seem confused in the matter, the most frequent error being to use accuracy in place of precision.

GENERAL APPRAISAL OF STOP-WATCH TIME STUDY

It is apparent that we are dealing with an objectionably erratic measurement system. Here are a few of the many sources of error, of which watch-reading itself is relatively insignificant and scarcely worth considering.

1. *The rating required to determine a normal time.* This judgment process by which the rater arrives at a rating factor is erratic and insensitive even in its simplest form, to say nothing of what it is when complicated by allowing the rater to take into account additional variables such as the volume of production, the fatiguing nature of the work involved, and deviations in method. Because of its critical role, this process of rating will be discussed at considerable length later in the chapter.

2. *The fatigue allowance.* At present, this can be estimated only by relatively crude means.

3. *The delay allowance.* There are reasonable means of estimation here, but in general, practice lags far behind theory so that relatively crude methods still seem to predominate.

4. *Overlap in consideration of certain factors.* For example, the fatiguing nature of an operation is often taken into account in rating and in the allowance.

5. *Observed times arbitrarily discarded.* In the opinion of the observer some recorded times are "not typical," and thus are discarded. This practice adds another subjective element to the time-study process.

6. *"Blind spots."* An example of a blind spot is the very short yet often occurring delay that is ignored in both the time study and the prolonged studies of delays like the ratio-delay study.

7. *"Working backward."* A time study is often conducted in such a manner as to substantiate an answer that the time-study man arrived at before starting to time the operation in question, as the result of experience on his part or of his taking desired take-home pay into account.

8. *Other flaws too numerous to mention.*

Yet in viewing these imperfections, the student, the scholar, and the practitioner must retain a realistic, middle-of-the-road perspective by remembering that this is a difficult measurement field, characterized by:

1. Severe opposing pressures exerted on the measurement system by representatives of management and the work force. The management exerts pressure to decrease labor costs, labor exerts pressure to increase earnings. The unfortunate time-study technician is caught in the middle. It is understandably difficult to always remain

objective and uninfluenced under the circumstances. This is no ordinary routine measurement situation to say the least.

2. The inability to apply a "safety factor" or its equivalent to the measurements produced. There are few parallel situations to be found in this respect. Other engineering specialties have their liberal "safety factors" which they apply to their carefully calculated results and with which they conceal their blunders and miscalculations. But this is not feasible in work measurement. There is someone to "call his shot" if the time-study man makes an error in either direction. William Gomberg described this situation quaintly when he wrote:

> Other engineers may have their problems, but nothing like these. If the civil engineer wants to understand what it's like to be a time-study engineer, let him visualize a dark, eerie Halloween night on which spirits, animate and inanimate, are abroad. Along comes the bridge, which is his pride and joy, spanning a majestic river and it addresses him in these accents, "Hey, jerk, do you know that I could have remained standing and carried just as big a load if you used one-quarter the tonnage of steel that my poor piles must hold up?" This is just an everyday experience for the time-study man.[2]

3. Intentional and virtually automatic changes in work rate and method whenever measurement is attempted.

4. Much more variability than is ordinarily encountered in the measurement of physical phenomena.

5. An elusive and difficult to define quantity to be measured.

6. A measurement unit—the standard minute—that is difficult to communicate and elastic in use, remindful of an elastic ruler.

7. Severe restrictions imposed by precedence and company-union agreements on what time-study procedures may be used and what results can be changed once they are established.

Remember, too, that time estimates must be provided regardless of the imperfections in methods of obtaining them. The need exists; time standards are indispensable to the successful operation of a manufacturing enterprise, so they *will* be obtained in some manner. The choice, however, is not one between perfection and imperfection, but of the best procedure from a set of imperfect alternatives. The techniques available all have flaws, all leave something to be desired, but one of them *will* be used, the one that seems best under the circumstances.

The casual observer of the field must be wary of two extreme points of view. It is apparently somewhat easy to be overly academic and unrealistically condemn the whole field because of its imperfections. It

[2] William Gomberg, *A Trade Union Analysis of Time Study,* 2nd ed., Prentice-Hall, Englewood Cliffs, N.J., 1955. By permission of the publisher.

appears just as easy to be overly receptive and naive, and remain relatively unaware of all the controversy, flaws, pitfalls, and questionable assumptions that abound beneath the surface. The student should resist being swayed too far in either direction. It can and does happen, even among writers and spokesmen in the field. Some go so far as to claim or make the pretense that the procedure is "scientific" (although few probably really believe this). Others, especially writers from fringe fields, go just as far in the hypercritical direction.

SOME MEANS OF IMPROVING THE STOP-WATCH TIME-STUDY TECHNIQUE

There is no denying that the results ordinarily obtained from stop-watch time study leave much to be desired. The weaknesses and objections to the technique have been thoroughly documented[3] and need not be belabored here. Our purpose is not to prove that errors exist; certainly the ordinary practitioner can testify to this. Rather, it is to suggest measures that may lead to improvements admitting at the outset that there is room for improvement in all phases of time study.

Improvements in Timing Methods

It seems somewhat of a paradox that stop-watch time study itself should have such a high labor content. In general, it is mostly a manual process, and, in fact, a rather costly process at that. In addition, it is difficult to find a time-study department that isn't "pushed" and does not have a sizeable backlog of standards to get out. The resulting pressures on the department leads to poorer results and undesirable delays in securing standards for operations. It would seem then that the time-study department might justifiably devote some attention to increasing its own productivity and thereby improve the cost and quality of the service it provides. What are some possible means of improving the productivity of the time-study department? One major possibility is in the increased use of clerical and other nontechnical help to relieve the higher-skilled observers as much as possible, from routine computational and paperwork tasks. Another is the use of various "computational aids," such as tables, graphs, alignment charts, and similar devices, as means of short-

[3] See, for example, William Gomberg, *A Trade Union Analysis of Time Study,* 2nd ed., Prentice-Hall, Englewood Cliffs, N.J., 1955; or H. O. Davidson, *The Functions and Bases of Time Standards,* A.I.I.E., New York, 1957.

cutting routine hand computations. Still another possible means of improving time-study department productivity is through mechanization of part or all of the time-study process. This possibility deserves further discussion.

Mechanization of stop-watch time study. Several of the time consuming portions of stop-watch time study can be performed automatically. Table 9 contrasts an electronic means with the ordinary

TABLE 9. Comparison of the Conventional Manual Method of Collecting and Processing Time Data with an Electronic Method

Phase	The Manual Method (Conventional Stop-Watch Time Study)	Automatic* (Electronic)
Collection of Times		
1. *Sensing* of the beginning and ending of the activity being timed.	Visual	Microswitch, photo-electrical cell, etc.
2. *Timing*	Stop-watch	Electronic timer (a time base, "gate," and counter).
3. *Readout* (transfer of the observations to some memorization medium).	Visual reading of watch, manual recording on observation sheet.	Punched card, punched tape, or magnetic tape.
Processing of Times	Manual	By computer, directly from tape or cards, or by punched card tabulating equipment.

* Not automatic, unfortunately, insofar as rating of the operator's performance.

"manual" method of collecting and processing the data in a time study. Note that under the "automatic" system the data is produced by the timer in a form that can be fed directly into an automatic data-processing device. This is the primary objective of an automatic timing system of the type described. Two feasible forms of mechanization of the time-study process will be illustrated. One

Figure 85. WETARFAC, the time-study device used by R. R. Donnelley and Sons Company, showing manual input unit and portable timing and readout unit in background. (Courtesy of R. R. Donnelley and Sons Company.)

type is predominantly electromechanical in nature, the other is primarily electronic.

The electromechanical system is illustrated by a unique device now used by R. R. Donnelley and Sons Company of Chicago, for time-study purposes. This device, pictured in Figure 85, is referred to as WETARFAC (Work Element Timer and Recorder for Automatic Computing). It consists of a manual input unit containing a keyboard with which the observer can identify the element being timed and enter his rating factor for that element, and a button he depresses to signal the timing unit that the element is complete. The main unit to which the input device is connected by cable is a console on wheels. This unit consists of the timer, a paper-tape punching mechanism, a direct-printing readout device, and a number of other useful features. When the time study is complete the recorded times and associated rating factors in punched paper-tape form are converted by standard equipment to punched-card form. From these cards it is possible to compute the mean time for each element, the standard deviation, a confidence interval, the normal time, and the standard time, on commercial punched-card processing equipment. Instead of the manual input device shown, an automatic sensing device such as a microswitch can be used to provide the input to the timer.

The main features of an electronic time-study device are shown in Figure 86 in the form of a simplified block diagram. A system based on the principles indicated can be constructed from commercially available components and has been done so at several educational institutions.[4]

The foregoing systems represent mechanization of a major part of the time-study process. In some instances it may be desirable to mechanize one particular phase of the system. For example, time-lapse photography can be used to replace manual observation in the case of the longer studies of delays. Photographic equipment can be programmed to automatically take pictures at prespecified regular intervals, say one per minute, one per 15 minutes, and the like, or at random intervals. In addition, there are numerous mechanical and

[4] "SEMTAR" at the University of Minnesota and "Automatic Motion-Time Recorder" at the University of California.

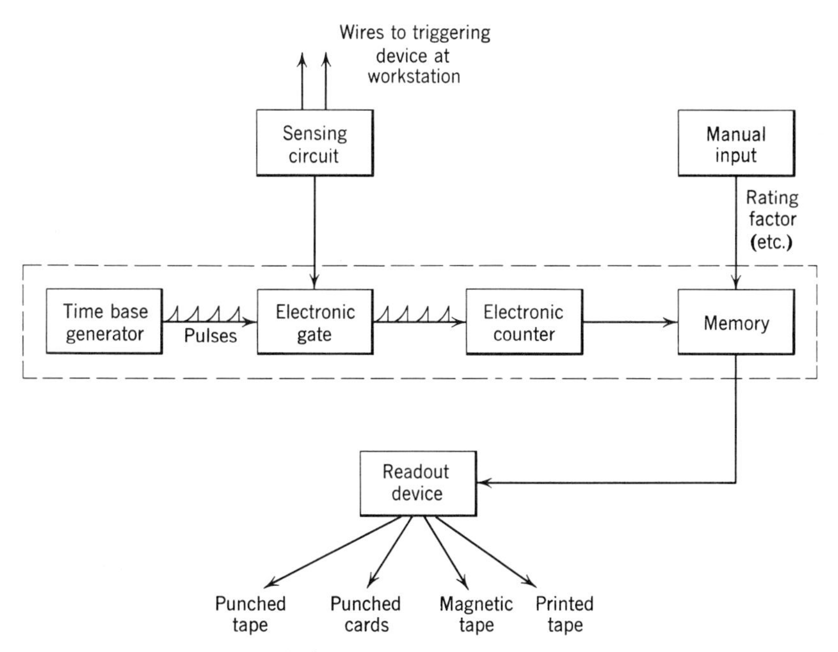

Figure 86. Block diagram of electronic time-study system.

electromechanical production recording devices on the market that can be used to automatically record running time and downtime of a piece of equipment.

It is important that the practitioner understand the primary purpose of the introduction of mechanization into the time-study system. The intent is not to automate for the sake of automation, which seems like a real temptation. It is easy to become intrigued by gimmicks and hardware. The initial cost of such devices is high and the usual consequences of increased mechanization, such as longer set up time, increased downtime, repair, maintenance, and the like, are inevitable.

The main intent of employing these mechanical aids is to increase the productivity of the time-study staff and thereby reduce the cost of setting standards and provide an improved service to the company. As long as the observer must accompany the timing device during the data collection phase in order to provide a rating factor and identify irregularities, the main saving through mechanization will be in the data-processing phase. The objective of employing a device like WETERFAC to assist in collecting the required data is to directly put that data in a form (tape, cards, etc.) that automatic data-processing equipment can handle. Thus the long periods of tedious, routine computation that the time-study man must perform after he has made his observations can be bypassed. Therefore, mechanization of the data collection phase, if it is worthwhile, will be so primarily because of its effect in reducing the data-processing cost. Only in prolonged studies of a grosser nature, such as downtime studies, for example, is mechanization likely to be economical for what it saves in observer time. Some additional advantages of mechanization of the collection of time-study data are the fact that the observer may be able to concentrate more on obtaining the rating factor and on the possibilities for job improvement, the fact that the relatively unattractive job of the time-study technician is made more appealing and that man-hours are made available for improvement of the measurement system and for better maintenance of time standards.

The Use of Formal Statistical Procedures to Estimate the Number of Cycles to Time

Since the textbooks and journals have recently publicized and promoted statistical procedures similar to those outlined earlier for estimation of the number of observations to make, some evaluation of the appropriateness of these procedures and the theory underlying them is desirable.

Under these statistical procedures, for a given sampling error as specified in terms of C and I, the number of cycles to time depends only on the amount of variation exhibited by the "raw" (unrated) watch times. Yet this variation in raw times appears to be a minor contributor to the error of a standard time, in contrast to error added by the rating process. An important limitation of these procedures, then, is the fact that the sample-size decision is based on a relatively insignificant source of error, with no recognition of the major determinant of an adequate observational period—rating. To elaborate on this argument, note that time study suffers from the two types of error mentioned earlier, a bias and a chance error. By way of the very nature of these two types of error, it can only be the chance error in a time standard to which the proposed sample size estimating procedures relate. This chance error of a time-study system, as measured by the variance of the standards set, σ_{ST}^2, is equal to[5] $\sigma_{BR}^2 + \sigma_{WR}^2 + \sigma_{DM}^2 + \sigma_A^2 + \sigma_{WT}^2$, where: σ_{BR}^2 is a measure of variation between raters; σ_{WR}^2 is a measure of variation within raters, that is, variation with passage of time; σ_{DM}^2 is a measure of the variation contributed by deviations in work method; σ_A^2 is a measure of the chance error contributed by the allowance estimating procedure; σ_{WT}^2 is a measure of the sampling error of the unrated watch times.

It is the chance error of the standard time, as measured by σ_{ST}^2, that is of ultimate concern in these sample size deliberations. The decision as to number of cycles to observe can justifiably be based on variation in unrated watch times *if* the resulting sampling error of these watch times is a major determinant of the chance error of a standard time. However, it seems very unlikely that the variance of the unrated watch times will be a major contributor to the variance of standard time, in view of the magnitude of errors within and between raters. It would appear then that the variance associated with the more troublesome phases of time study, especially the rating process, are the primary factors to consider in estimation of the length of time to observe. Stated in a different manner, in a stop-watch time study the performance is simultaneously timed and rated. Ultimately, the watch time and the associated rating factor are combined to obtain an estimate of the normal time. What constitutes an adequate

[5] The straightforward addition of these component variances to obtain σ_{ST}^2 presumes that there is no significant interdependency among these errors.

observational period depends on how long the operation must be observed to reduce the error of the normal time—the *combination* of the watch time and rating factor—to a desired minimum. Thus, the observer should continue to time and rate the operation until normal time error has been reduced to a satisfactory level, recognizing that this in turn depends primarily on when the rating error has been satisfactorily reduced by continued observation. At present, however, the manner in which the rating error and thus the error of the normal time estimate diminish with continued observation is a matter of conjecture.

It seems more logical and of greater practical significance to time a series of cycles, rate each cycle independently, and then apply a statistical method of estimating sample size to the resulting series of estimates of normal time. That is, instead of applying the statistical procedure to $T_1, T_2, T_3, \cdots T_N$, it would be applied to $(NT)_1$, $(NT)_2$, $(NT)_3, \cdots (NT)_N$, where

$$
\begin{aligned}
T_1 \times (RF)_1 &= (NT)_1 \\
T_2 \times (RF)_2 &= (NT)_2 \\
T_3 \times (RF)_3 &= (NT)_3 \\
\vdots \qquad \vdots \quad &\qquad \vdots \\
T_N \times (RF)_N &= (NT)_N
\end{aligned}
$$

However, it is very unlikely, in fact virtually impossible, for an observer to independently rate two or more cycles of an operator's performance of a task, especially if there is little or no time lapse between these cycles. There is little reason to suspect that $(RF)_2$, $(RF)_3$, etc., will be much or any different than $(RF)_1$ as long as these ratings are by the same observer, for the rating process is not that sensitive. Therefore, rating of successive cycles may not materially reduce the error of the normal time because of the likelihood of a strong correlation between ratings, yet interspersing these estimates with intervals of time sufficient to achieve independence of ratings is impractical.

A second limitation of some of the proposed statistical procedures is the time required to make the computations and arrive at a decision as to whether or not a sufficient number of observations have been taken. Under some of the procedures proposed, in the time that is required to decide whether or not M cycles is adequate, the observer could have been timing another M cycles. To make these statistical procedures practical, short-cut measures such as the use of R to estimate the sample standard deviation must usually be employed. Even then, there is some question as to whether or not the computational time is justified. Furthermore, the reasonably short procedures often are in error to an uncertain and controversial degree. It does not seem economical or even necessary to apply a statistical sample size estimation procedure in all or even most time studies. The wisest use of such a procedure appears to be to apply it to representative jobs throughout the plant to estimate for each *type* of operation what constitutes a reasonable sample size.

There are some misconceptions and poor practices prevalent in the textbooks and journals with respect to statistical procedures for sample size estimation. For example, some authors specify or imply that these procedures are to be applied to individual elements of the operation, instead of simply to the complete cycle. Aside from the prohibitive amount of time required if this practice is followed is the fact that *if* this recommendation is adhered to, the confidence interval demanded for individual elements need not be as severe as that required for the total cycle time. To insist on a plus-or-minus 5 per cent confidence interval for the time for each element of an operation means that a considerably more severe confidence interval is being demanded for the cycle time of that operation. The larger the number of elements, the greater the disparity.

Use of the range to estimate the standard deviation presumes that the underlying population of cycle times is normally distributed. The bulk of research on this matter indicates that the expected distribution of performance times for a manual task exhibits a moderate to severe positive skew. The resulting error in the estimate of s from R is apparently not of great practical significance and, in general, should not deter usage of this short-cut method.

Some readers may be disturbed by the fact that the author recommends use of the range of the whole sample as the basis for estimating s, whereas most authors recommend use of the average range ($\bar{R}$) for subgroups of four observations each. Thus: $s = \bar{R}/d_2$, which for subgroups of four becomes $\bar{R}/2.059$. However, Grubbs and Weaver have shown that for total sample sizes of $N = 2$ through $N = 11$, the best estimate of the population standard deviation is provided by the range of the total sample divided by the appropriate d_2 factor, and that any subgrouping as traditionally recommended provides a less efficient estimate.[6] For sample sizes of greater than $N = 11$, subgrouping is only slightly more efficient than use of the range of the whole sample, not sufficiently so to warrant the extra computation time required.[7]

Most authors recommend use of the normal rather than the t-distribution to characterize the distribution of sample means. However, since the population standard deviation must be estimated from the sample standard deviation, the t-distribution is appropriate. The resulting error is moderate. When an estimate based on the t-distribution indicates that a sample size of ten is necessary, basing the estimate on the normal distribution instead would indicate a sample size of eight as sufficient.

Another common, almost universal, recommendation in the textbooks and journals is use of a 95 per cent confidence level. It appears that many writers and practitioners adopt a value of $C = 0.95$ because this is what is used in other fields, and not as the result of a careful analysis of the needs in the ordinary practical time-study situation. A confidence level of 90 per cent appears adequate for most time-study applications. To insist on a value of $C = 0.95$ instead of $C = 0.90$ results in a 40 per cent increase in number of observations required. This is a disproportionate price to pay for a small and in fact unnecessary increase in confidence.

A number of authors insist on specifying the confidence interval as a percentage rather than as an absolute value. Thus it seems common to quote the confidence interval as "plus or minus 5 per cent," meaning that $I = 2 \times 0.05\,\overline{T}$. The main objection to this practice is the relative ease with which some percentage, 5 per cent for example, becomes universally and automatically employed in all applications of the sample size estimation procedure. Yet this is to be avoided. It is highly desirable that the practitioner decide *for each individual situation*, on the basis of the importance of the measurement being made, as to an appropriate confidence interval.

Some authors state that these statistical procedures are means of controlling accuracy of the average observed time. This is not the case. Such procedures are concerned only with sampling error. Un-

[6] Grubbs and Weaver, "The Best Unbiased Estimate of Population Standard Deviation Based on Group Ranges," *Journal of the American Statistical Association*, vol. 42, no. 238, June 1947.

[7] The best subgroup size being between six and ten, depending on N. For example, for $N = 15$, two subgroups, one of seven and the other of eight observations, are superior to other possible subgroupings.

like sampling error, accuracy of a series of measurements is unaffected by the number of observations taken.

Misconceptions and malpractices are prevalent in the applications of statistics to time-study problems. The field has much to learn insofar as the intelligent applications of statistics is concerned. It appears relatively tempting to use statistics for the sake of using statistics—to apply statistics blindly and pointlessly. Statistical ritual and rigor can give the impression of objectivity, of accuracy and precision, and of improved performance, when in reality these qualities are unaffected. These comments should by no means be construed as a condemnation of the use of statistics in work measurement. Actually the potential for the appropriate application of statistical procedures in various phases of work measurement is great. In fact, the penetration of statistics into work measurement methodology is long overdue. However, it is true that many writers and practitioners are inadequately prepared to intelligently appraise and apply statistical procedures. Some writers initiate incorrect and inadvisable practices, others perpetuate them. Practices are patterned after those employed in other fields. The general level of discrimination is low. Thus, it seems wise to obtain advice from a qualified, practical statistician when considering the appropriateness and mechanics of statistical procedures applied to time-study problems.

SOME PROPOSALS FOR IMPROVEMENT OF TIME-STUDY RATING

The practicing methods engineer who is not keenly aware of the deficiencies and difficulties associated with rating on the basis of first-hand experience alone, must be a rather rare individual. And those few rare ones are surely informed in the matter by way of the ample documentation this problem has received in the literature.

Some of the ideas that follow have been used in practice for some time, notably the use of rating films. Others are new proposals, some of which are directly applicable to current time-study practice. Others of these are yet to be tested. Presentation of the latter is intended to stimulate new avenues of thought, to stimulate research along positive lines, and eventually, to bring about some significant improvements in rating practices. With these purposes in mind the following means of reducing rating error are suggested.

More Specific Definition of the Rating Process

Just what this process of rating involves, specifically what is expected, what is and what is not being rated, and similar questions, are in most textbooks and at many companies rather vague and elusive matters. It seems that the newcomer to the art often receives only a very general and evasive explanation of rating, apparently with the expectation that he will learn the specifics of what he is to be doing and obtain a feel for the company's concept of normal performance by indirect (and usually unreliable) means. Under such procedures, diversity of interpretation is inevitable. Therefore, it is highly desirable that authors for the benefit of readers and managements for the benefit of the time-study staff, clearly and specifically spell out their conception of rating whatever it might be. Being explicit concerning what is to be rated and how, will in itself alleviate some of the problems that arise and will be especially beneficial to the novice in the art.

Careful Selection and Training of Raters

It behooves the administrator to give considerable attention to selection of the persons employed for stop-watch time-study work. One generalization that is pertinent to selection is that persons with experience as shop employees are likely to be superior raters. As urged earlier, it is very important that the initial training of the novice be thorough, explicit, and of sufficient duration, utilizing rating films, simultaneous on-the-job studies, and the like.

It is important that continuing attempts be made beyond the initial training period to keep the rater "in line" with other raters and the company concept of normal. With the passage of time a rater's concept of normal has a tendency to drift and become inconsistent with the concepts of others. To effectively control this tendency, extensive efforts should be made to provide a rater with knowledge of results (feedback) to facilitate reduction of rating errors. A blindfolded person can continue throwing darts at a target indefinitely and display no improvement in performance whatsoever if he receives no knowledge of the results of his attempts. This feedback is equally essential to improvement in the time-study practitioner's ability to rate. Yet many raters are in a position similar to that of the blindfolded dart-thrower.

Some of this desired feedback can be obtained through the periodic

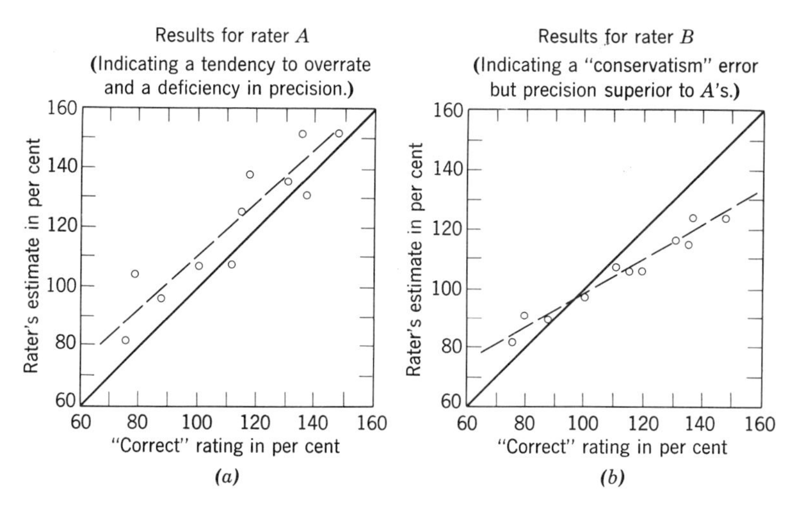

Figure 87. A common and very useful method of plotting the ratings of a series of filmed performances in order to detect various types of errors.

and intelligent use of a good set of rating films. A set of these films should provide scenes of a representative group of plant operations being performed at a variety of paces, along with the "correct" ratings for these scenes against which the raters participating in the practice sessions can compare their ratings and thereby profit. The usual practice is to project a series of performance scenes and allow the participants to rate each, then at the conclusion of the series, to give them the "correct" ratings for these. The results are then plotted and otherwise analyzed in a variety of ways in order to detect and inform each rater of the various types and degrees of error committed. One popular method of plotting the results of a series of rating trials is shown in Figure 87. A tendency to overrate is indicated in Figure 87*a*. A "conservatism" error, a reluctance to deviate appreciably from normal or 100 per cent, is indicated in Figure 87*b*. Figure 87*a* indicates that rater *A* is deficient with respect to precision, relative to what rater *B* has achieved. It is profitable to plot the results of a series of rating practice sessions for each rater to reveal any long-term trends in concept of normal, as illustrated in Figure 88. Over the 2-year period covered by the trials, rater *C* appears to have experienced a gradual tightening of his concept of normal. There are many other acceptable and profitable ways of processing the results of rating practice sessions to gain insight concerning rating

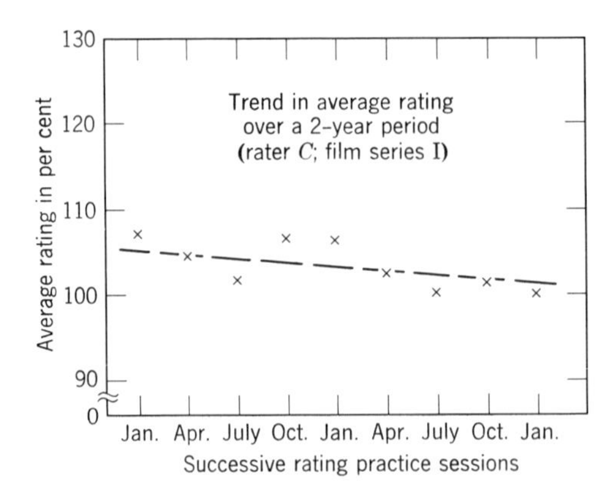

Figure 88. A chronological plot of the practice ratings produced by a given rater, rating the same set of filmed performances at 3-month intervals.

error, including a variety of accepted statistical procedures, use of control charts, and the like.

There are some precautions to be observed in the conduct of these rating practice sessions and in the use of rating films. The sequencing, spacing, etc., should be such as to minimize the chances that participants can memorize their previous ratings or the "correct" rating of a given scene. Furthermore, unless extreme care is taken in selection and planning of the use of a series of rating films so that they closely represent what the company holds as a concept of normal, the use of such films might do more harm than good by throwing the raters off and actually shifting the whole time standard and wage-payment structure.

Rating films may be purchased or rented from a variety of sources.[8] However, much is to be said in favor of a set of "homemade" films which presumably will show representative operations actually found in the plant involved. Purchased or rented films usually show a wide variety of types of jobs, only a small fraction of which are likely to be found at any one company. There is little point in having the time-study men at a textile mill practice and become proficient at rating the task of making molds for metal castings. If at all possible, practice should be with filmed performances of operations the participants are actually going to set standards for.

Multi-image rating films have come into vogue in recent years. With such films it is possible to project two or more performances simultaneously, side by side, for direct comparison. This seems especially useful for training novices to rate, because it is advantageous to simultaneously show normal and some other rate of performance

[8] The most elaborate and probably most widely used set of rating films is that produced and sold by the Society for the Advancement of Management (S.A.M.).

to demonstrate to them what they must ordinarily accomplish mentally.

The benefits offered by the use of rating films through more effective initial training of raters and through continuing control of their errors over the long run appear to make the investment in films and equipment and raters' time well worthwhile. Evaluation of the ultimate effectiveness of this technique in the reduction of actual on-the-job rating errors is left primarily to judgment, for just how much of the improvement that a rater experiences in the rating of films carries over to his performance out in the shop is unknown.

Another means of obtaining error feedback is through the use of occasional group time studies, where several or more time-study men simultaneously rate a number of different performances about the plant and compare and discuss results. This seems like a very effective method of controlling consistency among raters yet apparently is not widely used in practice. The relatively small amount of time required should prove well spent.

A third method of obtaining rating-error feedback is through the analysis of errors found in time standards actually set. A time-study department should effectively and persistently follow up on the time standards that it sets, not only to keep these standards abreast of change in manaufacturing methods and conditions, but to learn of and provide the bases for correction of errors in the measurement system itself. Thus, standards that have been set and put into use should be checked on periodically so that they may be brought up to date if a change in method or conditions has taken place, and so that erratically set standards can be isolated and improvements in the system made. The aggregate of (apparently) erratic standards detected over a period of time can be profitably analyzed in a variety of ways. For example, these standards can be classified according to the observer responsible as illustrated in Figure 84, page 262, according to type of operation, labor grade, operator studied, and so forth, in such a manner as to yield useful bases for correction of error. There are a number of sources of feedback of this type, there for the seeking, yet which are generally ignored or underutilized in practice. It makes sense to obtain and utilize knowledge of results achieved by a measurement system, at least if improvement is desired.

Exclusion of Complicating Factors

Rating is erratic even in its simplest form, to say nothing of what it is when we complicate it by expecting or allowing the rater to take one or probably several other variables into account in addition to and simultaneously with work pace itself. Among these factors are difficulty of the job, expected production volume, the fatiguing nature of the work, deviations in method, and the relationship between base wage (base rate) and desired take-home pay. It seems almost certain that a significant reduction in rating error could be achieved if rating were made as simple as possible by excluding many or all of the following complicating factors from the process. Thus the following proposals.

1. *That job difficulty be recognized in the time standard, but that no attempt be made to directly adjust for this variable in the process of rating.* The fact that some make an adjustment for job difficulty, that some do not, that some do not know whether to adjust for it or not, and that some have never faced the issue, is one of the matters that contributes to a general haze surrounding the subject of rating. It does appear, however, that it is desirable to recognize job difficulty in the time standard, at least if that standard is to be used as a basis for incentive wage payment. Failure to do so is likely to cause inequality in incentive earnings opportunity offered by the standards on different operations.[9] However, it seems preferable to let rating be simply an adjustment for speed of movement, leaving the adjustment for job difficulty to be made separately. Thus a procedure exemplified by Dr. Marvin Mundel's "objective" rating seems appropriate. Under Dr. Mundel's proposal, briefly described on page 237, work pace is judged; then independently and separately from this, an adjustment factor for job difficulty is selected on the basis of a table of values provided.

2. *That the rater ordinarily assume the same learning opportunity, the same production volume, for each job as he rates, and that volume be taken into account by appropriate adjustment of the standard on the basis of learning curves.* Apparently, in the course of arriving at a rating factor, most raters do and in fact must take into account the expected production volume for the operation under study. This is one more variable for the rater to consider as he rates, and one more he can and should be relieved of. Instead of the usual practice, the rater might better assume the same volume, and thus equal learning opportunity, for each operation he rates, for example a lot size that is

[9] The bases of this belief are the results of investigations that indicate that the greater the job difficulty the greater the degree of effort required to achieve a given incremental reduction in performance time. See, for example, H. A. Brea, "A Study of the Effects of Practice and Changes in Effort on the Performance Times of Different Motions in an Industrial Manual Operation," unpublished, M.S. thesis, Cornell University, 1955.

typical for the plant involved. Then in lieu of an adjustment via the rating factor, the rated time can be scaled up or down according to anticipated volume on the basis of learning curves or modified versions of them. Admittedly, this would require improved methods of predicting the effects of variation in volume on performance time, but this is certainly possible. (Once these improved predictive methods become available, it will be possible to create flexible standards that can be quickly and conveniently adjusted as significant changes in lot size or total volume occur. Furthermore, a satisfactory means of adjusting a time standard for different volume conditions would be an excellent adjunct to the standard data method, a procedure which ordinarily is insensitive to differences in volume.)

3. *That the fatiguing nature of the work not be considered in the course of arriving at a rating factor.* Although he is not ordinarily instructed to do so, it seems tempting for the rater to let the fatiguing nature, the heaviness, of the job under study influence his rating. The rater should ignore fatigue in arriving at a rating factor, for this factor is accounted for through the fatigue allowance, and often, in fact, in determination of the base wage through job evaluation.

4. *That the rater ignore the relationship between base wage and desired total incentive earnings for the job under study, and thus that consideration of money be excluded from the rating process.* Due to error sometimes made in the original determination of the base wage, shifts in supply and demand for some types of labor, and other reasons, the base wages for some jobs are noncompetitive or excessive with respect to those for other jobs within the company and at other plants. The time-study practitioner is usually aware of these out-of-line situations. In the opinion of the author, the observer is consciously or unconsciously prone to attempt to compensate for this unbalance as he goes about establishing time standards. If in his opinion the base wage for the job under study is unrealistically low, the observer is likely to rate more liberally than he otherwise would. If the base wage seems excessive, the observer is likely to be more severe in his rating than he would ordinarily be for the level of performance demonstrated. In the former case he is making the standard a little on the liberal side to permit higher incentive earnings to compensate for the tight base wage. In the latter instance he is attempting to cut back incentive earnings by setting the standard on the tight side. There is the inclination, opportunity, and often the pressure to take into account the relationship between the assigned base wage and what is felt to be a realistic total wage attainable through the incentive wage plan for the job in question, in the course of selecting a rating factor

for the performance observed. Time study would be better off in several respects if pay considerations were kept out of the process.

5. *That whenever possible the rater avoid attempting to adjust for deviations in work method in the course of arriving at a rating factor. That instead, if an adjustment must be made for variation in method, direct timing or predetermined motion times be used for this purpose.* In many instances it is impossible, or at least impractical, for the time-study man to eliminate deviations in method on the part of the operator(s) available for study. On the other hand, it is not wise to set the standard on a method that will not or should not be used. Consequently, as mentioned earlier, the observer is often forced into attempting to "rate out" deviations in method—a very erratic procedure at best. Here is one more variable the rater is expected to take into account. Repeating the suggestions made earlier, if it is not feasible to eliminate or ignore a deviation in method, it is proposed:

a. that the observer time and pace rate the method presented, making no attempt to adjust for the deviation through his rating factor, and

b. that he then adjust the resulting normal time on the basis of predetermined motion times (assuming the deviation is a manual one), or on the basis of elemental times recorded by stop-watch or other means, if possible.

Some writers and practitioners will object to this proposed use of predetermined motion times because of flaws they know of or suspect in existing predetermined motion time systems. It must be acknowledged that current predetermined action time systems have their imperfections. However, these limitations should not cause us to balk at this idea if we duly consider the primary known alternative, and that is to use pure judgment to estimate the effect of a difference in method as a part of the rating process. Furthermore, there is considerable potential and promise for improvement over currently available predetermined motion time systems.

Reduction of Rating Error Through a Basically Different Observational Scheme

The following two hypotheses support the appropriateness of a multi-observer, short interval sampling plan instead of the single observer, longer time study conventional in this field.

It is hypothesized that the moment an observer starts the process of arriving at a rating factor in making a time study of a given operator, the chance error of his rating will diminish with continued

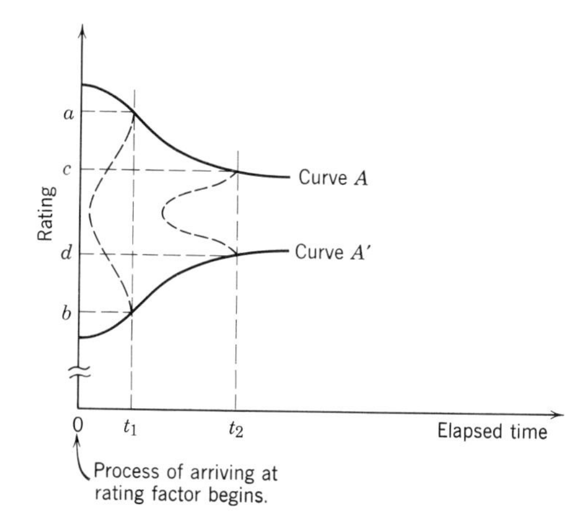

Figure 89. Curves indicating the expected improvement in precision of rating as a function of length of time observed.

observation in a fashion shown in Figure 89. Suppose the rater were forced to produce a rating factor at time t_1. At this point, his selection is likely to fall somewhere in the interval *ab*. If he were allowed time t_2, his selection is likely to fall somewhere within interval *cd*.[10] Thus, with continued observation the variability of ratings expected from the rater diminishes as postulated by curves A and A′, first at an accelerating and then at a decelerating rate, until a point is reached where continued observation accomplishes little in reduction of the variance of the rating or of the normal time estimate.

A second and equally important hypothesis concerns what the author refers to as the rater's decision point. It is postulated that at some time in the course of the time study there is an interval during which the observer concentrates on producing the rating factor, that after having given due consideration to the performance he has observed over that interval he decides upon the rating factor, and that this value remains fixed through the remainder of the observation period unless there is a radical change in level of performance. That is, once the decision has been made, it appears that the value selected is relatively inflexible to change. It appears unlikely that two or more independent, in fact, different ratings can be obtained from the same rater during the study of a given operator. If this is the case, there is little reason to continue timing beyond this decision point. A rating factor has been determined, and it applies *only* to those cycles timed

[10] These intervals and the curve comprised of them could be defined as containing y per cent (say 95 per cent) of the possible ratings that could be expected from that rater at time t_i.

over the period during which the observer was arriving at his rating. If additional cycles are timed after this decision point and these are not significantly different from those on which the rating factor is based, as pictured by the x's in Figure 90, what has been gained? If additional cycles are timed and are performed at a somewhat different effort level as indicated by the dots in Figure 90, and if, as expected, the rating factor remains fixed, the error has actually been increased by basing the rating factor on one sample of performance but actually applying it to another. This is also true of cycles timed before those on which the rating factor is based.

Implications of the foregoing hypotheses are numerous. One is that variation in or sampling error of unrated watch times is not ordinarily a realistic basis for estimating the most economical length of time to observe in a time study, as long as rating is involved. It is not a question of how many watch readings to take but primarily of how long to observe to reduce the error of the *rated* time to an economical minimum. It seems that this decision should be based primarily on the incremental reduction in the rating and normal time error contributed by added increments of observational time. Observation should be terminated when the cost of additional observational time no longer is justified by the resulting reduction in error. This period is probably relatively short, for the span of time the mind is capable of effectively encompassing in arriving at such a judgment is apparently rather

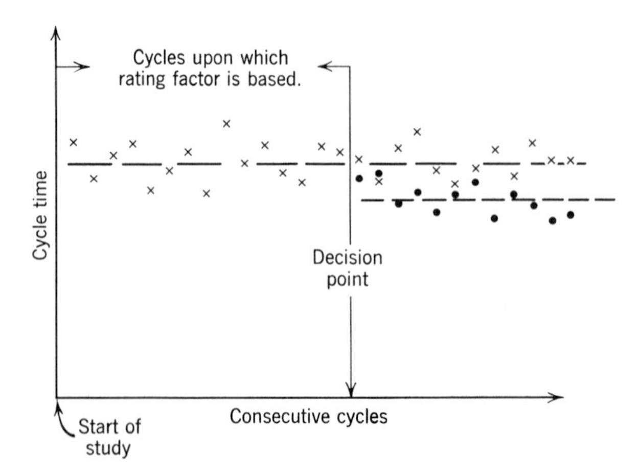

Figure 90. Graphical representation of a series of consecutive cycle times recorded by an observer, showing the postulated rating decision point.

brief. It may prove best, for example, to instruct the rater to observe for a relatively few minutes (or a minimum of one cycle if this is longer) and at that time choose the rating factor. The standard might be based on the normal time thus estimated, or the observer might return at a later time for another such sample (however, even with this time lapse the rating in the second sample is likely to be closely correlated with the first choice), *or similar estimates of several observers might be pooled.* It may prove best to have several observers each study a given operation for 5 minutes or so than to have one observer study it for 1 hour. It is quite conceivable that after 5 minutes the rater has done about as well as he is going to do with respect to his rating error. Why have the same person continue to observe if after he has chosen his rating factor it is unlikely that he is going to change his mind, either during that or future periods of observation? Or why have the rater continue to observe if additional observation adds virtually nothing to the precision of his normal time estimate? If any additional observing is to be done, it may as well be by different raters, especially since differences between raters contribute significantly to the errors in time standards. It makes sense statistically to sample more than one rater. The cost of having two or more observers study an operation in the course of establishing a time standard is ordinarily considered prohibitive in practice. However, this may be mainly because we are accustomed to thinking in terms of studies of traditional length. But as has been argued, this is probably not at all necessary.

Thus, it is proposed that a standard time be based on more than one rater's estimate of normal time, and that each estimate be based on an unconventionally short period of observation. Under this procedure the normal time for an operation would be based on a pooling of $(NT)_A$, $(NT)_B$, and $(NT)_C$, etc., where

Observer		Resulting Normal Time Estimate
A	$\bar{T}_A \times (RF)_A =$	$(NT)_A$
B	$\bar{T}_B \times (RF)_B =$	$(NT)_B$
C	$\bar{T}_C \times (RF)_C =$	$(NT)_C$

and where the different observers sampled at different times of the day and week, each for a relatively short interval, perhaps several minutes but not less than one complete cycle. *It is quite appropriate to apply conventional sample size estimation procedures to these relatively independent normal time estimates.*

Other writers have proposed the use of intermittent sampling in the course of time studying an operation.[11] Briefly, these proposals entail the collection of a series of randomly spaced subsamples each consisting of, say, five consecutive cycle times. Note, however, that the primary purpose of these procedures is to obtain a more representative estimate of the unrated mean cycle time, whereas the primary purpose of the procedure proposed herein is broader, namely, to obtain a more reliable estimate of the normal time, and is directed mainly to the more important problem of the rating error. The fact that the latter procedure likewise produces a more representative estimate of the average observed time is secondary.

A Selling Task for Time-Study Administrators

There is one important general observation concerning error and error reduction in work measurement that should not go without saying. It is this: management puts the pressure on the time-study department to get out the standards and to keep the cost of setting the company's standards to a minimum. Consequently, such departments generally seem understaffed. How can the time study department devote the attention that it should to training, error reduction, follow-up of the standards it sets, etc., under these circumstances? It cannot. In this author's opinion, too much emphasis is put on quantity, on cost of setting standards, and not enough on quality and the indirect costs incurred through erratic standards. The whole situation is somewhat of a paradox. Management wants low cost. In work measurement the only cost that is ordinarily apparent to management is the cost of operating the time-study department. But there is another cost that should be considered, the cost resulting from an unnecessarily erratic time standard structure. But since this cost is not so obvious, it is rarely duly recognized. So management puts all the emphasis on the cost that is evident and consequently and perhaps unwittingly greatly inflates the cost to which it is blind. Time-study administrators have a selling job to do here.

[11] John M. Allderige, "Statistical Procedures in Stop-Watch Work Measurement," *Journal of Industrial Engineering,* vol. VII, no. 4, July-August 1956; Adam Abruzzi, *Work, Workers, and Work Measurement,* Columbia University Press, New York, 1956.

REFERENCES

Barnes, R. M., *Work Measurement Manual,* William C. Brown Company, Dubuque, Iowa, 1945.

Buffa, Elwood S., "The Electronic Time Recorder; A New Instrument for Work Measurement Research," *Journal of Industrial Engineering,* vol. 9, no. 2, March-April 1958.

Davidson, Harold O., *Functions and Bases to Time Standards,* American Institute of Industrial Engineers, 32 West 40th Street, New York 18, N. Y., 1952.

Goldman, Jay and Gerald Nadler, "The UNOPAR," *Journal of Industrial Engineering,* vol. 9, no. 1, January-February 1958.

Gomberg, William, *A Trade Union Analysis of Time Study,* 2nd ed., Prentice-Hall, Englewood Cliffs, N. J., 1955.

Hoxie, Robert F., *Scientific Management and Labor,* D. Appleton and Company, New York, 1915.

Karger, Delmar W., "Instrumentation for Automatic Data Processing in Motion-Time Research," *Journal of Industrial Engineering,* vol. 9, no. 2, March-April 1958.

Rand, Christian, "A Hyper-Audio Pulse Triggered Input Device," *Journal of Industrial Engineering,* vol. 7, no. 2, March-April 1956.

Smirnoff, Michael V., "Significant Figures in Measurements," *Journal of Industrial Engineering,* vol. 10, no. 2, March-April 1959.

17

Work Sampling

The second of the two time-study methods involving direct observation is commonly referred to as work sampling. Work sampling involves intermittent sampling over a considerably longer period than is customary in a stop-watch time study. Note that in stop-watch time study a series of cycles are timed back-to-back over a period of time sufficiently long to obtain a reasonable estimate of the normal cycle time, but not long enough to provide a satisfactory estimate of the requirements of the job outside the straight work cycle. Work sampling fulfills the latter need.

Briefly, work sampling involves estimation of the proportion of time devoted to a given type of activity over a certain period of time by means of intermittent, randomly spaced, instantaneous observations. The following figure graphically depicts two states of activity, work-

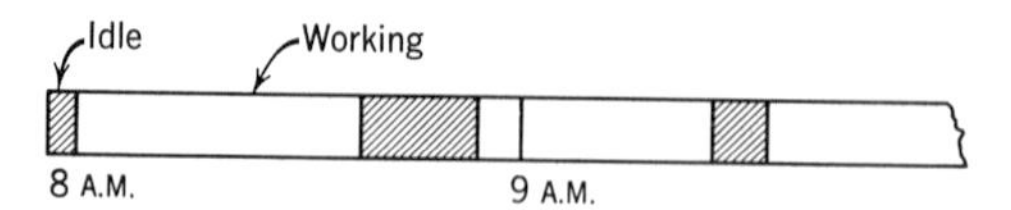

ing and idle, for a portion of an operator's workday. In conducting a conventional work-sampling study to estimate the proportion of time spent in an idle state, it is just as if we were to cut the preceding bar chart into very narrow strips representing instants of time, next collect and thoroughly jumble these strips in a bowl, and then select a certain number of these at random. After a reasonable sample has been drawn, the proportion of "idle strips" in the sample offers a practical estimate of the actual proportion of "idle strips" in the bowl and thus of the proportion of time actually spent in that state. Suppose one hundred strips were drawn and of these eighteen were idle strips. An estimate of the number of such strips in the bowl (the population) is

$$\frac{18 \text{ idle strips}}{100 \text{ strips drawn}} \times 100 = 18 \text{ per cent}$$

From this it is inferred that 18 per cent of the time is spent in an idle state.

In reality, of course, we do not have the advantage of the bar chart shown, so it is necessary to randomly select, in advance, the instants at which we wish to observe, then be there on the spot at those instants in time to observe what the operator is doing. So in practice the equivalent of this process of sampling strips is a series of randomly spaced, instantaneous, on-the-spot observations.

Applications of Work Sampling

In general, work sampling is used to estimate how time (man or equipment) is distributed over two or more types of activity, where it is not convenient, inexpensive, or possible to obtain this information from records or automatic recording devices. Following are typical applications.

1. Estimation of unavoidable delay time as a basis for establishing a delay allowance.
2. Estimation of the per cent of utilization of machine tools in a tool room, of cranes in a heavy machine shop, or of fork trucks in a warehouse.
3. Estimation of the per cent of time consumed by various job activities on the part of shop supervisors, engineers, repairmen, inspectors, nurses, schoolteachers, office personnel, etc.
4. Estimation of a time standard by combining rating with work sampling. For example, if a work sampling study shows that 20 per cent of a work week was consumed by avoidable delays; if each time a work-sampling observation was made the operator was also rated and the average of such ratings was 110 per cent; and if 1000 units were produced by the operator in that 40-hour period, the standard time would be

$$\frac{40 \text{ hrs.} \times 0.80 \times 1.10}{1000 \text{ units}}$$

or

$$0.32 \text{ hour/unit} \times 1.10$$

or

$$0.35 \text{ hour per unit}$$

Work Sampling Observation Sheet

Study of: *Utilization of Engineering Time*

Schedule of Observations. Observe at:	*Engr. to be observed*	Idle	Unaccounted for	*Nondelegable*											*Delegable*								
				Drawing	*Computation*	*Measurement*	*Construction*	*Paperwork*	*Acquiring info.*	*Thinking*	*Reading*	*Consultation*	*Supervision*	*Misc.*	*Drawing*	*Computation*	*Measurement*	*Construction*	*Paperwork*	*Acquiring info.*	*Acquiring mater.*	*Misc.*	
Mon. 8:02	*C*			✓																			
8:17	*M*										✓												
8:18	*B*	✓																					
8:31	*F*															✓							
8:48	*H*	✓																					
9:34	*C*																					✓	
9:36	*G*												✓										
9:57	*D*							✓															
10:07	*M*															✓							
10:41	*A*					✓																	
11:12	*C*					✓																	
11:25	*F*															✓							
1:09	*E*		✓																				
1:28	*B*											✓											
1:33	*H*			✓																			
2:28	*K*	✓																					
2:43	*G*		✓																				
2:57	*B*									✓													
3:10	*M*									✓													
3:18	*F*																	✓					
3:52	*D*																			✓			
3:55	*A*				✓																		
4:00	*F*							✓															
4:21	*E*												✓										
4:48	*B*			✓																			
Daily total		*3*	*2*	*3*	*1*	*2*	*0*	*2*	*0*	*2*	*1*	*1*	*2*	*0*	*0*	*3*	*0*	*1*	*0*	*1*	*0*	*1*	

Figure 91. Sample observation sheet for one day of a work-sampling study of an engineering staff. At the time designated in the first column, the observer made an instantaneous observation of the engineer listed. A check mark indicates that at the instant of observation the engineer observed was engaged in the activity identified at the head of the column.

Work sampling of an engineering staff—an illustrative case. The administrative staff of a large engineering organization has decided to conduct a work-sampling study of its personnel. The main objective of the study is to estimate the amount of the professional staff's workload that can be delegated to semitechnical and nontechnical personnel. The engineer designing the study has tentatively established the following activity categories to be observed.

Nondelegable	*Delegable*
Drawing	Drawing
Computation	Computation
Measurement	Measurement
Construction	Construction
Paperwork	Paperwork
Acquiring information	Acquiring information
Thinking	Acquiring materials
Reading	Miscellaneous
Consultation	
Supervision	*Idle*
Miscellaneous	*Unaccounted for*

The designer of the study has decided that 500 randomly spaced observations will be taken over a period of 20 working days. The 500 observations are to be distributed equally over the 20 days, so that 25 observations will be made each day. Thus, 25 randomly spaced observation times must be selected for each day and an observational schedule prepared. Each engineer will be observed an equal number of times during the study, the sequence to be random however. Upon preparation of the observation sheet illustrated in Figure 91, and specification of some additional observational details, the data collection period commences.

The manner of recording the observations is shown in Figure 91. When the 20-day observational period is completed, the observations in each category are totaled and a computation made of the proportion of observations in each. The results of the study are shown in Table 10. The study indicates, for example, that 5.6 per cent of the 20-day period was spent in idleness, that 36.4 per cent of this period was consumed by activities that could be delegated to personnel of lesser technical ability.

Outline of the Basic Work Sampling Procedure

The procedure for conduct of a work-sampling study, stripped of many of the refinements and niceties now available, is summarized by the following outline.

1. Preliminary steps, requiring
   a. definition of obectives, including specification of the states (categories) of activity to be observed;
   b. design of the sampling procedure, which involves:
      (1) estimation of a satisfactory number of observations to be made;
      (2) selection of the study length;
      (3) determination of sampling procedure details, such as the schedule of observations, exact method of observing, design of observation sheet, and routes to follow.
2. Data collection, by execution of the previously designed sampling plan.
3. Processing of data, including
   a. computation of $\bar{P}_i$, the proportion of the observations in the sample that were of activity i;
   b. analysis of the data in such other ways as seem useful and appropriate.
4. Presentation of results.

TABLE 10. Results of Work Sampling Study of Engineering Staff

Activity i	Number of Observations		Estimate of Proportion of Time Spent in Activity i ($\bar{P}_i$)	
Idle	28		0.056	
Unaccounted for	34		0.068	
Nondelegable	256		0.512	
Drawing		13		0.026
Computation		33		0.066
Measurement		24		0.048
Construction		11		0.022
Paperwork		23		0.046
Acquiring information		30		0.060
Thinking		31		0.062
Reading		17		0.034
Consultation		57		0.114
Supervision		10		0.020
Miscellaneous		7		0.014
Delegable	182		0.364	
Drawing		31		0.062
Computation		42		0.084
Measurement		13		0.026
Construction		37		0.074
Paperwork		25		0.050
Acquiring information		13		0.026
Acquiring material		6		0.012
Miscellaneous		15		0.030
Total	500		1.000	

Note that the major phases of this procedure are identical to those required in a stop-watch time study. A number of the details of this procedure will be described in detail. Before doing so it will be helpful to discuss three types of error that must be given careful attention in design of a work sampling study.

The sampling error. A sampling error arises when an inference is made concerning some characteristic of a population on the basis of measurements of less than the whole population, that is, on the basis of measurements of a sample from the population. The nature and behavior of this type of error may be illustrated by the assumption of a large urn of black and white beads. Even though the urn is known to consist of 20 per cent black beads, to demonstrate the sampling error a succession of samples of 100 beads were drawn. The proportion of black beads ($\bar{P}_b$) in the first sample was 0.23. After these were replaced, a second sample of 100 was drawn, yielding a $\bar{P}_b$ of 0.19. Other samples were drawn, each consisting of 100 beads each, yielding values of $\bar{P}_b$ of 0.24, 0.21, 0.15, 0.19, 0.27, 0.20, 0.22, and so on. Note the error that would result if any one of these sample proportions were to be used as our estimate of the proportion of black beads in the urn (P_b). This is what is meant by sampling error. If this bead-sampling process were continued for a long series of samples of 100 beads each, and the resulting $\bar{P}_b$'s plotted in the form of a frequency histogram, the result would be similar to that shown in Figure 92. This histogram is an approximation of the familiar binomial distribution.

This is clearly a sampling process; a chance error is obviously involved if a conclusion concerning the proportion of black beads in the population is to be based on a sample of 100 beads. Actually, by chance the sampler might well end up with a value of $\bar{P}_b$ anywhere

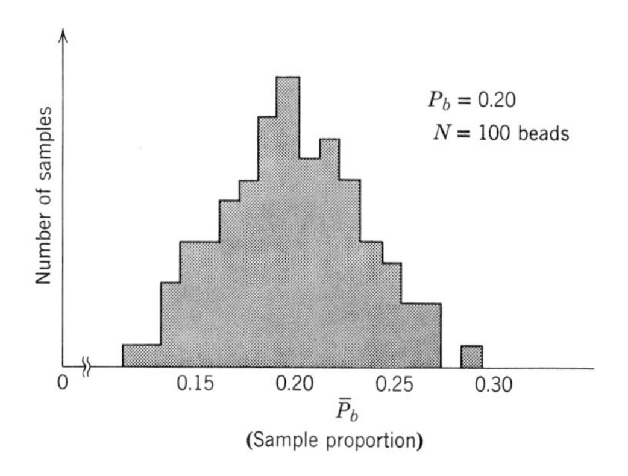

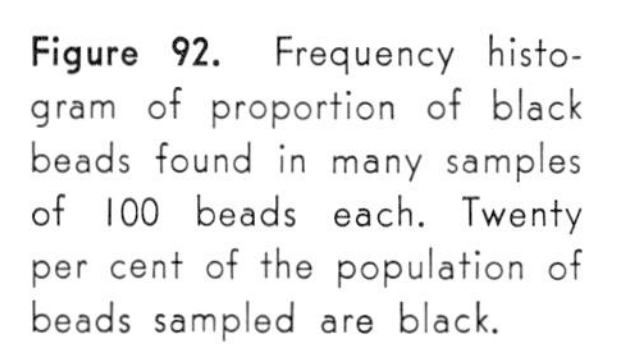
Figure 92. Frequency histogram of proportion of black beads found in many samples of 100 beads each. Twenty per cent of the population of beads sampled are black.

between 0.12 and 0.28 (see Figure 92) and draw a conclusion accordingly as to the magnitude of P_b. In the same fashion, work sampling is truly a sampling process. Instead of black and white beads we may have instants of time of two types—working and not working—where the population is all such instants in the period covered by the study. An inference as to the magnitude of P_w and P_{nw} is to be based on a sample of several hundred of these instantaneous observations. Thus, the designer of a work sampling study must contend with a sampling error. This error is controlled by the number of observations made. For example, if samples of 1000 instead of 100 beads each had been taken, most of the possible $\bar{P}_b$ values would have fallen between 0.17 and 0.23, instead of between 0.12 and 0.28. If the sample size were increased to 5000, most $\bar{P}_b$ values would be within 0.19 and 0.21, and so on. The choice of a sample size (N) may be made by a variety of means, ranging from pure judgment to the use of an objective statistical formula. An important advantage of the latter procedure is that it permits the designer of the study to numerically specify a tolerable sampling error in terms of a confidence interval and confidence coefficient, and to compute the sample size needed to control the sampling error to this level.

Bias in work sampling. A bias exists when the probability of observing a given state of activity, a delay for example, is different from the actual proportion of time devoted to that activity. Suppose the sampler who drew the successive samples of 100 beads from the urn was biased in his selections, so that even though the proportion of black beads was actually 0.20, because of the biased manner in which the sample was selected the probability of drawing a black bead was 0.30. If this were the case, the distribution shown in Figure 92 would converge upon 0.30 rather than 0.20. In the study of engineers the results indicated that 5.6 per cent of the time is spent in idleness. With this result the probability of observing idleness must be around 0.056, but it is very likely that the actual proportion of time spent in that state is at least several times this figure. As a consequence of bias there are two populations involved, the actual one and a fictitious one that exists only when an observation is made. In work sampling a bias may arise from three sources.

1. A nonrandom schedule of observations, say observing every half hour, so that any activity that is inclined to be periodic in its occurrence is likely to be observed more often or less often than it should on the basis of its true proportion, because the observations become in-phase or out-of-phase with that activity.

2. Bias on the part of the observer. If on-the-spot judgment is required of the observer because activity categories are ambiguously defined, or there are transitional states in which the worker is changing from one category to another and the observer can call it either way, or there is latitude as to the exact instant at which the observation should be made, there is opportunity for the observer's preconceived notions of what the results of the study should be to actually influence those results (and do not think that such preconceived notions are uncommon).

3. A change in behavior of the observed whenever observations are made, so that the actual and observed populations are not the same. In a work-sampling study of unavoidable production delays, it is not impossible that the probability of observing a delay would be around 0.30, whereas the actual proportion of delays is only half that, because the workers have been able to anticipate the moment of observation and adjust their performance accordingly. Whenever the opportunity to anticipate the observation exists, bias can be expected.

It is apparent from the numerous sources of bias that the designer of a work-sampling study must direct considerable attention to minimization of this type of error. Unfortunately, it is not possible to numerically specify and control bias as is the case with sampling error. The best that can be done with bias is to design and conduct the study in a manner that minimizes the opportunity for bias, and hope (but never really know for sure) that these precautions have been successful.

Nonrepresentativeness in work sampling. The third type of error to be dealt with depends on how closely the period sampled represents the future period to which the estimate will be applied. The estimate of P_i and the sampling error and bias associated with it apply only to the period sampled, which is past history at completion of the study. Nonrepresentativeness exists when the period studied is not typical—not characteristic of the circumstances that prevail in the long run.

Note the difference between an estimate and a forecast. An estimate per se is rarely of use until it becomes a forecast and useful for making decisions concerning future action. If there is a trend in the phenomenon or it is a cyclic phenomenon, and this fact is not recognized in the extrapolation, what might be a very accurate estimate could be a very inaccurate forecast. For example, delays are known to display a cycle throughout the week, often being at a minimum at midweek and a maximum on Monday and Friday. A sample taken only on

Wednesday might provide an accurate estimate of the proportion of delays for that day, but it hardly represents all days and, therefore, provides a less accurate forecast of the proportion of delays over the long run. If all days were the same with respect to proportion of delays, then there would be no point in extending a sample beyond one day. But days differ appreciably, so we cannot expect one day to be representative of all days. The same is possible of weeks, months, seasons, etc. General business activity, trends in the industry, company ups-and-downs, and the like, all cause random variations, cycles, and trends in production activity.

Thus, the designer of a work-sampling study must give careful attention to representativeness of the sample. Although it is not possible to numerically specify and control the error due to nonrepresentativeness, an attempt should be made to minimize this error by appropriate choice of when and how long to study. The period sampled should be typical of what is expected in the future. Periods of abnormal activity and unusual conditions should be avoided. The more pronounced cycles such as the workweek cycle should be spanned by the study.

Some information obtained during and at the conclusion of the observation period can be used to check on the representativeness of the sample. For example, during data collection a control chart (page 309) can be maintained and examined for evidences of trends and cycles that should be taken into account in order to obtain a representative sample. At the conclusion of the data-collection phase the production level of the group observed over the period studied should be compared with the customary production level of that group, to determine if the sampled period is typical.

In summary then, the errors associated with the estimate per se, that is, with measurement of the period studied, are sampling error and bias. When putting that estimate to practical use by applying it to the future, the three error sources are relevant. Note that each of the three types of error is controlled in a different manner.

1. Sampling error by altering N.
2. Bias by the manner in which the observations are made.
3. Nonrepresentativeness by appropriate selection of the period to be studied.

Design of the work-sampling study: selection of sample size. One method of selecting a sample size is to rely on the best judgment of the designer of the study. In some cases this is an adequate procedure. In others, however, a more reliable procedure is desirable.

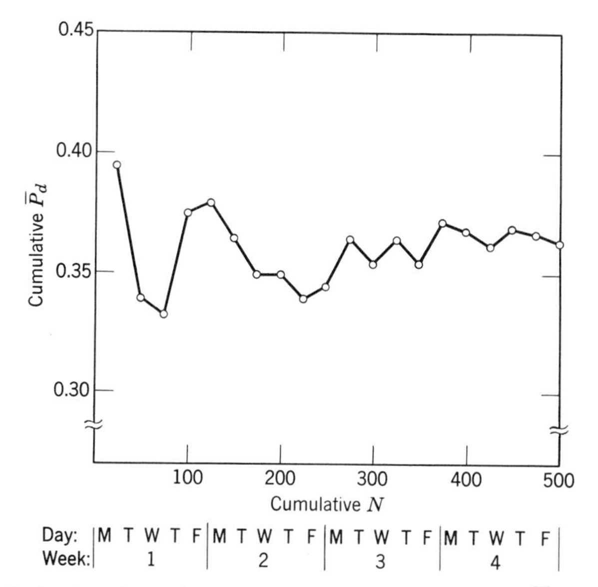

Figure 93. Daily plot of cumulative proportion of delegable time ($\bar{P}_d$) as a function of cumulative number of observations collected.

One such technique utilizes a plot of the cumulative estimate of P_i as illustrated in Figure 93. In this illustration, the observations collected during each successive day of the work-sampling study of engineers are pooled with those already accumulated to produce a cumulative and increasingly reliable estimate of the proportion of delegable time. Ordinarily, under this method the study is continued until the cumulative proportion appears to have "stabilized," and the accumulation of additional observations seems to have negligible effect on the level of $\bar{P}_i$. Of course, judgment must be exercised in deciding when the level of $\bar{P}_i$ has "stabilized."

A third and more objective procedure for sample size determination involves the use of elementary sampling theory, in the following manner.

1. Specify a maximum tolerable sampling error in terms of a confidence interval (I) and confidence coefficient (C), that is commensurate with the nature and importance of the decision to be based on the results of the study.

2. Obtain a preliminary estimate of the proportion of time devoted to the activity of primary interest in the study. This preliminary estimate can be based on judgment, or a portion of the study's observations can be taken and a preliminary estimate of P_i extracted from these.

3. Calculate the sample size required, in view of I, C, and $\bar{P}_i$, from the following expression:

$$I = 2\alpha \sqrt{\frac{\bar{P}_i(1 - \bar{P}_i)}{N}} \tag{1}$$

$$N = \frac{4\alpha^2 \bar{P}_i(1 - \bar{P}_i)}{I^2} \tag{2}$$

where $\bar{P}_i$ is an estimate of the proportion of time devoted to activity i, and α is a factor obtained from a table of probabilities for the normal distribution, for the value chosen for C.

For $C = 0.90$, $\alpha = 1.645$, and equation 2 reduces to

$$N = \frac{4(1.645)^2 \bar{P}_i(1 - \bar{P}_i)}{I^2} = \frac{10.8 \bar{P}_i(1 - \bar{P}_i)}{I^2} \tag{3}$$

Equations 2 and 3 provide an approximation of the sample size required to satisfy the sampling error specified.

As an illustration, let us examine how this procedure might have been applied in the work-sampling study of engineers. In this case, in view of the nature of the objective and the decisions to be based on the results of the study, values of $I = 0.06$ and $C = 0.90$ seem adequate. Since the proportion of time devoted to delegable activity (P_d) is of primary interest in this instance, N for the study should be determined by the N required to keep the sampling error of $\bar{P}_d$ to the level specified. To obtain the required preliminary estimate of P_d, the first 200 observations of the study might have been used. Application of equation 3 at that stage of the study, when the cumulative value of $\bar{P}_d$ was 0.35 (see Figure 93 at $N = 200$), would have indicated that

$$N = \frac{10.8\, \bar{P}_i(1 - \bar{P}_i)}{I^2} = \frac{10.8(0.35)(0.65)}{(0.06)^2} = 790$$

observations should be made to satisfy the sampling error requirement. (This is considerably greater than the 500 observations actually made. In retrospect then, it appears that 500 observations, a number the designer of the study arrived at on the basis of judgment, resulted in a larger sampling error than was deemed tolerable.) It is possible to use equation 1 to estimate the magnitude of the sampling error for $N = 500$. The confidence interval actually attained for $\bar{P}_d$ is

$$I = 2\alpha \sqrt{\frac{\bar{P}_i(1 - \bar{P}_i)}{N}} = 2(1.645) \sqrt{\frac{0.364(1 - 0.364)}{500}} = 0.071,$$

for $C = 0.90$.

Equations 1, 2, and 3 derive from elementary sampling theory. The sampling distribution for $\bar{P}_i$, similar to the distribution shown in

Figure 92, is adequately approximated by the binomial distribution, which in turn is satisfactorily approximated by the normal distribution under ordinary circumstances. The normal distribution is superimposed on the confidence interval as illustrated in Figure 94, indicating the source of the sample size estimating equations given. Recall that the standard deviation of the binomial sampling distribution is

$$\sqrt{\frac{P_i(1 - P_i)}{N}} \tag{4}$$

where P_i is the true proportion. In equation 1, $\bar{P}_i$ is substituted for P_i since in practice an estimate of the true proportion must be used. P_i is unknown. Yet $\bar{P}_i$ is obviously subject to a sampling error, so that the estimate of N by equation 2 is likewise subject to a sampling error, which explains why this equation can only provide an approximation. Thus, the reliability of the estimate provided by equation 2

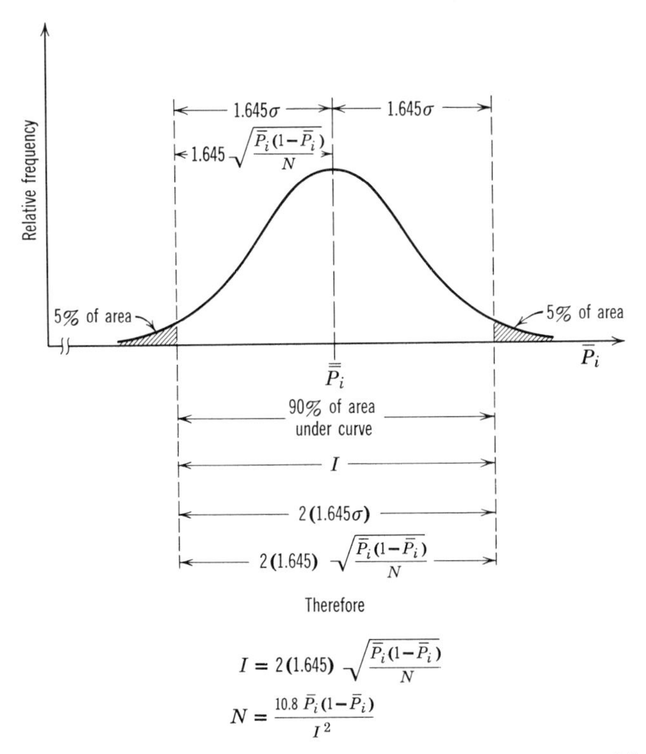

Figure 94. Illustration of the source of equations 2 and 3, assuming a confidence level of 90 per cent.

depends primarily on the reliability of $\bar{P}_i$. This being the case, N might well be periodically re-estimated during the collection of data phase of the study, as the estimate of P_i becomes more reliable as the result of a larger accumulation of observations. For example, N might be estimated early in a study, say after 100 observations have been accumulated. At that time equation 2 might indicate that 800 observations should be taken. After 500 of these have been accumulated, N might be re-estimated by equation 2 on the basis of the now more reliable $\bar{P}_i$. This process of convergence can be continued until the observer is satisfied.

Short-cut methods of estimating N, such as tables, graphs, and alignment charts, are readily available. These aids are based on and eliminate the need for actually using equation 2. These short-cut techniques will be introduced in a subsequent discussion of refinements of the basic work-sampling procedure.

An intelligent choice of values for I and C in use of these equations is crucial to the economy and competitiveness of the work-sampling technique. Selection of a confidence interval and confidence coefficient for a given situation is ultimately an economy decision, but such an involved one that judgment is ordinarily heavily relied on in practice. Naturally, these should be selected on the basis of the importance of the use of the estimate; they should not be inflexible from study to study. Furthermore, it is unwise to arbitrarily lift the value for a confidence interval or coefficient from some textbook, or automatically use values customarily used in some other field. For example, in some fields a 99 per cent degree of confidence is common; in others, 95 per cent is common. Certainly a value higher than 95 per cent is rarely justified in a work-sampling study. In fact, a 90 per cent confidence level seems adequate for most purposes. The consequences of requiring overly tight confidence intervals and coefficients are discussed in Chapter 18.

Design of the work-sampling study: details of the sampling procedure. Control of bias is a prime consideration and objective in determination of the detailed observational procedure. This type of error is minimized by designing and conducting the study so that the following conditions are assured.

1. Observations are made as close to randomly as possible,
   a. by avoiding ambiguous definitions of categories;
   b. by making observations at random intervals;
   c. by objective determination of the instant at which the observation should be made, for example, at the turn of the corner, at a certain point in the aisle;

  d. by objectively handling the transitional states that cannot be avoided by proper definition. Tossing a coin is an acceptable method.
2. The opportunity to anticipate the observation is minimized,
  a. by keeping observations randomly spaced;
  b. by making the observation as soon as the operation is in sight of the observer as he approaches;
  c. by randomizing the order in which individuals are observed when more than one observation is to be made per observer trip.

Another of the observational details to be determined in this phase of the study's design is the actual schedule of observations. Once the number of observations has been estimated and a representative period to study has been selected, it is necessary to distribute the N observations over this period. Ordinarily, the N observations are distributed equally over the total number of days selected. Thus if the preliminary estimate of N is 500 observations and a study period of ten work days has tentatively been selected, fifty observations would be taken per day. More often than not this practice is followed in order to avoid extremes in the number of observations to be taken on any one day. However, this practice is desirable for statistical as well as for convenience reasons, as explained later.

Observations are ordinarily distributed randomly over the day. There are a number of acceptable methods of achieving this. Probably the most suitable and most convenient is the use of a table of random numbers. Tables of this kind are commonly available in textbooks and handbooks. The appropriate type of table is one prepared from a rectangular distribution. The numbers 0 through 9 should have approximately the same probabilities of being selected. An extract from such a table might appear as follows:

31258
29541
01358
61440
71356

In such a table, then, the designer of a work-sampling study has a preselected series of random numbers. The next step is to convert these to a random time schedule. This can be accomplished by entering the table at a randomly chosen point, then allowing the first digit selected to represent the hour of the day, the next two to represent the minute, and, if desirable, the next two to represent the second. Under this procedure the above numbers yield the following observations.

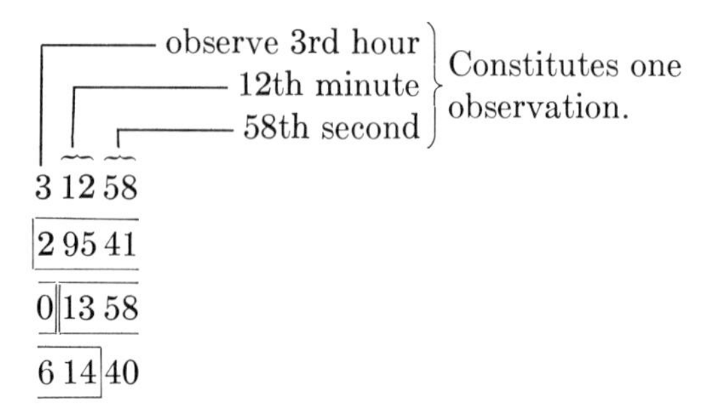

The next selection from this table would be the 2nd hour, 54th minute (the 9 is discarded), the 10th second. The next one would be the 1st hour, 35th minute, the 14th second (the 8 and 6 are discarded). These times are then chronologically ordered to complete the observation schedule.

After the observer has settled on the schedule of observations, design of the observation sheet, routes to follow, and other observational details, he is prepared for the data-collection phase of the study.

Processing the data. The data-processing phase of the study will consist of at least the computation of $\bar{P}_i$, and may well include the use of one or several of the many graphical and analytical procedures with which the data of a work-sampling study may be profitably treated. A number of these optional procedures will subsequently be described under the heading of refinements of the work-sampling technique.

Processing of the data may occur during as well as following the data collection phase. For example, as the data is collected, a cumulative plot of $\bar{P}_i$ might be posted daily, as illustrated in Figure 93. Or as suggested earlier, N might be periodically re-estimated from the increasingly reliable $\bar{P}_i$. Also, it is highly desirable that the value of $\bar{P}_i$ and the associated value of I computed from equation 1 be periodically reviewed during the collection of data phase. This review is desirable because in some cases it will become obvious that $\bar{P}_i$ and I are such that additional observations will not affect the ultimate decision and that there is no point in continuing the study. For example, suppose that a work-sampling study is being made to determine manpower needs in a certain situation. After the study has progressed and a substantial number of observations have been accumulated, the value of $\bar{P}_i$ indicates that 3.6 or, for practical purposes, four workers will be required in the crew being studied. Furthermore, it is apparent from the values of $\bar{P}_i$ and I that the probability is virtually nil that any number of additional observations would change the conclusion that four men are needed, by indicating that three or five men are required. Thus, inflexibility of the sampling error requirement during a study, as well as from one study to another, is undesirable.

Levels of Refinement in Application of Work Sampling

There is a wide range of levels of refinement with which work sampling may be applied. At one extreme is the strictly informal version, the basic work sampling process stripped of all optional features. Here there is no formally established schedule of observations. Observations are made when the course of routine activities happens to bring the observer near the subject of the study. No formulas are used; the only evidence of the study may be some strokes on a pad at the observer's desk, or on a slip of paper in his pocket. Work sampling may well be frequently used in this form to obtain a quick, rough, inexpensive estimate of time utilization. This version of the technique might be used to verify a claim that the office needs more adding machines, or more telephones, or to check on the suspicion that a certain person is only working about 50 per cent of the time, etc.

At the other extreme is the most sophisticated version of the technique, employing all the refinements and niceties known. Here the study is very carefully designed, employing relatively complicated "experimental designs," numerous special statistical and graphical methods of analysis, and the like. The procedure outlined and elaborated on in the preceding portion of this chapter lies approximately midway in this range of levels of refinement and formality. It is important that the designer of a work-sampling study select a level of application that is commensurate with the demands of the particular situation. In some instances the technique is "over-applied"; the formalities employed and the expense of the study are far in excess of what is warranted. In other situations the level of application falls far short of what could and should be employed considering the decisions involved and the complexity of the situation to be studied.

Refinements of work sampling—stratified sampling. Up to this point it has been assumed that the probability of observing a given state of activity is constant over the length of the study. When there are periods of time in which P_i is appreciably different, a random allocation of observations to these periods increases the error of the estimate of P_i for the interval studied. For example, suppose that the typical proportion of time consumed by delays (P_d) on different days of the week in a certain case are as shown in the table below, and that the number of work-sampling observations made each day (due to a random distribution of observations over the whole study period) are as indicated.

The variance of the estimate of P_d for the week consists of two components. One is purely the result of sampling. The other arises

	P_d	N
Monday	0.15	56
Tuesday	0.14	42
Wednesday	0.12	57
Thursday	0.14	44
Friday	0.16	51
		250

because in allocating observations to different days by chance, the P_d's for different days are given different weights. In the foregoing example, $P_d = 0.15$ for Monday receives a weight of 56/250 or 22 per cent in estimation of P_d for the week; $P_d = 0.14$ for Tuesday receives a weight of 42/250 or 17 per cent, and so on. If additional sets of 250 observations were each distributed randomly over the same 5 days, the number of observations per day and thus the weighting of P_d for each day would vary by chance from trial to trial, and thus inflate the variance of $\bar{P}_d$ for the week.

To avoid this and other undesirable consequences, whenever it is known or suspected that P_i is appreciably different during different periods covered by the study, such as the probability of delays being different on different days of the week or at different times of the day, the stratified sampling procedure to be described is preferred. To illustrate this procedure, suppose that for a proposed work-sampling study of machinist delays, it is estimated that the proportion of delays during the first and last hours of the typical workday are about the same, but appreciably greater than the proportion during the remaining 6 hours. Under these or similar circumstances the following procedure is recommended.

1. Segregate the observations made on each day studied.
2. Segregate the observations made in the first and last hours of the day, henceforth referred to as period A, from those made in the middle 6 hours, henceforth referred to as period B.
3. Allocate observations to these periods in proportion to the fraction of the week each represents. Thus, if 1000 observations are to be taken over a 5 day interval, each day will be allotted 200 observations, each

period A will receive $200 \times \frac{2}{8}$ observations, and each period B will receive $200 \times \frac{6}{8}$ observations. Within these intervals the observations should be randomly spaced.

4. In estimating the proportion of time consumed by delays for the week, the $\bar{P}_d$ for each of the segregated periods should be weighted according to the fraction of the work week each represents. Since the time represented by each period and the number of observations made in that period are proportional, the $\bar{P}_d$ of each period is automatically appropriately weighted when the delay observations in all periods are pooled and divided by 1000. The results of the one week study are as follows.

Work-Sampling Study of Machinist Delays

Summary of Results

	Period A		Period B	
Day	Delay Observations	Observations Taken	Delay Observations	Observations Taken
I	16	50	31	150
II	15	50	28	150
III	13	50	18	150
IV	13	50	22	150
V	15	50	29	150
	72	250	128	750

If the $\bar{P}_d$ of each period sampled is weighted by the number of observations allotted to that period (which in turn is proportional to the time involved), for the week

$$\bar{P}_d = \frac{(\bar{P}_{\text{I}A} \times N_{\text{I}A}) + (\bar{P}_{\text{I}B} \times N_{\text{I}B}) + (\bar{P}_{\text{II}A} \times N_{\text{II}A}) + \cdots (\bar{P}_{\text{V}B} \times N_{\text{V}B})}{N}$$

$$= \frac{(\frac{16}{50} \times 50) + (\frac{31}{150} \times 150) + (\frac{15}{50} \times 50) + \cdots (\frac{29}{150} \times 150)}{1000}$$

$$= \frac{16 + 31 + 15 + \cdots 29}{1000}$$

5. Segregation of the observations as illustrated in the table of results makes it possible to estimate $\bar{P}_d$ for different periods and to test $\bar{P}_d$ of different periods for significance of difference. Segregation of observations also makes it possible to apply a control chart to the daily $\bar{P}_d$, as described later.

On occasion stratification of another form is desirable. Assume that in the work-sampling study of the engineering staff it is suspected at the outset that P_d (the proportion of delegable time) is appreciably different for each of three engineering departments involved. If observations are allocated randomly to these departments and the data simply pooled to estimate P_d for the whole staff, the weighting of P_d for each department will vary by chance and thus inflate the variance of $\bar{P}_d$ for the population. Likewise, failure to segregate the observations made in each department will bar an analysis of the data for purposes of isolating significant differences in $\bar{P}_d$ among the departments involved. Therefore, it is desirable to stratify the population, using the staffs of departments A, B, and C as subpopulations, and to allocate observations to them in proportion to the number of engineers in each department. Thus, if 500 observations are to be taken, and the departments A, B, and C employ 25, 30, and 20 engineers, respectively, the 500 observations would be allocated as follows:

Department	Fraction of Total Staff	Number of Observations
A	$\frac{25}{75}$	167
B	$\frac{30}{75}$	200
C	$\frac{20}{75}$	133
		500

In turn, in estimating P_d for the staff, $\bar{P}_d$ in each department should be weighted by the fraction of total staff it represents.

In general then, when P_i is known or suspected of varying significantly during the period studied or of being significantly different for different portions of the population of men or machines under study, a stratified sampling plan is preferred. When stratifying, if:

T = total period of time encompassed by the study,
t_j = amount of time represented by the jth subperiod (stratum),
M = total number of men or machines being studied, and
M_j = number of men or machines in jth stratum (subpopulation, subgroup),
N = total number of observations taken over the period T,
N_j = number of observations allocated to the jth stratum,
P_i = proportion of time devoted to activity i over the period T,
N_i = number of observations of activity i over the period T,
$\bar{P}_i = N_i/N$ = the estimate of P_i,
p_{ij} = the proportion of time devoted to activity i in the jth stratum,
N_{ij} = number of observations of activity i in the jth stratum,
$\bar{p}_{ij} = N_{ij}/N_j$ = the estimate of p_{ij},

then the following procedure can be used in allocating the observations.

1. Select strata so that p_{ij} may be considered essentially constant in the jth interval and uniform in the jth subgroup. Observations made in each stratum should be segregated so that $\bar{p}_{ij}$ may be computed.

2. Proportionately allocate N to these strata, so that

$$N_j = N\left(\frac{t_j}{T}\right) \tag{5}$$

or

$$N_j = N\left(\frac{M_j}{M}\right) \tag{6}$$

whichever is appropriate. (The most efficient allocation of N would be on the basis of $\bar{p}_{ij}$. However, the resulting reduction in sampling error and/or N does not seem to warrant the extra computational effort in most instances.)

3. Distribute the N_j observations randomly within the strata.

4. Under these circumstances, $\bar{p}_{ij}$ is weighted according to the stratum length or size in determination of $\bar{P}_i$. Thus

$$\bar{P}_i = \sum_{j=1}^{n} \bar{p}_{ij}\left(\frac{t_j}{T}\right) \tag{7}$$

which reduces to the following if N_j is proportional to t_j (referred to generally as proportional stratified sampling):

$$\bar{P}_i = \sum_{j=1}^{n} \bar{p}_{ij}\left(\frac{N_j}{N}\right) = \sum_{j=1}^{n}\left(\frac{N_{ij}}{N_j}\right)\left(\frac{N_j}{N}\right) = \frac{\sum_{j=1}^{n} N_{ij}}{N} \tag{8}$$

If stratification is by men or machines,

$$\bar{P}_i = \sum_{j=1}^{n} \bar{p}_{ij}\left(\frac{M_j}{M}\right) \tag{9}$$

which reduces to

$$\bar{P}_i = \frac{\sum_{j=1}^{n} N_{ij}}{N} \tag{10}$$

if proportional stratified sampling is used.

5. The variance of $\bar{P}_i$ under these circumstances is

$$s_{\bar{P}_i}{}^2 = \sum_{j=1}^{n}\left(\frac{t_j}{T}\right)^2 \frac{\bar{p}_{ij}(1 - \bar{p}_{ij})}{N_j} \tag{11}$$

which reduces to

$$s_{\bar{P}_i}{}^2 = \frac{1}{N}\sum_{j=1}^{n}\left(\frac{N_j}{N}\right)\bar{p}_{ij}(1 - \bar{p}_{ij}) \tag{12}$$

if proportional stratified sampling is used. This is the appropriate formula for estimation of the variance of $\bar{P}_i$ when stratified sampling is used. Use of the expression $[\bar{P}_i(1 - \bar{P}_i)]/N$ to estimate $\sigma_{\bar{P}_i}{}^2$ when proportional stratified sampling is used is likely to result in an overstatement of the variance and of the number of observations needed to satisfy given confidence requirements.[1]

In general, stratified sampling with allocation of observations in proportion to t_j or M_j offers a means of reducing the sampling error, a more conveniently executed sampling schedule, a series of successive $\bar{p}_{ij}$'s to which the conventional control chart technique is readily applied, and an opportunity to test the results for significant variation of P_i with time and with subgroups of the population of subjects or objects studied. It would seem worthwhile to stratify to some extent in almost every work sampling study, at least to the extent of proportional stratified sampling of days or shifts.

Other sampling questions and refinements remain. A number of these are matters of some complexity and often are somewhat unique to the particular problem involved. On occasion the advice of a competent statistician is almost necessary if an efficient sampling plan is to evolve.

Refinements of work sampling—processing the data. Aside from computation of $\bar{p}_{ij}$ and $\bar{P}_i$, of the resulting confidence interval, and of the number of additional observations needed if this confidence interval is not adequate, processing of the data during and at the conclusion of the observational period might well include the following.

1. A plot of the cumulative value of $\bar{P}_i$ against the accumulated N as the study progresses, as discussed earlier and illustrated in Figure 93. A plot of this type reveals the progressive improvement and increasing stability of $\bar{P}_i$ as the accumulated N increases, and is therefore sometimes used as the basis for deciding when a sufficient number of observations has been accumulated, in lieu of or in conjunction with the use of equation 2 introduced earlier.

2. Application of the control chart technique, originally developed for quality sampling procedures, to $\bar{p}_{ij}$. Basically, the control chart as applied here is a time-ordered plot of successive $\bar{p}_{ij}$'s from which it is possible to detect trends and cycles in P_i as well as abnormality of a $\bar{p}_{ij}$. The results of a 15-day work-sampling study conducted to estimate percentage of time actually devoted to repair work in a utility company repair shop are as shown on the following page.

[1] This stratification procedure is based on ideas presented in "Some Statistical Considerations in Work Sampling," Richard W. Conway, *Journal of Industrial Engineering,* vol. 8, no. 2, March-April 1957.

t_j (day)	$\bar{p}_{ij}$	t_j (day)	$\bar{p}_{ij}$
I	0.74	IX	0.43
II	0.80	X	0.77
III	0.62	XI	0.84
IV	0.66	XII	0.83
V	0.76	XIII	0.63
VI	0.81	XIV	0.70
VII	0.70	XV	0.74
VIII	0.71		

Samples of fifty observations were taken on each of the 15 days. A control chart for the daily proportions ($\bar{p}_{ij}$) is shown in Figure 95. Three standard deviation "control limits," computed from

$$\bar{P}_i \pm 3\sqrt{\frac{\bar{P}_i(1-\bar{P}_i)}{N_j}} = 0.717 \pm 3\sqrt{\frac{0.717(1-0.717)}{50}} = 0.717 \pm 0.192$$

are used here because they are conventional in the field of quality

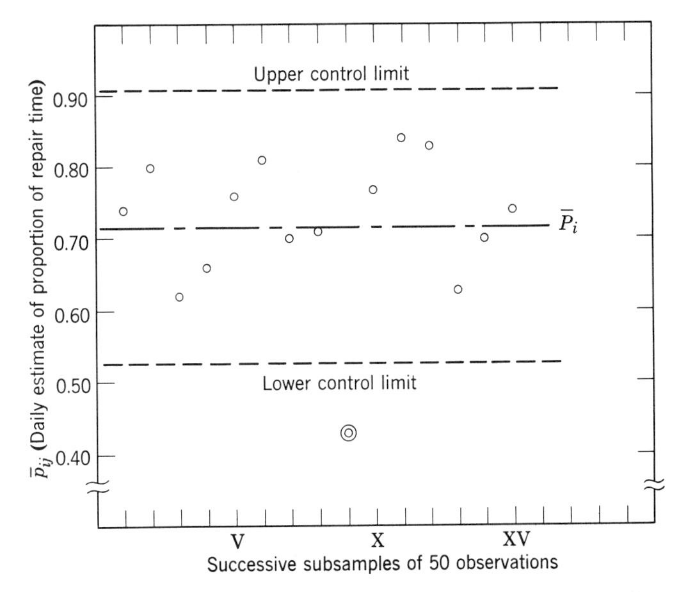

Figure 95. A control chart for daily estimates of the proportion of time devoted to productive work in a 15-day work-sampling study of a utility company repair shop. Daily sample size is constant at 50 observations.

control, and may or may not be equally appropriate for work sampling. The significance of these limits is this: if a $\bar{p}_{ij}$ falls outside of the control limits as occurred on day IX, there is only a very slim possibility that this event is the result of chance. In other words, it is quite likely that the $\bar{p}_{\text{IX}}$ of 0.43 is the result of some rather different circumstances on that day, and that the p_{ij}'s involved are not homogeneous.

3. When stratified sampling has been used, the resulting $\bar{p}_{ij}$'s may be tested for significance of difference by conventional statistical techniques. For example, in the work-sampling study of an engineering staff described earlier, the estimates of the proportions of idle time for the three departments are as follows:

Department	$\bar{p}_{ij}$
A	0.083
B	0.041
C	0.045

The administration might well be interested in knowing whether or not the 8.3 per cent is significantly higher.

Short-cut techniques. There are a number of short-cut procedures that substantially reduce the time for making various computations frequently required in designing a work-sampling study and processing the data. One such aid is the alignment chart shown in Figure 96 for estimation of sample size required, given $C = 0.90$, $\bar{P}_i$, and I as an absolute interval. Alignment charts may also be used to compute I, given $\bar{P}_i$, N, and C, or to compute limits for a control chart.

Graphs may be used to quickly solve for N, as illustrated in Figure 97 for $C = 0.90$ and several values of I. Figure 97 rather dramatically illustrates the consequence of demanding an unnecessarily severe confidence interval. Note the increase in sample size for $\bar{P}_i = 0.5$ if a confidence interval of 0.02 is used as opposed to 0.04. The graphical method of solving for N is recommended over the formula, table, and alignment chart, because it effectively illustrates the effects of employing different confidence interval values on sample size.

Summary

Certainly every time-study department should be equipped to apply this useful work measurement technique. Note, however, that work sampling is more competitive with and probably superior to ordinary stop-watch procedures only in instances where a more aggregative picture is sufficient. When refined measurements of work elements

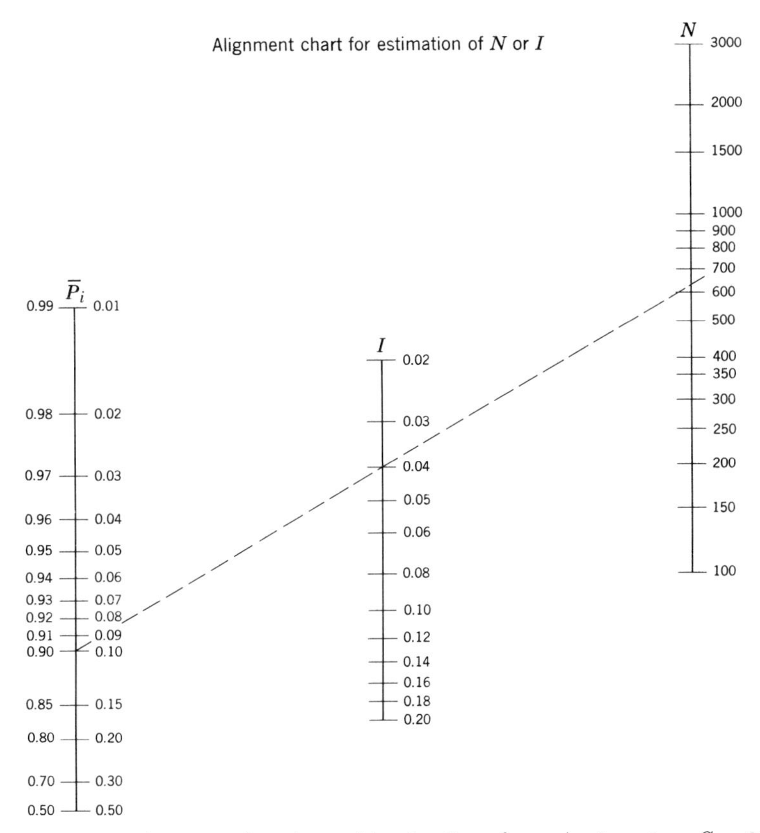

Figure 96. An alignment chart for rapid estimation of sample size, given $C = 0.90$, $\overline{P}_i$, and I, or for estimation of I given $C = 0.90$, $\overline{P}_i$, and N. For illustration, assume that $\overline{P}_i = 0.10$ and $I = 0.04$. N is found by projecting a straight line through these two points to the N scale. As indicated by the dashed line above, N is found to be 610 in this instance. This chart solves the expression $I = 2(1.645)\sqrt{\frac{\overline{P}_i(1 - \overline{P}_i)}{N}}$.

and cycles are not required, when behavior over a prolonged span of time is of special interest, when the behavior of aggregations of persons or machines is of concern, when gross impressions are desired, and in situations of similar character, this technique is especially adaptable.

EXERCISES

1. Design a work-sampling study to be made of the teaching staff to estimate the proportion of time devoted to activities that could be delegated to student assistants and secretarial help.

2. Two hundred observations of a work-sampling study of laborers at a utility company have already been made. Sixty-four of these observations were made when a laborer was idle. How many additional observations should be made if the sampling error of the estimate of the proportion of time consumed by idleness has been specified at $I = 0.06$ and $C = 0.90$? What is the confidence interval for the current estimate of the proportion of time consumed by idleness when $N = 200$ and $C = 0.90$?

3. Prepare a graph showing the relationship between $\bar{P}$ and N for a confidence interval of 0.05 and confidence coefficient of 0.90.

4. Prepare a graph showing the relationship between I and N, for a $\bar{P}$ of 0.1 and 0.5, with $C = 0.90$.

5. A situation that often arises in the field of methods design and work measurement is the desire to simultaneously observe a number of people working at concurrent and interdependent activities, for example a crew of men in operation. Suppose an engineer wishes to observe a ground crew in action at an air terminal, or the crew associated with a large printing press,

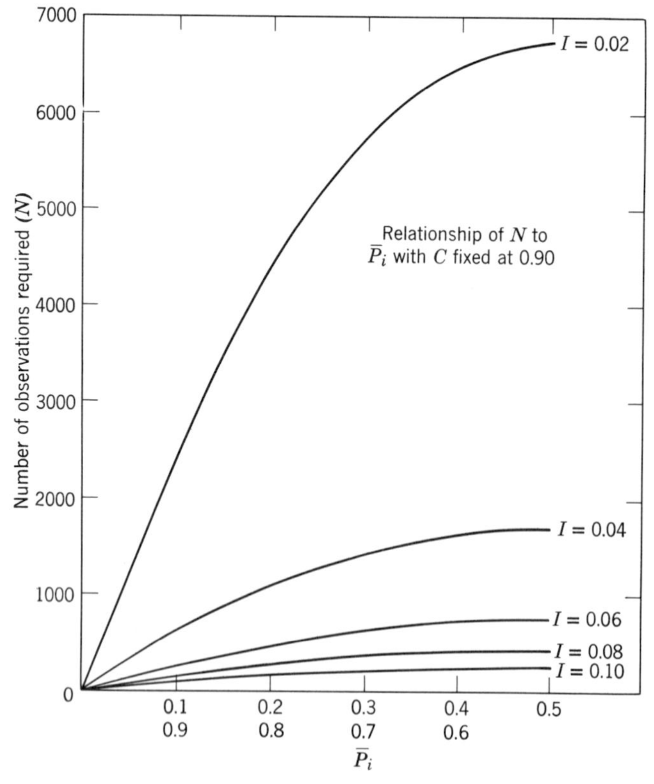

Figure 97. Curves for estimation of N for $C = 0.90$ and values of I of 0.02, 0.04, 0.06, 0.08, and 0.10.

or a surgeon and his crew of assistants, or a plane-overhauling crew, and in each instance wishes to learn what each crew member does as well as the duration and interrelationship of all activities. Suggest several means of achieving this and discuss the relative merits of these.

6. A device often used in complex data measurement, recording, and transmission systems to obtain a maximum of information with a minimum of equipment is called a data sampler. This mechanism, pictured in the accompanying sketch, consists of a rapidly revolving "wiper" and a series of

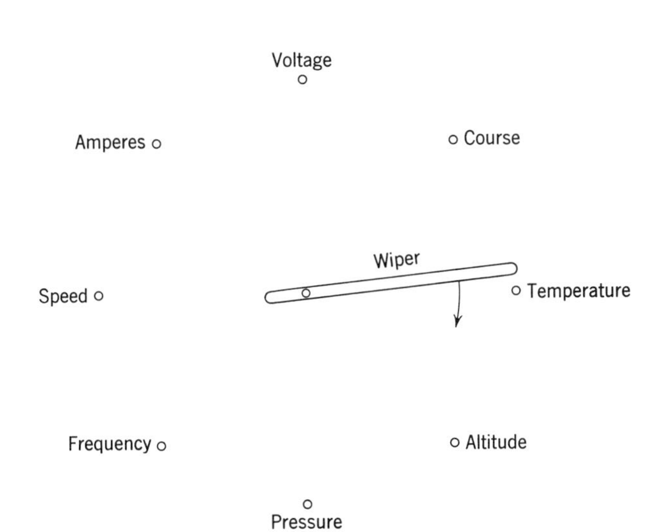

contacts. Each contact presents continuous information with respect to some characteristic of the system. As the wiper revolves the status of the information at each contact is sampled very briefly but very frequently, to be directly recorded or more likely transmitted to a distant recording point where the information is unscrambled by a similar device. By this means, numerous types of information can be handled by a single channel transmission or recording device.

The same principle might be profitably employed by the methods engineer to collect data. However, here the recording "channel" might be a single human observer. How might this scheme be profitably applied to the type of situation described in problem 5?

7. Work sampling might be used on a continuing basis as a means of estimating and controlling man or machine utilization or costs of various types. Design such a scheme for providing a running estimate of (and thus a means of controlling) the man hours required for materials handling, and the distribution of these hours over various types of endeavor, at a large factory. Handling of materials at this plant is responsible for a large percentage of the total manufacturing cost and therefore is of special and continuing interest to management.

REFERENCES

Allderige, J. M., "Work Sampling Without Formulas," *Factory,* March 1954.

Barnes, R. M., *Work Sampling,* 2nd ed., John Wiley and Sons, New York, 1957.

Halsey, John J., "A New Model for Work Sampling—The GREDS Theory," *Journal of Industrial Engineering,* vol. 11, no. 6, November-December 1960.

Hansen, Bertrand L., *Work Sampling for Modern Management,* Prentice-Hall, Englewood Cliffs, N.J., 1960.

Heiland, R., and W. Richardson, *Work Sampling,* McGraw-Hill Book Co., New York, 1957.

Isherwood, J. D., "Labor Cost Analysis by Work Sampling in the Small Business," *Journal of Industrial Engineering,* vol. 11, no. 5, September-October, 1960.

Malcolm, D. G., and L. L. Sammet, "Work Sampling Studies: Guides to Analysis and Accuracy Criteria," *Journal of Industrial Engineering,* vol. 5, no. 4, July 1954.

McAllister, G. E., "Random Ratio Delay," *Journal of Industrial Engineering,* vol. 4, no. 3, August 1953.

18

Evaluation and Improvement of Work Sampling

Popular Claims Concerning Cost and Error

Work sampling is relatively new to the work measurement field and has still to be adopted at many companies, even though there have been many profitable applications of the technique to a wide variety of problems. As yet, its potential as well as its limitations are not generally understood or fully appreciated. Many descriptions of work sampling, especially those appearing in its introductory years, harbor misconceptions, cite unjustified claims, present biased appraisals of its value and applicability. Particularly glaring among the unwarranted assertions are the common yet incompatible claims of "lower cost" and "less error" made for work sampling. Some authors claim that a work-sampling study can be conducted at a fraction of the cost (presumably of an equivalent study involving continuous observation). Such claims are made in spite of the following circumstances.

1. Work sampling involves what is inherently a relatively inefficient method of measurement. Ordinarily an appreciable amount of observer time and travel is required in order to obtain only a *qualitative* measurement, for example: working or not working.
2. Engineers frequently make these observations.
3. In spite of what some of the textbooks and articles naively claim, the time between these randomly spaced observations is not readily used for other work the observer may have to do. It is wishful thinking to expect that the intervals between trips will be effectively used, in view of the fact that the work-sampling observations ordinarily represent random and frequent interruptions of the "other work" and

that numerous time-consuming distractions present themselves while the observer is enroute to and from his observation point.

4. Relatively tight confidence intervals and confidence coefficients are commonly used, resulting in very large sample sizes.

A contention of "less error" is ordinarily made along with the "lower cost" claim. However, as it turns out, all writers who have made this claim are actually referring only to sampling error. They make no mention of, and in fact have not recognized, the other types of error—bias and nonrepresentativeness. Furthermore, to attain the small sampling error implied, a very large sample size is required. The latter, of course, hardly contributes to low measurement cost. For example, assume $P_i = 0.2$ and that a confidence interval of 0.04 and confidence coefficient of 95 per cent are specified. According to equation 2 this would require $[4(2)^2 0.2(1 - 0.2)]/(0.04)^2$ or 1600 observations. If each observation requires an average of 10 minutes of the observer's time, it will require approximately (1600 obs. $\times$ 10 min./obs.)/(480 min./day) or 33 days of observation time to obtain the 1600 observations. If we assume that 5 observations are obtained per trip and that doing so increases the average time per trip to 15 minutes, the total time required would be 10 days.

Although not impossible, it seems unlikely that *both* advantages, lower cost *and* lower error, can be realized simultaneously in a given study. The use of a sample size sufficiently large to obtain the lower (sampling) error inflates the measurement cost, whereas an attempt to attain the lower cost ordinarily involves smaller sample sizes and/or observing more than one person or machine per trip. The former inflates the sampling error, the latter increases the susceptibility of the estimate to bias.

Error in work sampling. In spite of the impression given by most authors, sampling error is not the only error to be dealt with in work sampling. In addition to ignoring the remaining two types of error, a majority of writers speak of accuracy when it is obvious from the text that it is sampling error they are discussing. This situation can cause confusion in the minds of readers who are aware of the fact that altering sample size does affect sampling error but *not* accuracy.

Probably because of the influence of the fields of experimentation and quality control where high levels of significance are customary, confidence coefficients often used in work sampling are unrealistically and uneconomically tight. Certainly a 99 and even a 95 per cent confidence level seem unjustified for most practical purposes, in light of the resulting study cost and of the fact that the errors attributable to bias and nonrepresentativeness are ignored, unknown, and quite possibly appreciable in magnitude. It would probably be more profitable to allocate a greater portion of the total cost of the study to minimizing and controlling bias and correspondingly less on holding the sampling error to a disproportionate level.

A similar argument holds against unrealistically tight confidence intervals. Note, in the formula for estimation of N, that N is a function of $1/I^2$, so that relatively small increases in the confidence interval result in appreciable reductions in N. To illustrate the consequences of relatively small changes in C and I, assume that in a work-sampling study the preliminary estimate of P_i is 0.25, and that a confidence interval of 0.05 and confidence level of .90 are deemed satisfactory. The estimated sample size required from equation 3, page 298, is

$$N = \frac{10.8(0.25)(0.75)}{(0.05)^2} = 810 \text{ observations}$$

Now suppose the value of C is increased from 0.90 to 0.95. The estimated sample size required under these circumstances using equation 2, page 298, is

$$N = \frac{4(2^2)(0.25)(0.75)}{(0.05)^2} = 1200 \text{ observations}$$

Now suppose the rather common practice of stating and applying the confidence interval as a percentage of $\bar{P}_i$ is adopted, and that the designer of the study wishes P_i to be within plus or minus 5 per cent of $\bar{P}_i$. In this case I is $[2 \times 0.25 \times 0.05]$ or 0.025, and the estimate of sample size required, keeping $C = 0.90$ for the moment, becomes by equation 3

$$N = \frac{10.8(0.25)(0.75)}{(0.025)^2} = 3240 \text{ observations}$$

Now suppose two common practices are applied, a confidence level of 95 per cent and a confidence interval of "plus or minus 5 per cent" as it is so often quoted. Under these circumstances the sample size becomes

$$N = \frac{4(2^2)(0.25)(0.75)}{(0.025)^2} = 4800 \text{ observations}$$

The results of these changes in C and I are summarized in the following table.

$C \backslash I$	0.05	0.025
0.90	$N = 810$	$N = 3240$
0.95	$N = 1200$	$N = 4800$

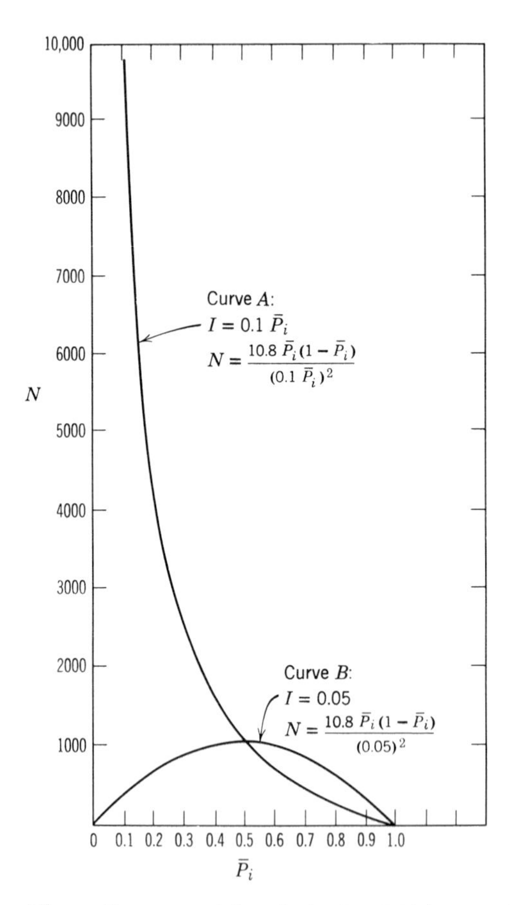

Figure 98. N as a function of $\bar{P}_i$ for two methods of expressing the confidence interval. The method yielding curve B is preferred.

The rather surprising fact about this sampling error matter is that the confidence level of 95 per cent and the confidence interval of "plus or minus 5 per cent" are commonly recommended in the textbooks and journals. The engineer in the case just related began with a very sensible $C = 0.90$ and $I = 0.05$. Perhaps in the course of examining some book or journal he came upon the usual recommendations, took them seriously, made what seemed like a very minor change of C to 0.95 and I to 0.025, and thereby increased the sample size of his study from a reasonable 810 to an absurd 4800 observations.

An unwise practice commented on earlier and illustrated in the case just cited, is that of expressing the confidence interval as a percentage of $\bar{P}_i$, for example $I = 0.10\bar{P}_i$. The effect of this practice on sample size is indicated in Figure 98 by curve A, which shows the rather drastic consequence if $\bar{P}_i$ should happen to be between zero and approximately 0.3, a not so rare occurrence. When the confidence interval is expressed as an absolute value and is thus independent of the magnitude of $\bar{P}_i$, N still varies with $\bar{P}_i$ as indicated by curve B, but not so drastically. In this case N is a maximum when $\bar{P}_i = 0.5$, for at that point $\bar{P}(1 - \bar{P})$ is a maximum. It is obvious from this graphical comparison that expression of the confidence interval as an absolute value is preferred over the percentage method.

Apparently, to facilitate more productive utilization of the periods between observations, some writers and practitioners have adopted the practice of observing at regular intervals, such as every half hour. This procedure makes it possible for observed individuals to anticipate the moment of observation and gives them ample opportunity to alter their behavior accordingly. Adherence to this practice is an open invitation to bias. Furthermore, if there is a tendency for any type of activity under observation to be periodic or cyclic in occurrence, for example a worker who likes to "take five" every hour for a cigarette, the use of regularly spaced observations will probably result in a gross over- or under-estimate of the proportion of time devoted to this cyclical activity. *If* the spacing and duration of an activity were random in nature, then equally spaced observations would not result in bias, but such randomness is not known to be common in practice. Therefore, to avoid what might well be a significant increase in bias, it seems desirable to use randomly spaced observations.

Another practice that has become common in an attempt to reduce the cost of the study is to observe more than one person (or machine as the case may be) per trip. Thus, on each trip he makes, the observer might observe each of the thirteen materials handlers in the shop (or in the study of engineers described on page 290, he might have observed say five engineers per trip to reduce the number of daily trips from twenty-five to a more reasonable five). Then the round of thirteen observations of thirteen different materials handlers will be treated as if these were thirteen different, randomly spaced observations of one materials handler. Unfortunately, the assumption of the equivalence of these two situations is ordinarily unjustified. When observations are taken in groups (rounds) two adverse consequences are likely.

1. The results of the study are easily biased, for although there may be an element of "surprise" in observing the first of the group of m individuals, $m - 1$ of them are probably aware of the observer's presence and thus have adequate opportunity to alter their behavior.

2. There is a likelihood that if in a given round of observations an individual is found engaged in a certain category of activity, others will

also be found engaged in that same activity. For example, some causes of idleness affect several or all of the persons in a group under observation. This correlation between observations, or lack of independence, ordinarily causes the binomial model to yield an underestimate of the variance of $\bar{P}_i$. One writer has proposed the following means of estimating the variance of $\bar{P}_i$ when nonindependence of observations is known or suspected.[1]

N = total number of observations;

m = number of individuals observed per trip or round;

$n = \dfrac{N}{m}$, the total number of trips;

N_{ik} = number of observations of activity i in the kth round of observations; thus $N_{ik} \leqq m$;

N_i = total number of observations of activity i during study;

$$s_{\bar{P}_i}^2 = \frac{m^2}{N^2} \sum_{k=1}^{n} \left(\frac{N_{ik}}{m} - \frac{N_i}{N} \right)^2$$

The preceding expression would replace the quantity $[\bar{P}_i(1 - \bar{P}_i)]/N$ ordinarily used to estimate $\sigma_{\bar{P}_i}^2$ when independence of observations may be assumed.

Some Misconceptions and Malpractices with Respect to Work-Sampling Procedure

One author has incorrectly stated that not more than one observation of a given delay should be counted in a work-sampling study. Thus, if an hour-long breakdown of an operation should happen to be observed three times, two of the three observations are to be disregarded under this recommendation. This contention is completely unfounded.

Another practice that has been proposed, presumably to reduce the observer travel in work sampling, is that of having the observer remain on the scene continuously, either for extended intermittent periods or for the length of the study. He then glances at the worker(s) at random times chosen in advance or by the observer on the spot. This is a rather odd proposal. If the observer is going to be on the scene continuously, why have him deliberately and literally turn his back on useful and readily obtainable information? He may as well take watch in hand and make continuous, quantitative observations while he is there.

[1] R. W. Conway, "Some Statistical Considerations in Work Sampling," *Journal of Industrial Engineering,* March-April 1957.

Making Work Sampling More Competitive as to Cost

There are a number of modifications that can be made in current work-sampling theory and practice in order to make the cost of the study more reasonable.

1. Adopt more liberal sampling error requirements than are now customary. In general, by doing so a better balance between measurement cost and error can be achieved. Above all, the practice of arbitrarily lifting a confidence coefficient and interval from a textbook or article, or adopting the values customarily used in another field should be avoided. Instead, the choice should be based on the particular demands of each situation, especially the purpose of the estimate being sought, the possible consequences of various sized sampling errors, and the like.
2. Reduce the cost of obtaining the observations by
   a. observing as many operators (or machines) per observer trip as apparent homogeneity of workers (or machines) will justify. It is the amount of observer time per observation that is so critical in determining the cost of a work-sampling study, and which on occasions becomes absurdly large. Reduction of this "unit observation cost" is crucial to the competitiveness of work sampling. Even though this will probably increase the bias of the study, this practice of grouping observations still seems desirable. Whenever observations are taken in multiple and independence cannot safely be assumed, the variance of $\bar{P}_i$ should be estimated by the method described earlier for this type of circumstance.
   b. making several work-sampling studies simultaneously. What might otherwise be independently conducted studies of different operations, activities, departments, etc., can sometimes be conducted in parallel, so that each observer trip will be reasonably productive of observations.
   c. using lower-paid, semitechnical personnel to collect the bulk of the observations required. Allowing engineers or other relatively skilled personnel to do the "leg work" involved in a work-sampling study inflates the cost of the study considerably. Not only is this an unnecessary expense, it is a paradoxical situation in that here is a specialty concerned with effective utilization of the human resource that is making grossly inefficient use of its own resources. Furthermore, this sort of practice certainly does not enhance the worth and prestige of the time-study department in the eyes of the rest of the organization. It can be especially detrimental to the attitudes of shop personnel toward time study and engineers.

   The collection of work-sampling observations is one of a number of such tasks that are well suited for and should be performed by a semitechnical man. However, design of the study, execution of a pilot study, supervision of the data collection phase, and interpretation of the results should be performed by a qualified expert (who, it is hoped, has had some training in statistics).

3. Reduce the cost of processing the data by
   a. directly recording the observations in a manner that will permit machine processing of the data. For example, there are special IBM cards that can be marked directly by the observer so that after that time the data can be tabulated, classified, and otherwise processed on IBM equipment.
   b. the use of short-cut computational procedures such as alignment charts, tables, or graphs, in order to minimize the time required to compute N, I, and control chart limits.

REFERENCES

Conway, Richard W., "Some Statistical Considerations in Work Sampling," *Journal of Industrial Engineering,* vol. 8, no. 2, March-April 1957.

Davidson, Harold O., "Work Sampling—Eleven Fallacies," *Journal of Industrial Engineering,* vol. 11, no. 5, September-October 1960.

19

Standard Data

The third major class of time-study procedures is referred to as the synthetic method (standard data method). By way of introduction, consider a factory in which a large number of drilling operations are performed on a wide variety of metal components. These parts differ in size, shape, weight, and hardness. The drilled holes vary in diameter, depth, number, location, and tolerance requirements. This company has been using stop-watch time study to establish time standards on all drilling operations. The management is dissatisfied with the number of man-hours consumed in setting these standards and they are anxious to have the time standards before jobs are actually run, for cost estimating, scheduling, and other decision-making purposes. In this case the company can and should take advantage of the large volume of time standards already on hand, by the following means.

1. Analyze these standards to determine if and in what manner the normal time for drilling depends on various characteristics of the part being operated on.[1] For drilling, the normal time (NT) is a function of most or all of the characteristics listed previously, thus

$$NT = f \text{ (size of part), (shape of part), (weight of part), (hardness), (number of holes), } \ldots$$

2. Then, from this time on, establish the normal time for any new drilling operation by substituting into the resulting formula the par-

[1] Theoretically, the normal time for an operation is a function only of the method and of the concept of normal performance in that plant. Conceptually, it is not a function of the operator, his condition, the time of year, or any other such variables that are affectors of *actual* but *not normal* time.

ticular characteristics of the part and calculating the normal time. Adding the appropriate allowance for delays and fatigue provides a standard time for the operation.

This is referred to as the standard data method of establishing time standards. Under this system it is not necessary to time study the operation. In fact, it is not necessary to see the operation in order to set the standard. It is necessary only to have the specifications of the part. This arrangement will permit the company to establish time standards for its drilling operations at a fraction of the cost required by individual time studies, and makes it possible to set a standard as soon as the part and tooling specifications are known.

Generally, standard data expresses the relationship between certain pertinent characteristics of a task and the normal time required to perform that task, in a form that permits synthesis of the latter from the former.

The most common form of standard data in use is referred to as the elemental type. Here the normal time for various elements E_1, E_2, etc., of the operation are individually expressed as functions of certain significant variables. The total normal time for the cycle is the sum of the elemental normal times, thus

$$NT = NT_{E_1} + NT_{E_2} + NT_{E_3} + \cdots NT_{E_N}$$

where

$$NT_{E_1} = f(V_1), (V_2), \cdots f(V_n)$$

$$NT_{E_2} = f(V_1), (V_2), \cdots f(V_n)$$

$$NT_{E_N} = f(V_1), (V_2), \cdots f(V_n)$$

and where V_1, V_2, $\cdots$ V_n are job variables affecting normal time, for example size, shape, weight, and hardness. If the elements E_1, E_2, $\cdots$ E_N are individual motions such as reach, move, and position, the standard data is commonly referred to as the predetermined motion-time technique (microscopic standard data). This form, introduced earlier, is exemplified by Methods-Time Measurement (Appendix A). The predetermined motion-time technique will be further discussed and appraised in Chapter 20.

If the elements E_1, E_2, $\cdots$ E_N consist of groups of motions, such as "pick up part and place in jig," "engage feed," "process part," and "remove part from jig and dispose," the data is commonly referred to simply as standard data (macroscopic standard data).

The use of elements is not an essential characteristic of standard data. It is possible and often desirable to bypass the elemental feature and derive a formula for computation of the *total* normal time directly. This type, suggested earlier for the company with many drilling operations, expresses the total normal time for a given type of operation as a function of certain variables (characteristics of the part), thus:

$$NT = f(V_1),\ (V_2),\ (V_3),\ \cdots\ (V_n)$$

In summary then, standard data commonly appears in several forms, namely, the elemental form, which may be of the "macroscopic" or "microscopic" type, and the nonelemental (formula) form.

Standard Data Derivation Procedure

The general procedure for derivation of standard data closely parallels that required for development of any predictive procedure. The same basic derivation procedure applies, whether it is a matter of predicting success on the job from certain characteristics of the prospective employee, or forecasting sales volume from certain business and market characteristics, or synthesizing the normal time for an operation from certain characteristics of the job to be performed. It is a process of determining what independent variables are significant affectors (determinants) of a dependent variable like job success, sales volume, or normal time for an operation, and of determining the nature of the relationship between dependent and independent variables so that the former may henceforth be computed, given specific values for the independent variables. Thus, in the derivation of standard data it is necessary to first hypothesize what job variables significantly affect normal time for a given type of operation, then collect times on a number and variety of jobs of that type, then use this data to determine the relationship (if any) between normal time and each of the variables suspected of being significant affectors of normal time.

For example, standard data has been derived for establishing standard times for cleaning offices, to be used by a firm which performs this service on a contract basis in many company office buildings. Standard data is useful to this firm for arriving at fees for potential new customers, for scheduling, and for establishing standards for incentive pay purposes. In deriving standard data for this purpose, it was initially hypothesized that such variables as floor area, type of floor, number of movable obstacles, number of stationary obstacles, and the like, are significant variables affecting the normal time to clean an office. To test these hypotheses a number of time studies were made of the cleaning of a variety of offices differing with respect to area, type of floor, number of movable objects, etc. With this data it was then

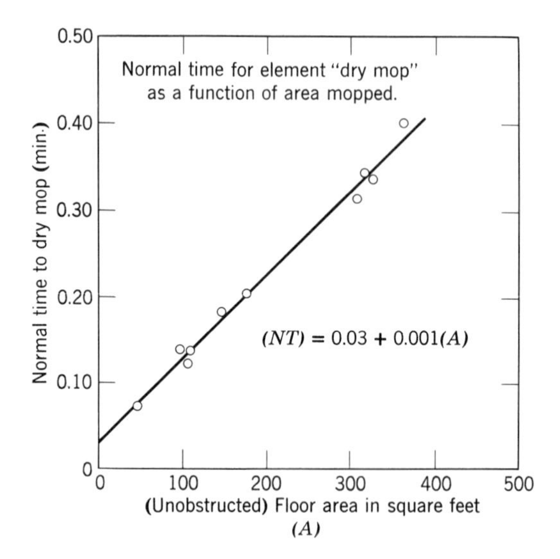

Figure 99. Plot made during investigation of the hypothesis that floor area significantly affects the normal time to dry mop.

possible to investigate the effect that each variable has on normal time. For example, normal time to dry mop was plotted against area, as illustrated in Figure 99, to determine in what manner normal time for that element varies with floor area. Similar investigations were made of other variables until the effect of each was known. A simplified version of the formula that finally evolved is as follows:

$$NT \text{ to clean office} = 0.85 + 0.001V_1 + 0.013V_2 + 0.032V_3 + 0.065V_4 + 0.03V_5 + 0.011V_6$$

where

V_1 = square feet of bare floor, wood or tile;
V_2 = square feet of rug;
V_3 = number of movable objects, for example chairs;
V_4 = number of stationary objects, floor accessible, for example table or desk;
V_5 = number of stationary objects, floor inaccessible, for example file cabinet;
V_6 = square feet of area to be dusted.

From this formula the normal time for a given office may be synthesized, given the characteristics of that office insofar as the foregoing variables are concerned, and to this the necessary allowance may be added to obtain the standard time. To illustrate, the normal time for a 30- by 35-foot office containing 4 desks, 10 chairs, 5 standard file cabinets, and 540 square feet of rug area, is computed as follows:

$$\begin{aligned} NT &= 0.85 + 0.001(1050 - 540) + 0.013(540) + 0.032(10) \\ &\quad + 0.065(4) + 0.03(5) + 0.011(95) \\ &= 9.86 \text{ minutes.} \end{aligned}$$

The addition of a 16 per cent allowance for fatigue and delays yields a standard time of 11.44 minutes.

The discussion thus far has dealt with the derivation procedure in general. The specific procedure suggested for derivation of standard data is as follows:

1. Preliminary steps, involving:
   a. determination at the outset, as to whether standard data is economically and practically feasible;
   b. improvement and standardization of the method used on the operation for which the standard data is to be derived;
   c. hypothesizing of the variables with which normal time is significantly correlated;
   d. decision as to what information must be collected and how, in order to establish the relationship between normal time and the variables hypothesized. Specifically:
      (1) What source(s) of time data should be used in investigating the effects of variables? Is it possible to use previously made time studies now in the files? Must special time studies be conducted? Can predetermined motion times or work sampling be used? Or is some combination of these preferable?
      (2) What information concerning the part, the method, the quality, should be collected as time studies are made for purposes of developing standard data? If the correlation between normal time and certain variables is subsequently to be tested, it will be necessary to know the particular value for each of these variables that underlies each normal time.
      (3) If elemental time studies are to be used to collect the data, what elemental breakdown should be used? Standardization of elemental breakpoints is usually important because collection of the necessary data frequently takes place over a fairly long period of time and involves a number of different observers. Failure to adhere to a standard elemental breakdown may render a large volume of time-study data useless insofar as deriving standard data from it.
      (4) If special time studies must be made to obtain the necessary time data, what jobs should be observed? In other words, what jobs

should the sample consist of in order that sufficient data be available over a satisfactory range of each variable?

2. Collection of data, which involves actual execution of the necessary time studies, collection of records on previously conducted time studies, and the like.
3. Processing of the data collected to establish the nature of the relationship between normal time and each of the variables hypothesized. This phase ordinarily involves plotting of graphs, fitting curves to sets of data points, and use of a variety of statistical procedures.
4. Presentation of results in a form that minimizes application time, difficulty, and possibility of misapplication. The final results should include:
   a. the "working data," the table, graph, formula, alignment chart, or combination of these, from which estimates of normal time will be made. The working data should be supplemented by the following.
      (1) Carefully prepared instructions as to how the applier of the data should proceed to prepare an estimate. This is important because ordinarily the applier of the standard data is not the same person who derived it. It usually requires a person with considerable training and experience to derive such data, but should require no more than a clerk to apply it.
      (2) A definite and obvious declaration of the limitations of the data provided. This should cover the method, equipment, range of variables, etc., to which the data applies. This should be done to minimize the chances that the data will be applied to operations it was never intended to cover.
   b. a complete file that thoroughly explains how the working data was derived. This file should include all supporting data, computations, analyses, graphs, and the like. Such information is vital to future adjustments, checks, and revisions of the data.

Sources of input data. Time data to support the standard data, that is, with which to determine the relationship between normal time and significant variables, must come from the time-study files, special time studies, predetermined motion times, work sampling, or most likely, some combination of these.

The accumulation of records of time studies taken over an extended period of time in the ordinary course of setting time standards is commonly referred to as the time-study file. Ideally, all that should be necessary to get the time data needed to derive standard data would be to go to the time-study file and pull out the studies on a variety of jobs sufficient to investigate the variables selected. The advantages are immediate availability and negligible cost. The disadvantages are the vulnerability of such data collected over a relatively long period of time to changes in method, quality, equipment, and conditions, and to variations in elemental breakpoints. The time-study file is certainly the preferred source of time data, but it is often unusable, unavailable, or inadequate, and thus must be supplanted or supplemented by another source.

If the time-study file cannot be used or must be supplemented, it is often necessary to conduct special time studies, solely for the purpose of developing the standard data desired. The obvious disadvantage is the extra cost involved. The main advantage is the likelihood that the data represents current shop methods, quality levels, equipment, and conditions.

To a limited extent, the predetermined motion-time technique may be used to synthesize standard data. For example, the number of screws to be inserted is a variable to be recognized in standard data being derived for a company's assembly operations. A predetermined motion-time system is used to synthesize the time to place a screw, 0.046 in this case. Then in the final formula the time for this phase of the assembly operation is $0.046N$, where N is the number of screws to be placed. The use of the predetermined motion-time technique for this purpose presumes that the analyst has prior knowledge of the significant variables and their effect, for the analyst is in effect synthesizing the effect of variables. Thus, variables the analyst does not anticipate on an intuitive basis and variables not recognized in the particular predetermined motion-time system used, probably will not be taken into account in the standard data that evolves. This technique is especially useful in filling in gaps in the data provided by time studies.

The results of a work-sampling study can sometimes be used as a basis for developing standard data. However, the standard data that evolves will not be of a refined nature and thus may not be adequate for many purposes.

Some of the best jobs of standard data development represent the combined use of the time-study files, special time studies, and predetermined motion times, in such a manner as to minimize the cost of derivation and maximize quality and coverage.

Balancing Cost of Derivation and Application Against Cost of Error

Suppose that for a certain type of operation variables, $V_1, V_2, \cdots V_7$ are suspected of contributing to determination of normal time. Time data is collected and the practical and statistical significance of the correlation between normal time and each variable is tested. As a result some variables are found to have a strong influence on normal time, others appear to have little or no effect. Assume that in the case in point the degree of influence is as follows, with variables listed in order of decreasing effect on normal time:

$$NT = f(V_4, V_7, V_2, V_1, V_3, V_6, V_5)$$

At this point in the analysis of the results it is necessary to decide how many of these variables should be recognized in the standard data. It does not necessarily follow that all variables having some effect on normal time should be taken into account, in fact, it is almost certain that for economy reasons some variables will be ignored. The issue at stake is as follows: The more variables recognized in the standard data, the more time that must be spent in derivation and the more unwieldy and costly it becomes to set a standard from the data. By attempting to recognize too many variables the cost of the data can thereby be inflated to the point where a primary purpose of standard data—reduced cost of setting time standards—is defeated. However, there is an adverse consequence to ignoring too many of the variables that are affectors of normal time. Granted, the fewer the variables taken into account the cheaper the data to develop and apply, but the greater the error of the standard times produced. Reduction of the number of variables recognized ordinarily inflates the error of the standards provided by standard data. This increases the costs that result from errors in the use of these standards for planning and evaluation purposes. Therefore, in developing standard data an attempt should be made to strike an optimum balance between simplicity of the data and the error of the estimates produced. In the foregoing hypothetical case, it is possible, but rather unlikely, that only V_4 need be recognized in the standard data in order to obtain precise enough standards for the situation at hand. It is quite likely that with only one variable taken into account the data would be very simple and quick to apply, but also that the standards would be rather inconsistent as a result of ignoring the remaining significant variables. If V_7, the next most influential variable were also recognized so that now $NT = f(V_4, V_7)$, the standard data would probably become more elaborate and time consuming to apply, but the consistency of the standards would improve. This process of increasing the number of variables recognized should continue until the cost of deriving and applying the data and the cost of the error of estimation appear balanced for the situation involved. That is, the process should continue until a point is reached where the added precision contributed by additional variables is not justified by the resulting higher cost of making the estimates. In most practical cases there will be some variables

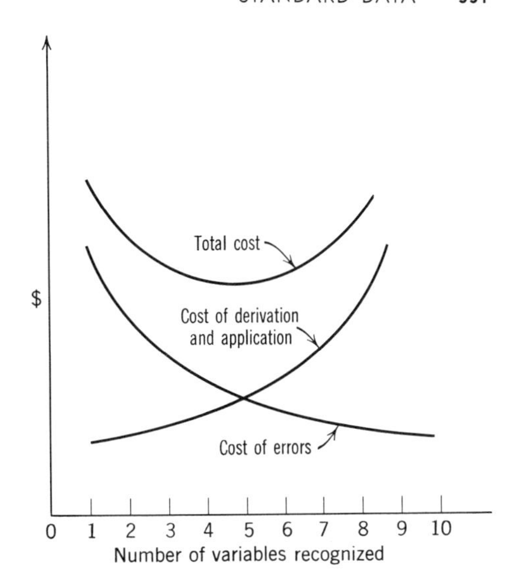

Figure 100. Expected relationship of the relevant costs with number of variables taken into account by standard data.

that, although they have an effect on normal time, simply are not economical to recognize in standard data. These variables are "averaged in" and contribute to the constant term in the final formula.

This balancing process can be depicted graphically. As indicated in Figure 100, the cost of deriving and applying the data is expected to increase at an accelerating rate as the number of variables recognized in the data is increased. The cost associated with the errors of the standards produced is expected to increase at an accelerating rate as the number of variables recognized is reduced. In each situation there is some number of variables for which total cost is a minimum.

A New and Promising Technique for Standard Data Derivation

It is not an easy matter to satisfactorily determine the relationship between a dependent variable and a series of independent continuous variables when the number of independent variables is more than two or three. The ordinary practitioner in the field of time study, if he could cope with the task at all, would require a very large volume of data in order to determine the effect of each variable if the number of independent continuous variables were four or more. Under these circumstances, he ordinarily finds it necessary to "hold the other variables constant" while investigating the effect of any one of them, a process that requires a relatively large amount of time-study data.

As a result, important variables are sometimes ignored and/or extra derivation costs incurred.

The inability to conveniently deal with more than a few variables simultaneously gives rise to the rather general practice of splitting the job cycle into elements and processing each one of these individually. In effect, by doing so the practitioner is developing separate standard data for each of the elements, then later aggregating these. If he tried to cope with the total cycle time and its various determinants simultaneously, he would probably have to deal with a prohibitively large number of variables at one time. For example, in developing the standard data for office cleaning, the analyst collected time data for such elements as the following, given along with the variables investigated for each.

1. Setup—room size.
2. Empty waste—room size.
3. Empty ash trays—room size.
4. Dry mop floor—floor area, floor type, obstacles.
5. Vacuum rugs—rug area, type, obstacles.
6. Dust furniture—surface area.

Note that in going about the task in this fractionated manner the largest number of variables he had to cope with simultaneously is three. If he had collected total cycle times only, and attempted to investigate the effect of these variables in aggregate on normal time for the cycle, he would be attempting to determine the relationship between a dependent variable and six independent variables simultaneously, namely:

$$NT \text{ to clean office} = f\,(\text{bare floor area, floor type, rug area, rug type, obstacles, furniture surface area}).$$

Without a large amount of time-study data, the task would be beyond the ordinary practitioner equipped as he usually is.

Yet the statistician has been effectively dealing with problems of this type for decades. In fact, since high-speed computers have become commonplace, solution of problems of this type, in which the relationship between a dependent variable and more than several independent variables must be estimated from a body of empirical data, have become routine matters in such fields as sales forecasting, industrial psychology, agriculture, and a number of others. The procedure that has been used in these fields is directly applicable to the problem of standard data development, with the same effectiveness, speed, and economy achieved elsewhere.

The technique commonly used by statisticians to estimate the relationship between a dependent variable and one or more independent variables is commonly referred to as regression analysis. This technique has great potential in the field of work measurement. Its use will be described and illustrated.

The regression technique. Regression analysis is an objective and very useful method of fitting a basic equation form to empirical data. This procedure yields an estimating equation (model) ordinarily referred to as a regression equation. To illustrate this method, the simplest case of one independent variable will be used and a linear relationship between the two variables assumed.

Standard data is to be developed at a certain print shop for the purpose of setting time standards on the operation of assembling booklets of varying numbers of pages. Special time studies have been made of booklets being assembled, yielding the results shown in Figure 101. Since linearity of the relationship apparently may be assumed, an equation of the form $y = a + bx$ fits the data shown. Solution of simultaneous equations 1 and 2 provides the best estimates of a and b, and thus the equation of the straight line that best fits the points given.

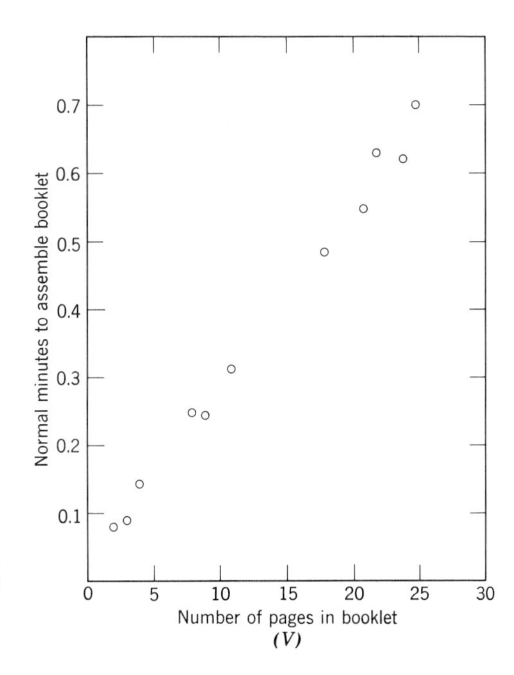

Figure 101. Plot of time-study data collected for derivation of standard data for booklet assembly operation.

$$\sum(NT) = na + b\sum V \quad (1)$$

$$\sum V(NT) = a\sum V + b\sum V^2 \quad (2)$$

where

n = the number of points available,
$\sum(NT)$ = the sum of the n normal times,
$\sum V$ = the sum of the values of the independent variable,
$\sum V(NT)$ = the sum of the cross products of the n points, and
$\sum V^2$ = the sum of the squared values of the independent variable.

Table 11 includes the values of the points plotted in Figure 101 and

TABLE 11. Computation of Linear Regression Equation for Booklet Assembly Data

Booklet Type	(NT)	V (Number of Pages)	V^2	$(NT)V$
$C4$	0.080	2	4	0.16
$A9$	0.090	3	9	0.27
$C2$	0.145	4	16	0.58
$C3$	0.247	8	64	1.98
$B1$	0.245	9	81	2.21
$B4$	0.314	11	121	3.45
$B9$	0.485	18	324	8.73
$A7$	0.548	21	441	11.50
$A4$	0.630	22	484	13.90
$C1$	0.620	24	576	14.90
$C8$	0.700	25	625	17.50
	$\sum(NT) = 4.104$	$\sum V = 147$	$\sum V^2 = 2745$	$\sum(NT)V = 75.18$

Substituting into equations 1 and 2:

$$\left.\begin{cases} 4.104 = 11a + 147b \\ 75.18 = 147a + 2745b \end{cases}\right\} \text{Solved to obtain values of } a \text{ and } b.$$

$$a = 0.0247$$

$$b = 0.0261$$

$$\therefore (NT)' = 0.025 + 0.026V$$

summarizes the computations made to obtain the values of the constants a and b. The resulting equation is

$$(NT)' = 0.025 + 0.026V$$

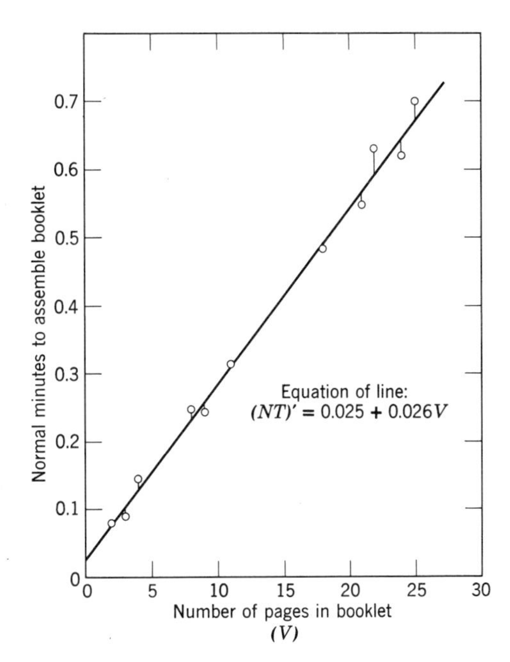

Figure 102. Time-study data collected for derivation of standard data for booklet assembly operation, shown with line fitted by the regression technique.

where $(NT)'$ is the synthesized normal time and V is the number of pages in the booklet to be assembled. Figure 102 shows this line along with the data points in the original sample. The vertical distance between each of the original points and the regression line has been computed and the results summarized in Table 12. For example, the normal time for booklet $A9$ according to the original time study is 0.090 minutes. According to the equation derived, the normal time for a booklet of three pages would be computed as

$$(NT)' = 0.025 + 0.026(3) = 0.103 \text{ minute}$$

The difference between the normal time for this booklet in the original sample and the synthesized normal time is $0.090 - 0.103 = -0.013$. This value squared is 0.00017.[2] As indicated in Table 12, similar computations have been made for each of the original points to obtain the total of the squared deviations. The sum of these squared deviations for the sample may be used to estimate the standard deviation of the original data points about the regression line in the following manner:

$$s_{(NT)} = \sqrt{\frac{\Sigma[(NT) - (NT)']^2}{n}}$$

[2] The deviations are squared to convert all to positive values.

where

$s_{(NT)}$ = the standard deviation of the original points about the regression line,
(NT) = a normal time from the original sample,
$(NT)'$ = the equivalent of (NT), but synthesized from the equation, and
n = the number of data points available in the sample of time studies.

Knowing $s_{(NT)}$, it is possible for us to estimate a confidence interval about the regression line from the following expression:

$$I = 2t_C s_{(NT)} \sqrt{\frac{n}{n-2}}$$

where

I = a confidence interval straddling the regression line,[3]
t_C = the appropriate factor selected from a table of values for the t-distribution (Appendix D) for $n - 2$ degrees of freedom and a confidence coefficient C,

TABLE 12. Computation of the Residual Sum of Squares and Standard Deviation of Data Points about the Regression Line; Booklet Assembly Data

Booklet Type	(NT) (From Original Sample)	$(NT)'$ (Synthesized from Standard Data)	$(NT) - (NT)'$	$[(NT) - (NT)']^2$
$C4$	0.080	0.077	0.003	0.00000
$A9$	0.090	0.103	−0.013	0.00017
$C2$	0.145	0.129	0.016	0.00026
$C3$	0.247	0.233	0.014	0.00020
$B1$	0.245	0.259	−0.014	0.00020
$B4$	0.314	0.311	0.003	0.00000
$B9$	0.485	0.493	−0.008	0.00006
$A7$	0.548	0.571	−0.023	0.00053
$A4$	0.630	0.597	0.033	0.00109
$C1$	0.620	0.649	−0.029	0.00084
$C8$	0.700	0.675	0.025	0.00063

$$\text{Residual Sum of Squares} = \sum[(NT) - (NT)']^2 = 0.00398$$

$$s_{(NT)} = \sqrt{\frac{\sum[(NT) - (NT)']^2}{n}} = \sqrt{\frac{0.00398}{11}} = 0.019 \text{ minute}$$

[3] This expression provides a rough approximation.

$s_{(NT)}$ = the standard deviation of the data points in the original sample about the regression line.

For the current example $s_{(NT)} = 0.019$ minute, so that for a confidence coefficient of 0.90,

$$I = 2(1.83)0.019 \sqrt{\tfrac{11}{9}} = 0.077 \text{ minute}$$

On the basis of the variability of the points in the original sample of time studies, if the normal time were estimated by time study, it would be expected to be within plus or minus $I/2 = 0.077/2 = 0.039$ minute of the time synthesized by the regression equation, with approximately 90-per cent confidence.

The sum of squares, $s_{(NT)}$, and the confidence interval estimated therefrom offer indices of the "goodness of fit" and thus the closeness with which the normal time that would ordinarily be set by direct time study can be estimated from the regression equation. A property of the regression equation is that the sum of the squares of the vertical deviations of the original data points from the curve[4] is a minimum, a feature that gives rise to the common title "least squares technique" applied to this procedure.

Thus, within the limits of the basic equation form selected and the empirical input data given, the regression (least squares) equation is the best estimate of the relationship between a dependent variable and one or more independent variables. Therefore, this equation offers the best estimate (in that the sum of the squared residuals is a minimum) of the dependent variable given specific values of the independent variables.

The procedure that has been discussed thus far does in a more objective manner what the time-study practitioner ordinarily accomplishes "by eye" in much less time and reasonably satisfactorily. So that when dealing with only one independent variable and a linear relationship, the regression technique offers no strong advantage over the common "fit by eye" procedure, unless the demand for objectivity is unusual. However, when the relationship between a dependent variable and two or more independent variables must be estimated, the regression technique assumes definite superiority.

The case of a dependent variable and two independent variables V_1 and V_2 where the relationships are linear, is graphically characterized by a plane in three-dimensional coordinates. In this instance the equation is of the form

$$(NT)' = a + bV_1 + cV_2$$

[4] Often referred to as the residual sum of squares.

The constants a, b, and c in this equation can be determined by solution of the following simultaneous equations:

$$\Sigma(NT) = a + b\Sigma V_1 + c\Sigma V_2$$

$$\Sigma V_1(NT) = a\Sigma V_1 + b\Sigma V_1^2 + c\Sigma V_1 V_2$$

$$\Sigma V_2(NT) = a\Sigma V_2 + b\Sigma V_1 V_2 + c\Sigma V_2^2$$

A desk calculator is of considerable assistance in obtaining the squared and cross-product members of these equations. The linear regression equation that results is a least-squares solution, and graphically is the plane that best fits the data provided.

To obtain the constants for the case of three independent variables V_1, V_2, and V_3, where

$$(NT)' = a + bV_1 + cV_2 + dV_3$$

by this technique requires the solution of four simultaneous equations and a considerable amount of laborious preliminary calculations. A desk calculator is very helpful if not essential.

Cases involving more than three independent variables are unwieldy and laborious to solve. Fortunately, digital computer programs for solution of such problems are commonplace and are frequently used in a variety of basically similar problem situations

Derivation of standard data by the combined use of multiple regression and a digital computer. Standard data is to be derived for the operation of applying special paints to large, flat surfaces, such as pictured in Figure 103. The sheet to be painted comes to and leaves the workstation on roller conveyor, and remains on the conveyor while being painted. After pushing the finished piece along on the conveyor, the operator pulls the next piece into painting position. He brush paints that sheet, then cleans along the border. After attaching a tag he shoves the finished piece along and repeats the cycle. On some occasions, masking tape must be used along the edges. Time studies are available for a number of different sized and shaped surfaces. A summary of the time-study data available from the files is given in Table 13; the shape and size of the parts studied are indicated in Figure 103.

Ordinarily, the practitioner would attack a problem like this on an elemental basis. In this instance he would probably investigate the effect of area of the sheet on the element "shove finished piece along conveyor"; the effect of area on the element "pull next piece to workstation"; the effects of perimeter and area on "paint"; the effect of

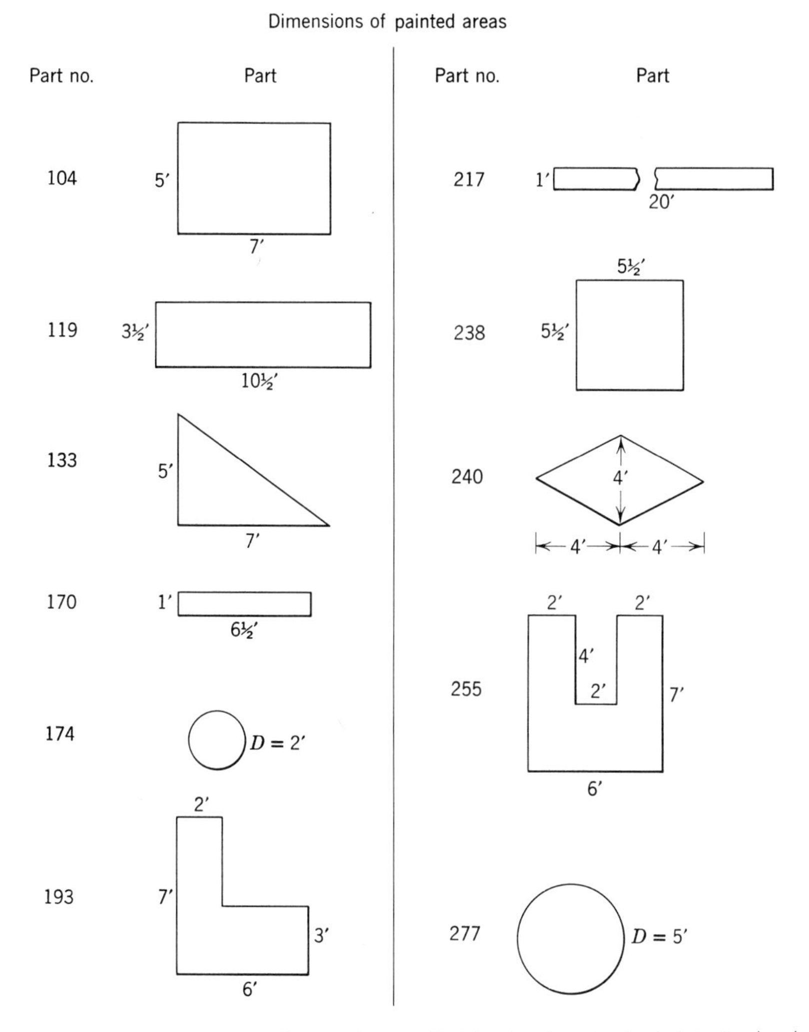

Figure 103. Size and shape of parts time studied to develop standard data for brush painting. Results of time studies are given in Table 13.

perimeter on "clean along edge"; and the effect of perimeter on "apply and remove masking tape."[5] However, use of multiple regression

[5] The element "attach tag" would probably be treated as what is commonly referred to in the trade as a "constant element," meaning that it is an element the normal time for which is apparently unaffected by any job variables, so that a fixed amount of time may be allowed regardless of size, shape, etc.

TABLE 13. Summary of Time Studies Available for Brush Painting Operation*

Element	Part Number										
	104	119	133	170	174	193	217	238	240	255	277
Shove finished piece along conveyor	0.18	0.16	0.17	0.16	0.17	0.15	0.18	0.17	0.16	0.17	0.18
Pull next piece to workstation	0.15	0.14	0.13	0.15	0.14	0.13	0.15	0.14	0.13	0.14	0.15
Paint	10.84	7.43	6.35	2.93	1.36	9.25	9.43	6.34	6.13	12.10	6.51
Clean along edge	4.31	—	4.13	3.50	1.34	5.44	8.32	—	3.35	6.88	3.12
Attach tag	0.08	0.08	0.07	0.08	0.08	0.07	0.09	0.08	0.07	0.08	0.08
Apply and remove masking tape	—	4.37	—	—	—	—	—	3.42	—	—	—
Normal Cycle Time	15.56	12.18	10.85	6.82	3.09	15.04	18.17	10.15	9.84	19.37	10.04
V_1 (Perimeter of unmasked part in feet)	24	0	20.6	15	6.3	26	42	0	18	34	15.7
V_2 (Area to be painted in square feet)	35	30	17.5	6.5	3.1	26	20	27.5	16	34	19.5
V_3 (Perimeter of masked part in feet)	0	28	0	0	0	0	0	22	0	0	0

* All times are in normal minutes.

analysis[6] eliminates the need for this elemental approach. This technique permits the practitioner to investigate the aggregate effect of the previously mentioned variables on total cycle time, an ability that has a number of advantages. Under the assumption of a linear relationship between normal time and each variable, the following basic equation applies:

$$(NT)' = a + bV_1 + cV_2 + dV_3$$

where

V_1 = perimeter of an unmasked part,
V_2 = area of part, and
V_3 = perimeter of masked part.

To obtain the values of the constants a, b, c, and d requires the solution of four simultaneous equations, a task performed in this instance by a digital computer. Note that the only data that must be given to the computer are the normal cycle time for each of the parts studied and the values of V_1, V_2, V_3 for each (the last four lines of data in Table 13). The computer handles the rest, providing among other useful types of information, the values of the constants in the regression equation. The constants and equation in this case are

$$(NT)' = 0.638 + 0.343V_1 + 0.190V_2 + 0.211V_3$$

In addition to providing the desired equation, the computer program can be designed to compute the residual sum of squares, $s_{(NT)}$, and a confidence interval for the regression equation. Thus, in paralleling the computations performed by hand for the booklet assembly regression equation, the computer will take the original values for V_1, V_2, and V_3 and compute the normal time from the equation provided, then determine the difference between this synthesized normal time and the normal time in the original time-study data (Table 13), then square this difference, then accumulate the residual sum of squares, and so on, as summarized in Table 14.

In addition to offering useful indices of the closeness of fit and ability to estimate, the residual sum of squares, $s_{(NT)}$ and I may be used very profitably as analytical aids in standard data development. If, after a series of variables have been investigated and the regression equation established, and the resulting confidence interval found to be objectionably or prohibitively large, improvement might be achieved by seeking out additional variables, adding these to the equation, recomputing the confidence interval, and repeating this process until

[6] Multiple regression analysis is a common term used to describe use of the basic regression technique in cases involving two or more independent variables.

TABLE 14. Computation of Residual Sum of Squares, Standard Deviation, and Confidence Interval for the Brush Paint Regression Equation

Part No.	(NT) (Original Time-Study Data)	$(NT)'$ (Synthesized from Standard Data)	$(NT) - (NT)'$	$[(NT) - (NT)']^2$
104	15.56	15.52	0.04	0.0016
119	12.18	12.26	−0.08	0.0064
133	10.85	11.17	−0.32	0.1024
170	6.82	7.02	−0.20	0.0400
174	3.09	3.39	−0.30	0.0900
193	15.04	14.50	0.54	0.2916
217	18.17	18.84	−0.67	0.4489
238	10.15	10.04	0.11	0.0121
240	9.84	9.85	−0.01	0.0001
255	19.37	18.77	0.60	0.3600
277	10.04	9.73	0.31	0.0961

$$\Sigma[(NT) - (NT)']^2 = 1.4492$$

$$s_{(NT)} = \sqrt{\frac{\Sigma[(NT) - (NT)']^2}{n}} = \sqrt{\frac{1.4492}{11}} = 0.362 \text{ minute}$$

$$I = 2(1.83)0.362\sqrt{\tfrac{11}{9}} = 1.47 \text{ minutes}$$

the engineer is satisfied. Similarly, it is possible to systematically eliminate various variables from the regression equation and note the effect of such omission on the residual sum of squares, standard deviation, and confidence interval. In this manner it is possible to *objectively* isolate the variables that are not carrying their weight in the estimating process, that is, the variables that do not contribute sufficiently to the precision of the final estimate to justify their recognition in the regression equation. This is a very useful feature of the regression technique.

Appraisal of the regression technique. Advantages offered by the regression technique in addition to greater objectivity in processing, analyzing, and interpreting the input data, are much reduced time required to derive standard data, reduction or even elimination of the need for elemental type time studies, and an improved means of isolating the effect of a variable on normal cycle time. The latter advantage requires some elaboration. In the traditional elemental approach to standard data development, each variable is treated in fragmented fashion, as if the effects of such variables were compartmentized. For example, variable X may have a major and obvious effect on the time for element A, but only a relatively small effect on the times for the remaining elements. Ordinarily, by the traditional approach, variable X will be recognized as affecting element A and ignored in the case of the remaining elements. When the regression approach is employed and total cycle time is used instead of individual element times, the *total* effect of the variable is taken into account, regardless of what elements it might affect and to what degree. In the regression approach the latter is irrelevant. Thus, in the brush painting example, the effect of area of the sheet on normal cycle time is estimated, rather than the separate effect of area on the time to "shove finished piece along conveyor," "pull next piece to workstation," and "paint." If there is a residual effect of area on the remaining elements it is ignored under the compartmentized elemental approach. The ability to isolate and take into account the total effect of a variable is a significant advantage of this approach to standard data development.

When regression analysis is used, elemental time-study data is unnecessary, since the standard data can be derived directly from normal cycle times. Therefore, insofar as standard data development is concerned, elemental time studies would no longer be necessary under this method of derivation. The regression technique is one of the most significant and promising tools to be introduced to the field of time study in decades. It has the potential for bringing about major improvements in time-study methodology in a variety of ways. Certainly, this technique should be given equal consideration with traditional means of deriving standard data.

There are, of course, some disadvantages in the use of this approach. For example, the technique can become rather complex. Aside from requiring some technical training which many practitioners have never had or have long since forgotten, the occasional assistance of an applied statistician may be necessary or certainly highly desirable. Furthermore, a computer is sometimes necessary or close to essential to perform the necessary computations and solve the more cumbersome sets of simultaneous equations.[7] Some assumptions must be made in using this technique, such as independence of the effects of variables, the restrictiveness of which have not been fully explored. A significant change in method of performing an operation for which standard data of the regression equation type has been developed makes recompu-

[7] Commercial computing centers are ordinarily equipped to handle such problems at a moderate fee.

tation of the equation almost inevitable, whereas in the case of elemental standard data there is a chance that the change may affect only one or several elements and that adjustments can therefore be made without a major overhaul of the standard data.

To be sure, this is not an exhaustive treatment of the regression technique. There are many details, refinements, and extensions beyond what has been described herein. For example, the technique can be extended to nonlinear relationships between variables, confidence intervals can be estimated for the constants, and so on. These extensions are described in most intermediate statistics textbooks.[8]

The Economics of Standard Data

The main virtue of the standard data method is the rapidity with which a time standard can be established by this technique, relative to the amount of time required by stop-watch time study, the predetermined motion-time technique, and work sampling. For example, in a situation in which it might previously have required several hours to set a time standard on the typical operation by stop-watch time study, with standard data it might well now require, say, 5 minutes to do an equivalent job and perhaps with only the services of a clerk rather than a person of higher technical skill.

Standard data is analogous in a number of ways to the special purpose production machine. The latter is ordinarily capable of producing parts at a much faster rate than the general purpose machine, but the initial cost is higher and the machine is readily rendered useless to its owner by major changes in product design. Standard data is likewise of special purpose for it is capable of setting time standards only on a certain type of operation. An extra expenditure is required for its development, but standard data is capable of producing standards at a faster rate. Similarly, standard data is readily rendered useless by a major change in method in the operation for which it can set standards.

The similarity does not end here. A primary determinant of the economic feasibility of employing a special purpose production machine is volume of production. When volume becomes sufficiently large, the savings in production costs make investment in the special purpose machine profitable. Similarly, as the number of standards to be set increases, a point is eventually reached where the saving in time required to set the standards justifies investment in the derivation of

[8] See, for example, Mordecai Ezekiel and K. A. Fox, *Methods of Correlation Analysis,* 3rd ed., John Wiley and Sons, New York, 1959.

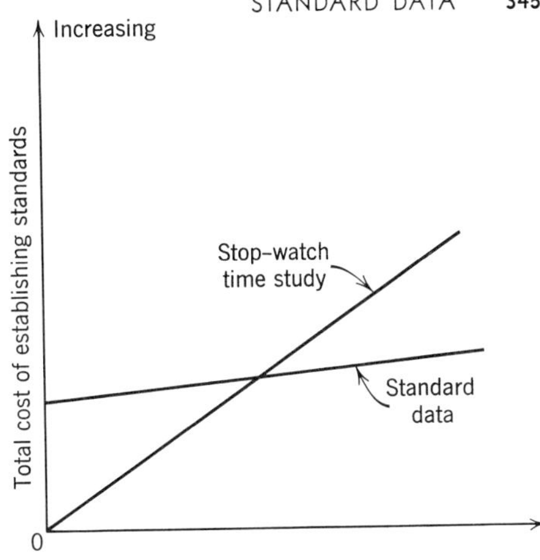

Figure 104. Expected relationship between total cost of setting time standards and number of standards to be set, for two major work measurement methods.

standard data. This situation is pictured graphically in Figure 104, in which the total cost of setting standards is shown as a function of the total number of standards to be set, for standard data (the special purpose tool) and for stop-watch time study (the general purpose tool).

Although volume of standards to be established is a primary determinant of the economic feasibility of employing standard data, it is not the only factor. The expected cost of deriving the data is another. This cost is a function of the type and quality of input data on which the standard data can be based, and will vary from a negligible sum if usable time studies are already in the files, to an appreciable extra expenditure if time studies must be made especially for this purpose. Cost of derivation is also a function of the complexity of the operation insofar as the number of variables and the ease of quantifying and isolating them are concerned.

In summary, the economic feasibility of employing standard data to establish standards depends primarily on the following.

1. The volume of standards that must be set, a function of:
   a. the frequency with which standards must be set and reset on the type of operation in question;
   b. the stability, the expected life of the method for the operation in question.
2. The magnitude of the derivation cost, which is primarily a function of:
   a. the type, quality, and availability of the input data that must be relied on;

b. the relative complexity of the operation insofar as variables are concerned.

Thus, standard data is *likely* to be economically worthwhile where there are a number of similar pieces of equipment performing similar operations on a substantial number of different component parts, and where major changes in method are infrequent. These characteristics are often true, for example, of a press, milling, drilling, or lathe department. Another situation in which standard data is likely to be profitable is the so called "job shop," an establishment performing a large variety of different short-run jobs on one or several basic types of production equipment. In this type of situation, a stop-watch time study on each job would be economically out of the question, to say nothing of the fact that many jobs would be started and finished before the time study could be made. Standard data may well be the answer under such circumstances.

Standard data is not likely to be economical in a situation in which changes in method are relatively frequent, as is often the case with assembly operations for a product subject to frequent major redesigns. Contrast this with a punchpress department, for example, where the method has been and probably will continue to be basically the same for decades.

Standard Data in Summary

If standard data of a more refined nature is desired, as it surely would be if the standards produced are to be used as bases for wage payment, there are two competitive methods of derivation. One is the newer regression technique applied to data from stop-watch time studies. The other is the traditional elemental approach applied to stop-watch time-study data or predetermined motion times. In some instances, standard data of a less refined nature will suffice for the purposes at hand, as for example in the situation in which the resulting time estimates are used for cost estimating, manpower allocation, and other less demanding uses, but not for wage payment purposes. Under these circumstances work sampling might well provide satisfactory input data upon which to base the standard data. In fact, work sampling and regression analysis may be profitably combined to derive a grosser form of standard data fairly inexpensively.

To some extent the derivation of standard data is a pooling process in which a considerable volume of individual time-study results are combined in such a manner that each standard contributes to the reliability of the others. Thus dilution and compensation of errors would tend to make standard data a more reliable method of setting time standards. However, a loss is suffered because of the synthetic feature, for almost certainly some influential variables are neglected, and too, any deviation in method actually used in the shop from that assumed in the standard data results in error. So there are some factors contributing to reduced error and other factors contributing to increased error, and what the net effect is seems generally unknown. It must be, however, that this net effect is not too great whatever its direction, or it would be noticeable. These statements presume that an effective job of derivation has been achieved. In the absence of a thorough analytical approach in the derivation of a standard-data installation, the results are likely to be noticeably and objectionably erratic.

EXERCISES

1. Develop standard data with which time standards may be synthesized for sweeping classrooms of varying size, type, and contents.

2. The normal time for a messenger to mount his bicycle, ride to a destination, dismount from his bicycle, and park it, is given below for a number of trip lengths. Obtain the least squares line for this data and compute the residual sum-of-squares about the fitted line.

Normal Time	Distance Traveled
1.20 min.	600
1.39 min.	700
1.38 min.	800
3.14 min.	2400
4.78 min.	4300
5.09 min.	4500
5.88 min.	5200
6.42 min.	5900

3. An executive of a large corporation has learned that there is a complete handbook of standard data on the market, available for a nominal price. This volume provides standard data for a majority of metalworking operations ordinarily encountered. He has asked the supervisor of the methods engineering department if it is possible and in fact desirable for the department to use this handbook to set their time standards.

What investigations would you make in order to intelligently answer this executive's question? What is your eventual recommendation likely to be?

REFERENCES

Auburn, Travers, and Joseph Motycka, "Electronic Data Processing Comes to Time Study," *Journal of Industrial Engineering,* vol. 8, no. 1, January-February 1957.

Ezekiel, Mordecai, and K. A. Fox, *Methods of Correlation Analysis,* 3rd ed., John Wiley and Sons, New York, 1959.

Ladd, George W. and Keith L. McRoberts, "A Regression Method for Determining Visual Inspection Times," *Journal of Industrial Engineering,* vol. 11, no. 5, September-October 1960.

20

The Predetermined Motion-Time Technique

The predetermined motion-time technique, introduced on page 167 and exemplified in detail in Appendix A, is of course an appropriate and competitive method of establishing time standards for manual operations. The sum of the predetermined motion times for a given task constitutes an estimate of the normal time for that job. The addition of an allowance to this provides an estimate of the standard time.

Advantages Offered by the Predetermined Motion-Time Technique

The use and usefulness of predetermined motion times as a means of establishing time standards for manual operations are attributable primarily to the following characteristics of the technique.

1. No timing is required in studying manual operations by this technique so that there is little need for the observer to stand over the worker with a watch. This feature has certain desirable aspects that are rather attractive to some companies.
2. The use of predetermined motion times eliminates the need for the very troublesome performance rating required in stop-watch time study. Some companies are rather eager to avoid rating and turn to this technique for this reason alone. The application of predetermined motion times may require no less judgment in total than time study rating, but this judgment takes place in a different, not so obvious, less objectionable form.
3. The synthetic feature, the ability to estimate the normal time for a task before the operation actually comes into existence, is a very

important reason for use of the technique. This feature makes the procedure very useful in methods design work, cost forecasting, equipment selection and design, assembly line balancing, and other situations in which more than a subjective estimate of performance time is wanted and yet there is no opportunity to observe the task being performed.

4. The use of predetermined motion times *forces* the person who is setting the time standard on an operation to give detailed consideration to the method being used. In view of the fact that it is extremely poor practice to establish a time standard for an operation without attempting to improve the method beforehand, and that the temptations and pressures are to bypass this crucial step, a procedure for setting time standards that automatically forces the user to give consideration to the method employed on that operation is an attractive one to many. As a consequence of this forced consideration, the analyst can hardly avoid coming up with more improvements, even though he may not be actively seeking them, than are likely to result if ordinary stop-watch time study is used without a methods analysis preceding. Just as much consideration of method *could* take place under stop-watch time study or any other standards-setting procedure, but the fact of the matter is that it just is not commonly done.

5. A by-product of the use of predetermined motion times is a detailed record of the method on which the time standard is based. This is useful for installation of the method, for instructional purposes, and for detection and verification of changes in the method in the future. This aid that the technique offers in keeping standards abreast of changes in method is considered an important advantage by some companies.

Although there are other factors in this technique's favor, the foregoing features appear to be the most influential.

Some Factors to Consider in the Selection of a Particular Predetermined Motion-Time System

A number of specific versions of the predetermined motion-time technique are on the market. Among the more competitive ones are Methods-Time Measurement (MTM), Work Factor, and Basic-Motion Time Study (BMT). With these and others to choose from, what are some criteria to consider in making a selection?

1. Cost of installation, consisting mainly of the cost of obtaining analysts proficient at applying the system in question.

2. Operating cost, which is determined primarily by the length of time required to set a standard by the system in question.

3. Level of performance embodied in the system. It is quite likely that the level of performance embodied in a system under consideration will differ to some degree from normal performance in the plant involved. This is almost to be expected, and is easily isolated and corrected for by adjusting the times given in the tables or by application of an appropriate allowance. This is simply a process of calibration. Therefore, a spread between the level of performance embodied in the system and the level desired is not an important factor to consider, unless the disparity is extreme.

4. Consistency of the standards set by the system under consideration. In Chapter 21 there is a discussion of a number of errors associated with currently available predetermined motion-time systems. From such an analysis it becomes apparent that existing systems have their imperfections, a number of which lead to inconsistency in the time standards thereby set. The degree of inconsistency in the resulting standards is not expected to be the same for competing systems. In contrast to the level of performance the system represents, consistency among the standards set by a system *is* an important factor to consider. This, however, is a difficult matter to judge, and probably can be best accomplished by a trial application of the system to a series of operations in the plant and examination of these for internal consistency.

5. The type of operations in the plant involved in contrast to the source of time data used to derive the system under consideration. One of the existing systems is based primarily on studies of light metal-working operations, another is based mainly on operations typical of the electronics industry. As a consequence, these systems seem more adaptable to some types of operations than to others. It seems advisable to take this into account in selecting a system, and to avoid extending a system beyond the range justified by the input data if this can be avoided.

Some Precautions to Observe in Making Comparisons of Predetermined Motion-Time Systems

Sometimes, as part of the selection process, practitioners attempt to compare two or more existing predetermined motion-time systems at the motion level or at the total estimate level. Performance times for what appear to be equivalent motions in two or more systems are compared, or total estimates for certain tasks are compared. However, these systems differ structurally in a number of respects such that differences in motion and total task times will result between systems even if said systems are errorless. To illustrate, existing systems differ with respect to performance level embodied in the time data, by as much as an estimated 35 per cent. In addition, different element breakpoints were used in deriving the motion times. Furthermore, note that the times for manipulative motions, for example position and grasp, are relatively unaffected by changes in pace, while the times for travel motions, for example reach and move, are drastically affected by pace changes. Therefore, if predetermined motion time system *A* embodies one level of performance, and system *B* embodies a level of performance 30 per cent above this, we should find that the times for these manipulative and travel motions in the two systems compare quite differently. The manipulative motion times in the two systems should be about the same (other things being equal), whereas the percentage difference between the travel motion times should be considerably greater than 30 per cent (again, other things being equal). This and other circumstances complicate a motion by motion comparison of systems.

Finally, to make a satisfactory comparison between systems, the person making the comparison should be fully qualified to apply each of the systems. Such individuals are rare indeed.

Some Precautions to Observe in the Installation and Use of a Predetermined Motion-Time System

A small percentage of companies have adopted a predetermined motion-time system and at some later date expressed dissatisfaction with the idea. In a number of these cases the difficulties and dissatisfaction seem more the result of malpractice and misapplication by the users than of faults in the technique itself. In view of these unfortunate experiences and the current prevalence of certain inadvisable practices, the following precautions are advanced.

1. Have analysts trained by competent instructors. Avoid, if at all possible, the second-hand training that seems common. Under this system a small group of men are trained by an outside consultant and subsequently become instructors for others in the company. Eventually this original group of company instructors, as a consequence of promotion, transfer, and the like, is displaced by a "second generation" of instructors. This process continues over a period of time until eventually the current instructors are about the sixth generation. Each time an instructor moves on and one of his students replaces him, a certain percentage of the original

"know-how" is lost, and eventually this shrinkage becomes appreciable enough to have a major and adverse effect on quality of application.

2. Carefully select the persons who will apply the system. In general, younger men are preferred for several reasons.
   a. Learning to successfully apply one of these predetermined motion-time systems involves learning a tremendous number of details, a task that many older men find difficult and perhaps impossible.
   b. Older men tend to resist and abuse the technique because it threatens to reduce their value to the company. Many of these older men who are skilled stop-watch time study practitioners fear that they will never be able to master the new technique and become as skilled and as valuable to the company as they formerly were as stop-watch time study men.
   c. The older men who have had years of experience in stop-watch time study find it difficult to resist "working backwards" in applying predetermined motion times. There is a tendency on the part of the man with such experience to attempt to make the synthesized estimate agree with what his experience tells him the standard should be for the operation under study. The objection to this is that if the analyst is going to use his experience to set the standard and predetermined motion time simply to support this subjective estimate, this is a rather expensive way to mask an opinion.

   The most suitable type of individual to equip this technique with apparently is a younger person who is grounded in the fundamentals of time study and methods design and who has had some experience in each.
3. Maintain control over the judgment required by analysts in application of the system. Considerable judgment must be exercised in the application of a predetermined motion-time system, partly because of certain weaknesses in existing systems, and partly because an element of judgment is inherent in the technique. (The need for judgment in the application of a predetermined motion-time system is discussed in detail in Chapter 21.) Although the role of judgment cannot be eliminated, it certainly can and should be controlled to a minimum level. Recall that rating practice sessions are used to improve the uniformity of that form of judgment. The same policy should be adopted in the application of a predetermined motion-time system in order to maintain uniformity in the selection and classification of motions. As new motions and groups of motions are encountered, and as the system is extended to different types of operations, interpretation and extrapolation of the system is required. The company's analysts should agree on how newly encountered patterns should be analyzed, instead of each applier developing his own set of rules.
4. The analyst should be thoroughly familiar with the operations to which he will apply the technique, since he will be constantly forced to decide what motions are required for those operations. This knowledge is vital, for more often than not the worker presents a biased picture of the requirements of the job by introducing extra motions, making motions

appear more difficult than he really finds them, and the like. If the analyst is not able to successfully isolate these situations, the standards he produces will suffer accordingly.

5. Make certain the analysts recognize the limitations of the technique they are applying. These limitations pertain to such matters as:
   a. the types of operations and motions to which the system should not be applied. For example, some systems apparently were not meant to be applied to heavy work, such as that frequently encountered in the foundry. The same may be true of very complex, intricate, manipulative type operations, and of operations in which the method is subject to considerable variation from cycle to cycle.
   b. the important variables ignored or poorly accounted for by the system, so that if the system is to be applied to jobs in which these variables are influential, some compensating adjustment may be made by the analyst.
   c. the maximum length of cycle to which the system should be applied. Apparently one of the causes of dissatisfaction with the predetermined motion-time technique on the part of some users is the length of time required to establish a standard. This technique is likely to become prohibitively expensive if applied to longer cycle operations.
   This limitation could be effectively relaxed if the system had a rougher, faster-to-apply version to use on longer cycle operations. This simplified version might recognize fewer variables, have fewer categories and values, and deal in small *groups* of motions such as "gets and places." The latter is the most important feature of this simplification process. If there are just as many individual motions to isolate and classify, not much of a time saving will result even though there might be fewer variables and values recognized in an abbreviated version of a system. (If this type of simplified version is not provided as a part of the original system, the individual user can derive one for his own use.)
6. Familiarize operating personnel and union officials with the nature and workings of the system. Foremen, especially, should thoroughly understand the technique.

Summary

Since this procedure eliminates the need for the very troublesome performance rating required in stop-watch time study, and since no stop-watch need be held over the worker, and because of its synthetic nature which permits standards to be established before an operation is actually installed on the shop floor, the predetermined motion-time technique appears attractive to and is used by an appreciable number of companies for setting time standards. Although existing predetermined motion-time systems offer considerable room for improvement, even in their current underdeveloped state, these systems are definitely useful in time study and methods design work.

EXERCISE

1. Using the MTM system, synthesize the normal time for loading a stapler. See problem 7, page 120, for the general method to assume.

REFERENCES

Brady, P. J., "Work Factor Analysis Saves Money, Time, Complaints," *Factory,* vol. 109, no. 6, June 1951.

Lynch, H., "Basic Motion Timestudy," *Journal of Industrial Engineering,* vol. 4, no. 3, August 1953.

Maynard, H. B. et al., *Methods-Time Measurement,* McGraw-Hill Book Company, New York, 1948.

Mundel, M. E., and I. P. Lazarus, "Predetermined Time Standards in the Army Corps," *Journal of Industrial Engineering,* vol. V, no. 6, November 1954.

Quick, Shea, and Koehler, "Motion Time Standards," *Factory,* May 1945.

Segur, A. B., "Motion-Time Analysis," *Proceedings of National Time and Motion Study Clinic,* Chicago, 1948.

Stewart, T. C., "Work Factor Analysis Takes Stopwatches Out of Time Study," *Factory,* vol. 106, no. 5, May 1948.

"Timing a Fair Day's Work": *Fortune,* October 1949.

21

Evaluation and Improvement of the Predetermined Motion-Time Technique

The predetermined motion-time technique is the object of many exaggerated claims, both of an optimistic and pessimistic nature. A variety of these assertions will be discussed at some length because such unrealistic and unreliable claims often appear in work measurement literature. Furthermore, an analysis of them can be a very enlightening experience. An objective examination of these statements will reveal some glaring examples of complete disregard of professional objectivity, accepted rules of scientific inference, and facts of the business world, from which some very profitable lessons can be learned. First, it is desirable to introduce the more important errors associated with predetermined motion-time systems, for many of these flaws are the bases of unjustified claims.

Some Errors Associated with the Predetermined Motion-Time Technique

The sources of error known or suspected of being significant are as follows.

1. The effect of the context of a motion on its expected performance time. This is also referred to as the effect of interdependence or nonadditivity of motion times. This effect may be illustrated by the following hypothetical case. Suppose the motion C appears in two tasks as indicated. Is the expected (average) time for motion C in Task I the same as the expected time for motion C in the different motion context represented by Task II? Probably not, because of the subtle interrelationships that usually exist between the simultaneous and sequential actions of the living organism. For basically the same reason that a human being reacts differently to a given action, event, or person, in different contexts, so the

TASK I		TASK II	
LH Motions	RH Motions	LH Motions	RH Motions
M	*R*	*T*	*Q*
T	*B*	*B*	*Y*
(*C*)	*Y*	*M*	*B*
P	*D*	(*C*)	*Q*
B	*F*	*U*	*F*
		A	*D*
		Q	
		R	

character of a minute movement depends on the movements that surround it. Thus, *for no obvious or predictable reason,* the expected time for motion C is *likely* to be somewhat different when surrounded by different motions in different sequences and numbers.

There seems little doubt that the context of a motion has some effect on its expected performance time. However, whether or not this effect is of any practical significance is not yet known. Various researchers have attempted to investigate this matter, but the results have been inconclusive, contradictive, and in some cases incorrectly interpreted.[1] This source of error is used by psychologists and unionists as a basis for condemning the standard-data technique, but unjustifiably so in light of present lack of evidence.

2. Discreteness of motion categories and time values given. For example, time for the motion position is a continuous function of the clearance between the mating parts, as illustrated in Figure 105. In system X only three different position times T_a, T_b, and T_c are provided to cover the whole range of the continuous variable clearance. Thus, the time T_b is assigned to all cases falling in the range of clearances embraced by category b, and similarly for the remaining categories. Therefore, the single position time allowed for category b does not apply equally well to all motions assigned to that group. A bias results for most motions falling in a given category, an error that is considerable around the boundaries of the categories.

A similar situation exists for all variables and times in such a system, and is of course inevitable to some degree. It is not a matter of eliminating discreteness, but one of optimizing it. Therefore, an attempt should be made to obtain an economical balance between the number of categories and thus the time required to apply the data on the one hand,

[1] See, for example, the results and conclusions reported by Adam Abruzzi in *Work Measurement,* Columbia University Press, 1952, where interdependence of *average* motion times is confused with interdependence of individual motion times in a given cycle.

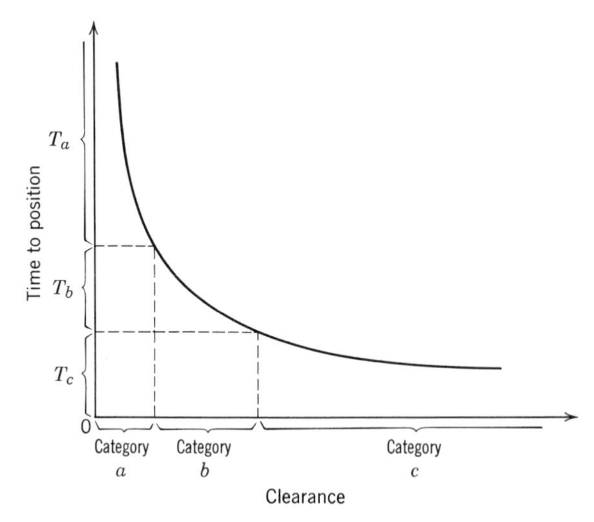

Figure 105. Position time as a function of clearance between the mating objects. Position times assigned in predetermined motion time system X are indicated along with the range of clearances covered by each time.

and the resulting error on the other. It appears that in many instances the data is overly discrete. Often the categories are too broad, the steps in time allowed are too large. For example, in one system the important variable control required by the motion is recognized by two categories: "control" or "no control," with a corresponding time allotted to each. In another system, left- and right-hand motions are recognized as either capable of being performed simultaneously or otherwise capable only of being performed successively, as indicated below. In reality, most motions are partially overlapped under the circumstances.

LH RH or LH RH

3. Errors contributed by the person applying the system. The analyst can affect the time ultimately allowed, primarily because liberal amounts of judgment are ordinarily required in
   a. deciding what motions the "average" operator will require to perform the task at hand, and
   b. deciding to what categories certain motions should be assigned. A number of motion categories are ambiguously defined. Furthermore, the categories are not mutually exclusive.

In synthesizing the normal time for a given task by means of the predetermined motion-time technique, it is necessary that the user decide what motions are normal for the task *in view of the expected production volume.* To illustrate, take the task of turning a nut down on a bolt. What method is natural: deliberate turning of the nut by the thumb and index finger, spinning with the two fingers, spinning with strokes of the index finger, "rubbing" it down by running it down the index finger, or perhaps others? Which of these is natural under low-volume conditions? Medium volume? High volume? This is a difficult prediction problem which requires liberal amounts of judgment with only experience and common sense to guide the analyst. The problem is amplified considerably by the fact that what is a normal, natural, or reasonable method depends upon the opportunity the worker has for practicing the job. This type of decision is especially difficult in the case of manipulative motions.

Once the analyst has decided what motions the normal operator will use for the job, these must be classified according to the motion categories of the particular predetermined motion-time system being used. But there are certainly many types of motions, yet only a relatively small number of categories to which these may be assigned, and these often are loosely defined categories at that. This is true, for example, of the element position in the MTM system, where there is a very vague distinction between a *P*1 and a *P*2, such that there are differences of opinion between different analysts and inconsistencies committed by an analyst over a period of time. The user of any of the existing systems is forced to use considerable judgment in classifying many of the motions he encounters.

The necessity for using judgment introduces random errors and bias (both of an intentional and unintentional type) into the results obtained from current predetermined motion-time systems, to a degree that seems both objectionable and unnecessary. This situation manifests itself in the different time estimates different users obtain in applying the same system to the very same task. The extent to which judgment is required and to which the analyst can affect the results means that it is *hardly satisfactory to evaluate a predetermined motion-time system apart from the person applying it.* The particular predetermined motion-time system and its applier are two closely interrelated components of a complete measurement system, such that the two should be evaluated in combination. One practical consequence of this situation is that some shortcoming in the system by itself, a poorly accounted for variable, for example, might well be compensated for by the analyst as he applies the data.

4. Error introduced by assuming a flawless performance. Regardless of the system used, the motion pattern synthesized is smooth, clean, free of all of the unavoidable miscues and irregularities so characteristic of human motor performance. If a number of repetitions of a manual task are carefully observed and timed, it will be noted that several different types of work cycles occur. The frequency distributions of times for these

different types of cycles are believed to appear approximately as pictured in Figure 106. These types of cycles are characterized as follows:

a. the miscue-free cycle, a repetition of the task in which there is no deviation from the manner in which the operator *intended* to perform the activity.

b. the miscue cycle, a repetition of the task in which one or more subtle miscues give rise to extra, discrete, corrective motions, minor enough so as to cause no disturbance of the operator's rhythm. Corrective action in this instance is automatic. Examples of these miscues are the simple fumble in grasping a part or a slight bounce off the edge of a hole as a plug is being positioned in it.

c. the irregular cycle, a repetition of the task in which a chance deviation in method is unusual or long enough to disturb the operator's rhythm, yet does not require cessation of work. Examples of these irregularities are dropping a part or tool on the floor, jamming of a fixture, severely tangled parts, and the like.

It is the miscue-free cycle that is synthesized in the course of using a predetermined motion-time system. Not only does this type of cycle appear to constitute a relatively small percentage of the total number of cycles, but also this percentage changes drastically with practice and differs considerably among different types of tasks. Failure to allow for these miscues and irregularities constitutes a source of sizeable error in current predetermined motion-time systems.

5. Errors made in assigning time values to different motions and their various subclassifications in originally setting up the system. This includes a sampling error, for a rather limited number of time observations were utilized to estimate the normal time for a given motion classification. In

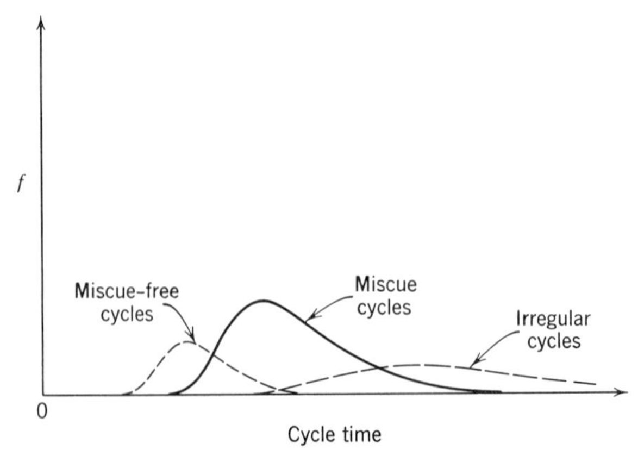

Figure 106. Postulated form and relationship of three types of cycle time distributions contributing to the total distribution of cycle times generated by repetition of a manual task.

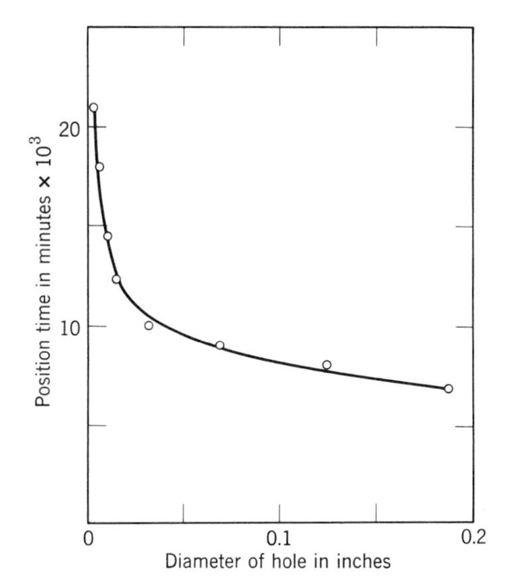

Figure 107. Position time as a function of hole diameter, with clearance between plug and hole held constant.

the case of current systems, it appears that a remarkably small amount of data actually underlies the times given in the tables so that an appreciable sampling error is likely.

Errors in the table values are likely also as a result of errors and malpractices in processing the time data underlying a given system. All available evidence indicates that the originators of existing systems used far from the best known statistical practices in treating the input data and arriving at time estimates for their tables.

6. Failure to recognize certain significant variables introduces another source of error. Each of the existing systems takes into account several important task variables that affect the normal time to perform a motion, such as distance and degree of control required. However, they do not recognize the same variables, and none of the systems recognizes all of the significant variables. In other words, not all of the relatively influential variables were exhausted in setting up existing systems. For example, in positioning, the absolute size of the hole has an appreciable effect on time to position. This is illustrated in Figure 107 which pictures the results of an investigation of the effect of hole diameter on position time with clearance between plug and hole held constant.[2] Yet this variable is ignored in most existing predetermined motion-time systems. The neglect of this and other task variables having an appreciable effect on normal time contributes to a system's error.

[2] Unpublished results of research performed by the Department of Industrial and Engineering Administration, Sibley School of Mechanical Engineering, Cornell University.

7. Other errors arise because of influential variables that are inadequately recognized. The most glaring example is the inferior manner in which the effects of practice are recognized in most existing systems. Learning opportunity, dictated by production volume and cycle length, has a strong effect both on the number and type of motions required and on the time that should be allowed for each. Yet in some systems this variable is virtually neglected; in others it is recognized in a very crude and unsatisfactory fashion.

Of the many sources of error associated with the predetermined motion-time technique, the preceding seven seem deserving of concern. In the literature some of these types of error are overworked, whereas others are scarcely or never mentioned. In summary, the estimate of normal performance time for a given motion provided by the ordinary predetermined motion-time system is

$$(NT)' = (NT) + c_1 + c_2 + v$$

where

$(NT)'$ = the estimate of the normal time;

(NT) = the "true" normal time;

c_1 = bias that arises because of a difference in the overall level of performance embodied in the system and the level of performance identified as "normal performance" at the particular company involved;

c_2 = a bias that results from a majority of the sources of error introduced previously, such as discreteness, bias introduced by the analyst, original errors in the table values, ignored and inadequately recognized variables, and the like; and

v = a random or chance error that arises because of chance errors made in classifying motions, measuring length of transport motions, etc., and because of the error due to differences in motion context.

The two types of bias differ as to seriousness and as to method of remedy. c_1 is a uniform percentage error for all motion times in the system and is easily compensated for by a process of calibration to suit the company's concept of normal. In contrast, c_2 differs for various motion times given in the tables. Fortunately for the technique, the c_2's in a given system differ in direction (+ and −) as well as in magnitude. Unfortunately though, this variation in magnitude and direction makes any correction of the c_2's in a system a major overhauling job.

The Error Reduction Potential in Current Predetermined Motion-Time Systems

With the possible exception of the effect of motion context, significant reductions seem possible in the various types of error associated with current predetermined motion-time systems. *If* changes in motion context are eventually shown to have an effect on motion time of practical significance, it is possible that the error could be reduced somewhat by certain provisions in the system, but in the main this error (whatever its magnitude may be) appears to be primarily inherent in the technique. The reduction potential is different, however, for the remaining sources of error. The discreteness error can profitably be reduced until a better balance between application time and error of the estimate is achieved. A certain degree of discreteness is inevitable, of course. In the case of the judgment required in applying current systems there seems to be appreciable error reduction potential. This can be achieved by:

1. providing improved guides for the selection of the normal method for a task under different volume conditions, especially for the bimanual case. There is a definite limit as to the amount of assistance that can be given in this respect, but certainly more can be done than is being done at present to guide the analyst in the course of deciding what motions are "natural" under various circumstances.
2. by reducing the ambiguities in definition of motion categories.

Judgment is far from eliminated from the standards-setting process through the use of current predetermined motion-time systems (as they now stand). And although judgment and the consequential effects of observer bias cannot be eliminated from such a system, there is certainly much room for improvement. Synthesis of a miscue-free performance is a source of error that can be remedied for practical purposes by the inclusion of a percentage adjustment factor based on practice opportunity and difficulty of the task. In fact, since reduction in relative frequency and severity of miscues is one of several effects of practice, an allowance for these miscues could be part of a general adjustment of the synthesized normal time for learning. The errors made in originally assigning time values to the motion categories in existing systems, if they are of the order of magnitude currently suspected, can be significantly reduced by the accumulation and proper statistical treatment of additional time data. Insofar as ignored and inadequately accounted for variables are concerned, redesign of current systems in this respect could result in a major improvement in the estimating ability of the technique.

In the opinion of this author, existing predetermined motion-time systems are considerably underdeveloped. Apparently a substantial portion of the error in estimates made by this technique can be eliminated through a relatively moderate investment in further development.

Why Use These Predetermined Motion-Time Systems If the Sources of Error Are So Numerous?

Why would anyone use a predetermined motion-time system if it suffers from all of the sources of error described? How can it be that many users claim at least moderate success and satisfaction, that some are very enthusiastic about the results they have obtained? The reasons for this are numerous.

1. The first is the fact that this technique adheres to what might be referred to as a cardinal principle of good estimating practice. Suppose you have a number of shopping errands to do and are asked to estimate the time of completion. One method of arriving at this estimate is to view the errands in aggregate and simply make an estimate in total. Another method would be to break down the total estimate into a number of smaller subestimates, then sum these individual estimates to obtain an estimate of completion time. Thus, you might estimate 5 minutes to pick up shoes at the repair shop, 10 minutes to pick up laundry, 15 minutes to get the groceries, etc., and in this fashion obtain a more precise estimate of completion time than the estimate-in-total method.

The same principle is applied in the predetermined motion-time technique. The total normal time estimate consists of a series of relatively crude estimates of normal time for very small parts of the task. The c_2 errors for individual motion-time estimates may be considerable, yet the fact that they vary in magnitude and direction and that many estimates go to make up the whole, ordinarily results in considerable compensation of error. As long as there are more than only a few motions involved, as long as the analyst is not overly biased, and as long as an estimate is obtained for each motion, the normal time estimate for the task will usually be considerably less in error than the rather large individual motion errors might lead you to suspect.

2. The fact that the alternatives to the predetermined motion-time technique likewise have their errors and other disadvantages. Com-

pared to perfection, this technique leaves much to be desired, but compared to the alternatives, it represents a superior, or at least competitive means of setting time standards in many situations.

3. The fact that in some instances the practitioners who apply the technique can, to some degree, make the results turn out favorably whereas without this help the outcome might not be so satisfactory. Analysts who have had time-study experience often have a preconceived notion of about what the standard time should be for the operation under study. If the synthesized time is not in reasonable agreement with this preconceived notion, the analyst can force a reasonable agreement by appropriately reselecting and reclassifying motions. This *is* done, and as a consequence the standards established by predetermined motion times can be *made* to show reasonable agreement with the standards that were or would have been set by the previously used work measurement method.

4. The fact that in spite of its imperfections, the technique has a number of unique and useful features that are attractive to a large number of practitioners. Some of these qualities, cited in the previous chapter, are reduced need for the stop watch, elimination of the need for rating, benefits attributable to its synthetic feature, and so on.

Thus, these systems are popular in the face of numerous sources of inaccuracy and inconsistency because the errors in standards that ultimately result apparently are not as severe as the many sources of error would lead us to expect; because the alternatives to the predetermined motion-time technique are likewise erratic; and because of certain unique advantages offered by this method of setting time standards that at least partially offset any error or cost disadvantage that may exist.

CLAIMS AND COUNTERCLAIMS

No other topic in the field has been more subject to unjustified claims and extreme views than that of the predetermined motion-time technique. Because of their prevalence and unreliability, many of these claims and views warrant discussion and appraisal.

Most authors on the topic of predetermined motion times may be classified on the basis of the points of view they express as either hyperenthusiastic or hypercritical. The hyperenthusiasts make liberal use of such words and phrases as "accurate," "they work," "provide true standards," "eliminate judgment," and similar assertions. Their evaluations of the technique usually lack objectivity in that they report a biased picture of their experience with predetermined motion times, or fail to cite important assumptions, failures, pitfalls, and known or potential sources of error. Their evaluations reflect a strong enthusiasm that appears highly subjective in origin.

Writings of the hypercritics may be identified by such quotes as "dangerous to use," "be wary of such standards," "meaningless," "an impossibility," "little predictive value," and "invalid." Their articles frequently describe one or several studies that indicate that errors (of an unestimated magnitude) will result if standards are set with a predetermined motion-time system, and then proceed to generalize that the technique should therefore be avoided. Their evaluations of the technique lack objectivity, but in this case the lack is in the form of failure to weigh the practical consequences of unsatisfied assumptions, failure to recognize certain unique advantages offered by the technique, comparison of the probable error with an abstract and unobtainable level of perfection rather than with that of alternative procedures, or failure to adhere to accepted methods of inference. Their evaluations reflect pessimism and impracticalism, and at times they become as subjective in their statements as their counterparts.

Some claims made by hyperenthusiasts. The term scientific is a byword among ardent proponents of the predetermined motion-time technique, and although the matter is not of serious practical consequence, the frequency with which this word is misused tempts some comment. Description of this technique as scientific is justified if the system has been developed according to a certain set of rules conventionally followed in the derivation of knowledge. These rules require such procedures as the formulation and testing of hypotheses, collection of sufficient amounts of data, the making of inferences, statement of assumptions, estimation and statement of experimental error, full disclosure of all of these, etc. No evidence is available to support the contention that any current predetermined motion-time system was so developed, and that any of them have been is highly doubtful. Of course, it is not necessary that this technique be scientific in order to produce satisfactory results.

Proponents frequently describe the technique or a particular system of same as accurate. Accuracy, however, is a matter of degree, not bilateral. What a particular writer may mean by the term accurate is any one of a large range of degrees of exactness, depending on his ability to judge this characteristic and on the stringency of the standard he uses for judging it. Without qualification such an expression should not be taken seriously by the reader.

Perhaps more important than this careless and misleading use of the term accurate is the problem of the measurability of accuracy. To estimate accuracy, it is necessary to know the true value, or at least have a technique for estimating it that is "more accurate" than the one being evaluated. If there is a technique that is *known* to be "more accurate" than predetermined motion times, what is it and how do you know that it is more accurate? In the answer to this question lies the reason why comparison of estimates derived from predetermined motion times with rated stop-watch times does not justify a statement about accuracy of either technique.

A statement or implication that the technique or a particular system "works well," "works," or "works satisfactorily" is quite indefinite and subjective. There are a number of reasons why a statement of this kind should not in general be taken seriously. For instance, industrial environments differ radically so that what "works" in one situation may not in another. Furthermore, we can hardly expect a salesman or user of predetermined motion times to write about the cases in which it didn't work! And, too, the statement "works satisfactorily" has little meaning unless we know the criteria used in arriving at that opinion. Since the decision involved here is primarily subjective and the factors are deceptive, it is quite possible that what is judged satisfactory by one user may not seem at all so to others. For example, some merely want something that works in the sense that there is a minimum of friction and that life is peaceful and simple, and so they are easily satisfied. Those few writers who do describe their criteria cite some rather unsound bases for reaching the conclusion they have. To illustrate:

1. Extensive savings through improved methods are claimed to result from installation of a predetermined motion-time system. Yet, some portion of this saving has resulted merely because attention has been given to work methods in the process of applying the system, attention that could be given without the benefit of a predetermined motion-time system. There is good reason to suspect that this portion is appreciable, especially in view of some of the examples that writers have described to illustrate how the technique has been "responsible" for better methods.

2. Standards set with a predetermined motion-time system are claimed to compare favorably with prevailing standards in the company or with a group of standards specially set by another method (usually stop-watch time study) for purposes of comparison. Yet, as stated earlier, it cannot be denied that considerable judgment and interpretation are required in the selection and classification

of motions under any of the current systems, and that it is quite possible, in fact very easy, to permit previous time-study experience to influence these selections and classifications to the point where favorable agreement between the results is obtained. Thus, although perhaps unintentionally, the results of the two methods can be forced to compare satisfactorily. The extent to which this practice takes place would probably come as a surprise to some. Even if the results of the two methods do actually agree closely, this means no more than just that. *But why expect agreement?* Stop-watch time study has an objectionable error associated with it, so how can any method that agrees with it be any less in error? When and if a more accurate and precise system does come along, will we reject it if the results do not agree with stop-watch time standards?

3. It is stated or implied that employees seem more satisfied with the new system. However, the absence of employee complaints is not necessarily an indication that the standards are accurate; in fact, we might conclude the opposite.

Another source of weakness in almost any statement about how well a work measurement method works is the difficulty involved in satisfactorily estimating the degree of error present in any standard or group of standards. The limitations of attempting to estimate this by comparing times set by two different methods, predetermined motion times and stop-watch time study, for example, have already been discussed. Another practical alternative is to use the actual performances of operators working under the standards as the basis for evaluating the error. This is commonly done by observing the ratios of "actual to standard" and noting evidence of tightness and particularly of looseness. Aside from the philosophical flaw in such a procedure is the fact that actual output under a time standard is strongly influenced by the standard itself. With loose standards, at least, employees tend to maintain a ratio between actual rate and standard rate of output that they believe is expected by the company, by judicious regulation of their output. Thus, loose standards are *made* to appear satisfactory. Therefore, basing a conclusion as to the satisfactoriness of a set of standards on the number of tight standard complaints and high earnings cases is an unreliable procedure.

The claim that use of predetermined motion times eliminates judgment from the process of setting time standards is completely unjustified. In fact, the total judgment demanded may not be any less than that exercised in stop-watch time study.

Some hypercritical claims. The point of view of these authors seems to be that the predetermined motion-time technique should be avoided, or that its usage should be drastically curtailed. Such conclusions are sometimes based on a comparison of systems or on some research concerning a particular assumption. A popular subject for these investigations is the effect of the context of a motion on its performance time. If the particular assumption is shown to be violated, the writer usually proceeds to make the generalization that such systems "are dangerous to use," "invalid," "meaningless," etc. There are several flaws in this generalization process.

First, these investigators usually study one task with one or several subjects. Statistical significance under these circumstances is not sufficient to justify the generalizations made. Before extending findings beyond the particular task and subjects used, other individuals and tasks must be studied because differences in results are expected among them.

Second, these investigators are usually content with testing a hypothesis, for example, that a motion time depends on its context. If this hypothesis is accepted, and it usually is, they proceed to make a drastic generalization, such as the claim "dangerous to use." But there is no doubt that the times for some motions depend on the nature of other motions in the task. The relevant question to be answered is *to what extent* do motion times depend on their contexts? We are interested in an estimate of the error resulting from the effect of context on motion times, not merely in whether or not there is an error. This technique cannot justifiably be condemned for violation of an assumption when the degree of violation is unknown.

Third, these investigators fail to estimate the practical consequences of violation of the assumption before they condemn the system. Granted, the context of a motion influences its performance time, but what error does this ultimately cause in time standards estimated by this technique? Even knowing to what extent context may be expected to influence a motion time does not tell us what the net effect is on the standards obtained, for two reasons.

1. As a consequence of this error, for a particular task it is probable that the standards for some *motions* will be tight and others loose. The probable net effect is that the *expected* error (due to this cause) of the cycle time estimate will be a smaller percentage than the typical percentage errors of the individual motions, because of some compensation. This should not be construed as a statement that the error is thus eliminated or rendered negligible. The point is that this

is a likelihood that authors are responsible for investigating before they make their condemnation.

2. It is quite possible that some of the effects of context may be accounted for in the structure of the system. This may make the system more cumbersome to apply, but the possibility certainly exists and is not unreasonable.

Fourth, and probably most obvious, is that these authors attempt to condemn the technique on the basis of its own error without evidencing recognition of the practical problem facing management, which must have time estimates to operate successfully. There are a number of methods of estimating time standards, each of which has a sizeable inherent error. Yet no method is arbitrarily rejected because of this. Clearly the problem involves comparison and selection from a number of imperfect alternatives. Just as clear is the absurdity of abstractly rejecting the predetermined motion-time technique or any other method, because it has been shown to be imperfect.

Furthermore, the hypercritical authors seem to ignore a worthy feature of the predetermined motion-time technique—its synthetic nature. If eventually the accuracy and precision of standards established by this technique should be demonstrated to be inferior in competition with actual observation methods, it may well be that it will be retained for use where the latter methods are ineffective, for example where a time standard is desired in the preproduction stage. Thus it seems unrealistic to speak of outright rejection, for this and other work-measurement methods have certain features that make them uniquely useful for different types of problems facing management.

Aside from this, it is not probable that any one work measurement method will prove superior for all types of operations, production volumes, products, industrial environments, etc. It seems likely that different techniques will prove superior under different circumstances.

In summary, this technique cannot be completely and arbitrarily rejected simply on the basis of imperfection, for several reasons.

1. The alternatives are imperfect also, and as yet the degrees of imperfection of various methods are unknown. Most will concede that stop-watch time study results contain an undesirable error but this does not stop people from using it.

2. It is unrealistic to talk of *complete* rejection, for the technique will at least be useful for some purposes because of the synthetic nature of its results.

3. Variation in characteristics of the tasks for which time standards

must be obtained and differences in conditions make it seem likely that each work-measurement technique will be superior under some circumstances.

In any event, the technique should not be judged on the basis of characteristics of the current versions, for these have faults not inherent to the method. Such defects are correctable, and either existing systems will be revised or superior ones will supplant them. This is one reason why attempts by some investigators to "prove" that the times allowed by various systems for certain motions differ significantly are of doubtful merit. Of what consequence is it to show that these differences in motion times exist, differences which are inevitable in light of the variation in performance level, motion definitions, and breakpoints among systems? Certainly the standards established by various systems will not *necessarily* differ appreciably because there are differences among the systems in individual motion times. Furthermore, standards estimated by different systems may differ appreciably and still be no cause for concern. That is, if the standards estimated by two predetermined motion-time systems (for the same group of operations) differ by a reasonably constant ratio, no problem exists. If the values arrived at by the two systems are not reasonably correlated, then one system is more precise than the other, and this *is* important in selecting a system. In summary, showing that motion-time values differ among current systems does not provide sufficient basis for condemning these systems, let alone the technique in general.

Predetermined Motion Times—Conclusions

This is a very useful technique and should be included among a company's alternative time-estimating procedures. Even if it is not to serve as the primary means of setting standards, the predetermined motion-time technique is an excellent adjunct to the other work-measurement techniques, as well as a useful tool in methods design. True, there are a number of important faults in each of the existing systems, not the least of which is the fact that the quality of results attained depends to a considerable degree on the person applying the system. And, too, in many respects the synthesized motion patterns generated by current systems are gross over-simplifications of what in reality takes place in manual activity. Yet it appears possible and economically worthwhile to significantly reduce the error and to otherwise render the technique a more economical and flexible work measurement procedure.

EXERCISE

1. Assume that you have been appointed by the factory manager to make recommendations concerning the feasibility and desirability of adopting the XYZ predetermined motion-time system to establish time standards at the company, in place of the currently used stop-watch time study.

(*a*) How would you proceed to gather the information on which to base your recommendations?

(*b*) What matters would your investigation, conclusions, and recommendations cover?

REFERENCES

Abruzzi, Adam, *Work, Workers, and Work Measurement,* Columbia University Press, New York, 1956.

Abruzzi, Adam, *Work Measurement,* Columbia University Press, New York, 1952.

Abruzzi, Adam, "Developing Standard Data for Predictive Purposes," *Journal of Industrial Engineering,* vol. 3, no. 3, November 1952.

Boyce, C. W., "How Good Is MTM?" *Factory,* vol. 108, no. 8, August 1950.

Buffa, Elwood S., "The Additivity of Universal Standard Data Elements," *Journal of Industrial Engineering,* vol. 8, no. 6, November-December 1957.

Davidson, H. O., *Functions and Bases of Time Standards,* American Institute of Industrial Engineers, 1956.

Gomberg, W., *A Trade Union Analysis of Time Study,* 2nd ed., Prentice-Hall, Englewood Cliffs, N.J., 1955.

Nadler, G., "Go Slow On Predetermined Motion Times," *Manufacturing and Industrial Engineering,* vol. 31, no. 4, April 1953.

Nadler, G., "Critical Analysis of Predetermined Motion Time Systems," *Proceedings of the National Time and Motion Study Clinic,* Chicago, 1952.

Nadler, G. and D. H. Denholm, "Therblig Relationships: I—Added Cycle Work and Context Therblig Effects," *Journal of Industrial Engineering,* vol. 6, no. 2, March-April 1955.

Reiss and Pigage, "The Incompatibility of Predetermined Motion Time Systems," American Society of Mechanical Engineers, Paper S4-F-5.

White, K. C., "Predetermined Elemental Motion Times," American Society of Mechanical Engineers, Paper 50-A-88.

part V

SPECIAL METHODS ENGINEERING PROBLEMS

22

Coping with Unbalance in the Manufacturing System

Design and operation of a manufacturing facility would be only a fraction of the problem it actually is if it were not for *unbalance* in the output capacities of men, machines, departments, and other components of the system, and *variability* in the performance of these components and in conditions surrounding the business. These two characteristics of the system are responsible for the idle or under-utilized capacity so costly to the ordinary manufacturing plant, and for the costly inventories of raw materials, work in process, and finished goods. Because of their widespread effects and the difficulties they cause, unbalance and variation and methods of coping with them are given special attention herein. Variation will be treated in Chapter 23.

Consequences of Unbalance in Productive Capacities

Six successive operations are required to manufacture product X. These operations along with their respective performance times are as follows.

1. Cast— 1.09 minutes per part
2. Mill— 2.40 minutes per part
3. Drill— 1.58 minutes per part
4. Clean— 0.97 minute per part
5. Paint— 1.54 minutes per part
6. Inspect— 1.30 minutes per part

The production rates for these operations are shown graphically in Figure 108. These production rates represent about the maximum

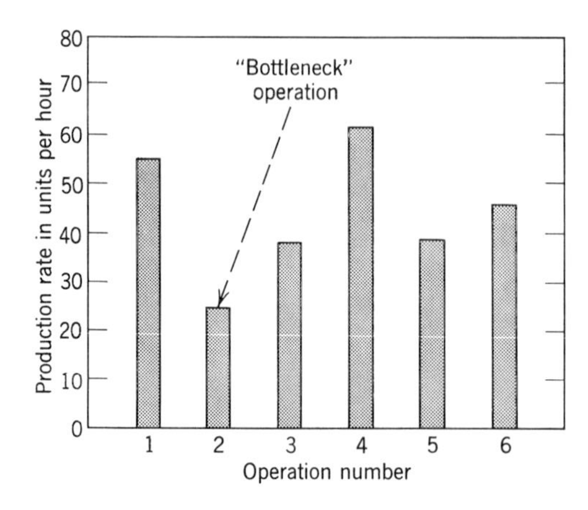

Figure 108. Average production rates for the six operations required to produce product X, illustrating unbalanced production capacities.

that can reasonably be expected from these machines and people under the existing circumstances. This series of production system components with unequal rates of consumption is an example of an unbalanced production situation.

What are some of the alternative courses of action in a situation like this? Obviously, unless a basic change is made in the system and/or a special operating procedure is adopted, an intolerable work-in-process inventory build-up will develop if an attempt is made to run all operations at their respective capacities. Here are some of the choices.

1. Physically alter the system to improve the balance by:
   a. making improvements in the bottleneck operation by alteration or replacement of equipment, improvement in procedure, etc.;
   b. redistributing the work between operations;
   c. duplicating facilities for the bottleneck operation to increase capacity.
2. Adopt some special operating procedure for the system, such as:
   a. running the parts through successive operations at capacity but in relatively large, intermittent batches;
   b. diverting some parts from the bottleneck operation to another similar machine in the plant if such is available, or to overtime or second-shift operation of the same machine, or to a subcontractor;
   c. and other less common special operating arrangements.
3. Operate all operations at the rate of the bottleneck operation.

In many situations a combination of the foregoing possibilities is adopted. It is virtually impossible to completely eliminate unbalance by the methods suggested in 1, primarily because of the discreteness associated with such remedies. So in most cases there will be some unbalance in the system that must be made the best of by means of the alternatives suggested in 2 and 3. Notice that the courses of action suggested in 2 and 3 incur a penalty in one form or another. It may be the costs associated with inventories, overtime, extra handling, extra equipment setups, or idle capacity.

The assembly case. When the assembly of a product is of choice or necessity to be divided into two or more successive operations, the balancing problem arises and the previously mentioned courses of action apply. Since assembly work is ordinarily comprised of a number of manual elements that can be combined in various sequences and divided in various ways, in contrast to the typical machining operation, "redistribution of work" is an effective means of achieving better balance of operations.

The balancing of assembly operations may be illustrated by the following example. Comb and brush sets are to be inspected, wrapped in tissue paper, placed in boxes, and then packed in shipping cartons. The set and packaging materials are pictured and identified in Figure 109. The specific elements involved and estimated performance times are as listed on page 378.

Figure 109. Comb and brush set to be inspected and packaged. Comb, wrapped brush, and tag are placed in the cardboard display tray (at upper left). Celluloid cover (upper center) is placed over tray and unit is placed in outer box (upper right). Boxed sets are then placed in large shipping carton (not shown).

1.	Get empty tray, remove celluloid cover, locate tray for packing.	0.09 minute
2.	Inspect brush.	0.08 minute
3.	Wrap brush in tissue paper.	0.18 minute
4.	Place brush in tray.	0.05 minute
2′.	Inspect second brush.	0.08 minute
3′.	Wrap second brush.	0.18 minute
4′.	Place second brush in tray.	0.05 minute
5.	Place comb in tray.	0.04 minute
6.	Place price tag in tray.	0.05 minute
7.	Place celluloid cover on tray.	0.07 minute
8.	Get outer box and disassemble.	0.05 minute
9.	Place tray in box.	0.04 minute
10.	Place cover on outer box.	0.04 minute
11.	Place box in large packing carton.	0.06 minute
	Total assembly time.	1.06 minutes

Note that it is possible for all of the foregoing elements to be performed by an operator at one workstation as one job. It is also possible to apply the division of labor principle and distribute the total task over two or more successive operators and workstations. In fact, it is possible to subdivide this task into as many as fourteen different "stations" along an assembly line, although this would surely be stretching the point. Increasing it beyond twelve would lead to ridiculous and impossible situations. But note the rather wide range of degrees of specialization that are within reason.

Except in the rare case in which equal division is possible, dividing a task into two or more successive subtasks results in unbalance. Furthermore, as the number of subdivisions is increased, the total time lost through unbalance shows a general increase. In the foregoing example, splitting the task into two successive operations, the first feeding directly to the second, as indicated in Table 15 results in 0.08 minute of idle time on the part of one operator and thus increases the total inspection and packaging time from 1.06 to 2×0.57 or 1.14 minutes. Distributing the same work over a three-station "assembly line" results in 0.14 minute of idle time, increasing the total time to inspect and package to 3×0.40 or 1.20 minutes. Increasing the number of stations beyond three results in additional (but not uniform) increases in idleness.

Discreteness

The primary cause of these increases is the discreteness of feasible subdivisions of a task and of their associated times. There is a limit as to the number of operations into which it is possible, at least for

TABLE 15. Work Assignments for a Two-Station and Three-Station "Assembly Line" for Comb and Brush Sets

Two Stations

Elements	Time
1. Prepare tray	0.09 min.
2. Inspect brush	0.08 min.
3. Wrap brush	0.18 min.
4. Place brush in tray	0.05 min.
5. Place comb in tray	0.04 min.
6. Place tag in tray	0.05 min.
Total, Station #1	0.49 min.
2′. Inspect brush	0.08 min.
3′. Wrap brush	0.18 min.
4′. Place brush in tray	0.05 min.
7. Place celluloid cover	0.07 min.
8. Prepare outer box	0.05 min.
9. Place tray in box	0.04 min.
10. Place cover on box	0.04 min.
11. Pack in carton	0.06 min.
Total, Station #2	0.57 min.

Idle time due to unbalance:

0.57 − 0.49 = 0.08 min. per set

Three Stations

Elements	Time
1. Prepare carton	0.09 min.
2. Inspect brush	0.08 min.
3. Wrap brush	0.18 min.
4. Place brush in tray	0.05 min.
Total, Station #1	0.40 min.
2′. Inspect brush	0.08 min.
3′. Wrap brush	0.18 min.
4′. Place brush in tray	0.05 min.
5. Place comb in tray	0.04 min.
Total, Station #2	0.35 min.
6. Place tag	0.05 min.
7. Place celluloid cover	0.07 min.
8. Prepare outer box	0.05 min.
9. Place tray in box	0.04 min.
10. Place cover on box	0.04 min.
11. Pack in carton	0.06 min.
Total, Station #3	0.31 min.

Idle time due to unbalance:

(0.40 − 0.35) + (0.40 − 0.31) = 0.14 min. per set

practical purposes, to subdivide a total task for balancing purposes. Take the element "wrap brush" for example. It is hardly possible for one person to pick up the brush and tissue paper and another to do the wrapping. Take another example involving the "wrap" and "place in box" elements. It is possible for one operator to wrap the brush and then place it in a tote pan or on a conveyor to be transported to another operator for placing in the tray. However, there is a time penalty associated with this procedure due to extra handling and, of more consequence in this case, to the fact that the brushes will become partially unwrapped. This is a common situation in a balancing problem; subdivisions that are not impossible, yet which are more time demanding because of extra motions made necessary by the separation. In practice, if this penalty is not too great, the separation is made if desirable, thus causing some increase in the total assembly time. Often, however, this penalty is such as to make it impractical and out of the question to further subdivide a given set of motions. Thus, because of impossibility or economic infeasibility, there are groups of motions that cannot be split for balancing purposes.

As a result, it is possible only to juggle what are the smallest practical or possible subdivisions of the total task, rather than fractions of seconds as we would prefer, to achieve better balance. The fact that these chunks of time we are at liberty to manipulate are relatively large, and that their magnitudes vary considerably as demonstrated in the preceding example, makes it virtually impossible to achieve and difficult to approach a perfect balance.

Precedence Requirements

Balancing is made difficult by a second characteristic of the ordinary task, and that is a limited number of sequences of elements that are physically possible or economically feasible. It is impossible, at least for practical purposes, to wrap the brush before inspecting it, or to place the brush, comb, or price tag inside the box before it is opened. Situations such as these must be taken into account in combining and recombining elements in an attempt to achieve improved balance. The existence of a limited number of feasible permutations of elements makes the limit on attainable balance due to discreteness more severe than would otherwise be the case. The precedence requirements for the case at hand are illustrated in Figure 110 in the form of a precedence diagram.

Procedures and Aids for Balancing Assembly Operations

Predetermined motion times are of considerable value in predicting the station times for different motion group assignments, especially when the balancing must be done while the product is still in the blueprint stage, as is usually the case in planning assembly lines for automobiles, appliances, television sets, etc. Thus, an engineer could have synthesized the times used in Table 15 from a predetermined motion-time system, and could have done this before the comb and brush set went into production. He could also use a predetermined motion-time system to estimate the time penalty incurred by separating certain elements that would best be performed in immediate succession, such as the time lost through an extra lay-down and extra pick-up if the brush is not placed in the tray immediately following wrapping.

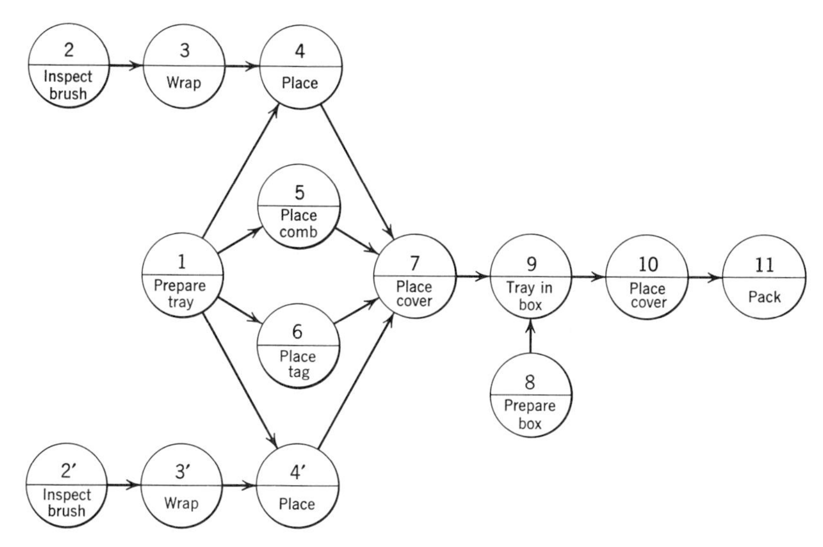

Figure 110. Precedence diagram for inspecting and packaging comb and brush sets.

As to specific balancing procedures, the most widely used method is strictly a shuffle and reshuffle process. Guided by good judgment, motion groups are distributed and redistributed in different combinations and sequences over the preferred number of assembly stations, until a satisfactory balance is obtained. In this juggling process, the precedence diagram is useful as a means of suggesting arrangements of elements and of verifying the precedence feasibility of various sequences. Figures 25 and 26 illustrate the usefulness of the multiple activity chart for balancing purposes.

Other and more rigorous solution procedures have been developed and are described elsewhere.[1] In some of the more complex situations, balancing has been done with the aid of a computer. A computer is especially helpful in a situation in which an assembly line must be rebalanced fairly frequently because of marked changes in rate of production.

REFERENCE

Salveson, M. E., "The Assembly Line Balancing Problem," *Journal of Industrial Engineering,* vol. 6, no. 3, May-June 1955.

[1] M. E. Salveson, "The Assembly Line Balancing Problem," *Journal of Industrial Engineering,* vol. 6, no. 3, May-June 1955.

23

Coping with Variation in the Manufacturing System

In most of the discussion so far the variation in behavior of men and machines that is so true of real life has been very conveniently assumed away. Neither man nor machine performs with perfect uniformity with respect to timing or quality. There is variation in the time required to perform an operation, especially if man partially or completely controls the rate of work. There is variation in the timing and type of interruptions of the production activity of man and machine. There is variation in the timing and type of demands for auxiliary services, such as services of the supervisor, materials handler, repairman, inspector, and the like. The variation in these and many other cases is the cause of a multitude of difficult problems for designer and administrator.

Types of Variation

It is useful to distinguish between two types of variation. One is the shift, instantaneous or gradual, of the *expected* value of a quantity. A gradual change, the trend, is illustrated by the common learning curve, a sample of which is shown in Figure 66, page 176. As a consequence of operator learning, a plot of successive cycle times for the job in question exhibits a gradual, persistent downward trend. In this case the trend is likely to continue for a substantial period of time, although at a gradually diminishing rate. Another example of the trend is the case in which dulling of the tool on a drilling operation causes a gradual increase in drilling time, continuing until the tool is replaced or sharpened and the expected drilling time returns to a minimum value once again. This case differs from the first example in that the trend is cyclical. Thus, a trend may be cyclical or non-cyclical in nature. In contrast to another type of variation—that dependent on chance—the trend is ordinarily much less troublesome primarily because of its relatively predictable nature.

As the *expected* cycle time gradually diminishes as the result of learning, and as the expected cycle time gradually increases because of tool dulling, *individual* cycle times vary noticeably about the average value, in what appears to be a random fashion. This chance variation is inherent in behavior of man and machine. It is a pronounced characteristic of the human being's behavior, and although

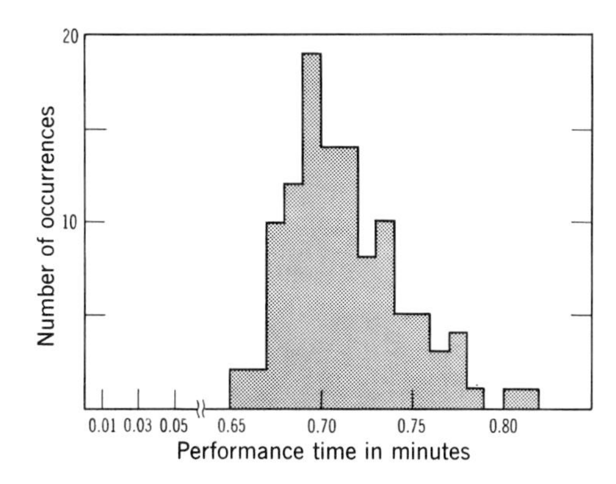

Figure 111. Frequency histogram for a series of consecutive repetitions of a semimanual task, showing the consequence of random variation in performance.

it has been reduced to a negligible degree in most instances, chance variation has not been eliminated from the behavior of machines.

Chance[1] variation is illustrated by the frequency histogram shown in Figure 111, constructed for a series of consecutive operation performance times. The same data is plotted in order of occurrence in Figure 112. Chance variation similar to that demonstrated by this operator and his machine is found throughout all phases and in all components of the manufacturing organism. It is the chance type of variation that is primarily responsible for the adverse consequences of variation to be described. Because of its random nature and the fact that it is impossible to eliminate or compensate for, and difficult to even reduce, chance variation is a real challenge (and headache) to the designer and manager of the business enterprise. The remainder of this chapter is directed primarily to this type of variation.

[1] What are for practical purposes commonly referred to as chance variations are not in most instances strictly random in nature.

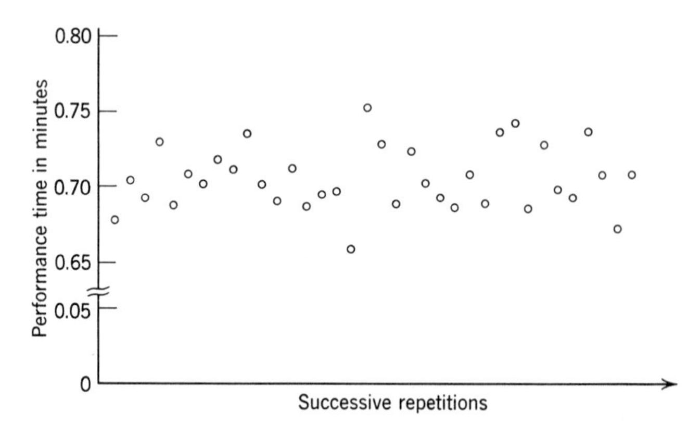

Figure 112. Times from Figure 111 plotted in order of occurrence.

Effects of Variation in the Timing of Events

The primary consequence of variation in the timing of events, such as in the arrival of customers, the completion of units, the breakdown of equipment, etc., in a production situation is idleness or waiting. Take the familiar case of the barber shop. If it took a barber exactly 15 minutes to process a customer, and if customers arrived at regular intervals, each arriving just as the previous one was finished, then the barber would be busy throughout the 8-hour day, no customer would find it necessary to wait, and the barber could process (60 min./hr.)/(15 min./cust.) $\times$ 8 hrs./day, or 32 customers per day. Actually, of course, the time to service a customer varies considerably, the general pattern of variation appearing similar to that demonstrated in Figures 111 and 112, so that even if the patrons arrive at 15-minute intervals, some of them will find it necessary to wait and others will find the barber waiting for them. If the interval between customer arrivals varies but the processing time is constant at fifteen minutes, a similar situation will result, unless, of course, demand is so great that customers are always on hand. Inevitable variation in process times and intervals between customer arrivals creates a typical barbershop situation familiar to all.

This same type of situation abounds in the typical manufacturing plant. Take the case of individual parts coming along on a belt conveyor to a manual inspection operation. The experiences of these parts and the inspector parallel those of the barbershop customers and the barber, for the same reason—variation. At times the inspector is waiting for a part to inspect, and at times parts pile up waiting to be inspected, even though the *average* interval between arrival of parts may be slightly greater than the *average* time to inspect a part. Or take the case of the overhead crane servicing a group of machines in a heavy machine shop. Because of variation in the interval between calls for the crane and variation in the time to service a machine, there are periods of crane idleness and periods during which one or more machines and their operators are waiting for the services of the crane. Or take the case of a lathe in a job machine shop. Because of the variation in the interval between and duration of jobs, there are periods during which one or more jobs are "in queue," waiting to be processed on the machine, and other periods during which the machine sits idle for lack of jobs.

The generality of variation and the idleness and queuing that result will be illustrated by other commonplace examples from the point of view of worker, product, machine, and customer. The worker might well be found waiting to punch in at the time clock, waiting for a work assignment, waiting to pick up tools at the toolroom or materials at the stockroom, waiting for the foreman's technical assistance, waiting for a materials handler to bring or remove materials, and so on. The product, still in the form of raw material, waits in the freight car to be unloaded, waits in the stockroom to be processed; in the form of work-in-process inventory, it waits to be processed at successive operations; and, in the form of finished-goods inventory, it waits to be sold. The individual machine might well be found waiting for work, for servicing by its operator, or for a repairman in the case of a breakdown. In many instances units of product are completed at irregular intervals. Similarly, customers ordinarily "arrive" at irregularly spaced times, so that a queue of finished units of product forms for customers or a queue of customers forms for completed products, depending on the particular circumstances.

The waiting described in these examples arises as a consequence of trends in demand or capacity (primarily "peaking," as experienced by the restaurant, supermarket, power company, toolroom near quitting time, the toy manufacturer, etc.) and especially as a consequence of chance variation. The generality, consequences, and stubbornness of the latter are such that a discussion of means of dealing with chance variation is warranted.

Coping with Chance Variation—Reducing Variability to a Minimum

There are two major courses of action available in dealing with chance variation. One is to attempt to minimize the variability itself. The other is to find ways and means of making the best of such variation. Certainly the former possibility should be considered first, for even though such variation cannot be eliminated there is a possibility that it can be reduced to a negligible level. Furthermore, any reduction that can be made, whatever the magnitude, means there is that much less to make the best of by the means suggested later, all of which are costly.

The most obvious and most effective means of reducing random variation is scheduling. The common appointment system used by physicians is an example of the use of scheduling to reduce chance variation. If the doctor were to depend on the chance arrival of patients, the average length of waiting time, the dissatisfaction of patrons, the size of waiting room required, etc., would be significantly greater than they ordinarily are under the common appointment system. Furthermore, the doctor risks the possibility of idle periods under the chance method. Although appointments do not eliminate, they significantly reduce the magnitude of chance variation in patient arrival times. Scheduling is extensively used in almost every plant to reduce the variation in demand for production equipment, in the same manner as the physician uses it.

Another example of the use of scheduling for this purpose is the periodic replacement of certain machine parts (bearings, cams, etc.) that are known from experience to have a limited life and to be frequent causes of machine breakdown. Instead of waiting for the failure to take place, the timing of which is strongly dictated by chance, the part is replaced at periodic, convenient times. This permits the replacement to be made when disturbance of production is at a minimum and assists in reducing some of the chance nature of the demand for repairmen.

There are numerous other cases in which scheduling is utilized in an effort to minimize chance variation, and still others in which it could be used to advantage for the same purpose. But of course in the production situation there are many instances in which scheduling is out of the question. For example, it is futile to attempt scheduling where variation in performance time is inherent to the human operator. In this and many other instances the only course of action is to determine how to minimize the adverse consequences of such variation, as described in the remainder of the chapter.

Coping with Chance Variation—Use of the Inventory

Generally speaking, the adverse effects of chance variation are minimized by the use of an appropriate sized inventory, a bank or reservoir of waiting objects, persons, machines, or whatever is receiving the service involved (henceforth referred to as "customers"). This inventory should be of such a size as to minimize the *total* cost of waiting customers and idle service facility. Some examples of this use of the inventory will be described.

Some physicians apparently use this scheme to protect themselves from idleness during office hours. Appointments are made in such a manner as to always keep a sufficient number of patients on hand in the waiting room to assure that the doctor will not be idle even though some patients arrive late for appointments and even though some require less than the expected amount of the physician's time. Thus, by the use of this reservoir of waiting patients the doctor minimizes the adverse effects (from his point of view) of chance variation in customer arrival times and in service times.

The second example pertains to what is commonly referred to as the finished goods inventory, which protects the customer from chance variation in rate of output of the finished product, and protects the plant from chance variation in the arrival of buyers. Note that the costs of space, invested capital, insurance, etc., increase as the quantity of finished product "in waiting" is increased, but that as the quantity is reduced the cost associated with dissatisfied customers increases. Thus, there is an optimum sized inventory, one that minimizes the combined costs of carrying the inventory and of dissatisfied customers.

The third example involves two successive manufacturing operations of a predominantly manual nature. As parts are completed by operator A they are transported directly to operator B on a short belt conveyor. Even though the mean production rates of the two operations are essentially equal, because of chance variation in the performance time for each there will be frequent instances in which the two operators "interfere" with each other. Interference occurs when operator A finishes a unit before operator B completes his, so that A must delay in releasing his unit onto the conveyor until B is ready. Interference also occurs if operator B completes his unit before A does, such that B cannot continue until he receives his next unit from A. The resulting idleness in both instances arises because of chance variation. The obvious remedy for this interference is to reduce the interdependency of the two operators by introducing an inventory of parts between them. The larger this work-in-process inventory, the less chance that a very long cycle or extended delay could cause one operation to interfere with the other. From this point of view, the larger the inventory the better. Opposed to this, however, are the costs of extra factory space required, invested capital, and so forth, associated with a work-in-process inventory. Therefore, there is an optimum sized inventory, one which represents the most favorable compromise in the conflict between the cost of carrying an inventory and the cost of operator idleness.

The fourth example concerns breakdowns in production equipment. Unfortunately the particular time at which a machine breaks down is strongly dictated by chance, so since repair crew size is relatively fixed there are times when there are a number of machines "down" and waiting for a repairman, and there are other times when part or all of the crew is idle. Just as in the foregoing instances, there is an optimum sized "inventory," which in this case consists of machines awaiting a repairman. Increasing the size of the plant's repair crew reduces the production time lost due to the unavailability of a repairman, and in fact such production delays could be eliminated by keeping a small army of repairmen on hand. But this would hardly be economical, especially in view of the high-priced nature of these individuals and of the restrictive regulations of various trades. Likewise, it is not economical to operate at the opposite extreme and eliminate all idleness of repairmen. At that point there would be a prohibitive amount of idle production equipment. Thus, there is some repair crew size and resulting amount of idle equipment time that represents an optimum balance between cost of the crew and cost of waiting machines.

The fifth example concerns a toolroom that supplies tools for the machinists in a shop. There is considerable chance variation in the intervals between demands for this service and in the time required for a toolroom clerk to obtain the tools needed by machinists. In this as well as in the previous examples, there is waiting, there is a cost conflict, and there is an optimum average number of waiting machinists. Figure 113 graphically depicts the general behavior of the relevant costs in a situation like this. Note that as the staff of the toolroom is increased, for each clerk added there is a progressively smaller reduction in amount and cost of machinist waiting, until additional clerks are of negligible benefit. Progressive reductions in the number of clerks increase the penalty for waiting machinists at an accelerating rate, until a point is reached where the system breaks down completely. The most favorable number of clerks, three in this case, is that for which cost of manning the toolroom plus cost of waiting machinists and their idle equipment is a minimum. Instead of toolroom clerks, the service units in Figure 113 could just as well be repairmen, fork lift trucks, toll booths, bank tellers, and so on.

Although all of the foregoing examples involve an inventory in the broadest sense of the term, it seems common to speak of such a

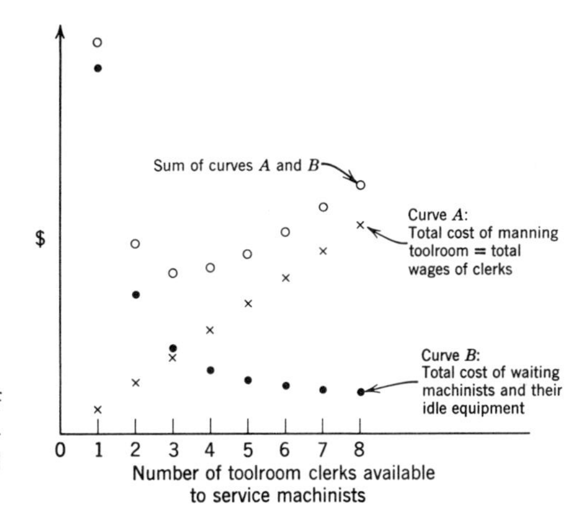

Figure 113. General behavior of the relevant costs in a waiting-line problem as number of parallel service units is varied.

situation as an inventory problem only if it involves materials, and to describe the situation as a waiting-line (queuing) problem if it concerns persons or pieces of equipment. Although the problem is basically the same, some of the factors to be considered and methods of solution are different for the two cases. The waiting line problem is of particular interest to the methods engineer, for he encounters it frequently in his design work and in establishing work loads and standards for those cases in which demand for the operator's services is subject to chance variation, as in repair work, materials handling, machine setup, and the like. Work standards established for such jobs as these can be very unrealistic if the effects of variation are not appropriately taken into account. For this reason the characteristics and methods of solving waiting line problems are to be discussed at length.

General Description of a Waiting-Line Problem

This is a very common type of problem; very few businesses escape it. Manufacturing, distribution, and service enterprises are equally plagued by various forms of the waiting line problem. A waiting line problem ordinarily arises when:

1. there is chance variation in the time required to provide a service to customers, or there is chance variation in the interval between demands for that service, or both; and
2. there is a cost associated with customer waiting *and* a cost involved in providing the service in question.

Under these circumstances there is a particular number of parallel service units (repairmen, toll booths, etc., sometimes referred to as service channels) and resulting amount of waiting that minimize the sum of the relevant costs involved. Note that there is a direct relationship between the capacity of the service facility to process customers and the total time spent by customers in waiting. In fact, customer waiting is controlled by adjusting the total service rate capacity. This is usually accomplished by changing the number of parallel service channels in the service facility. Ordinarily in dealing with a waiting-line problem, *after* an attempt has been made to minimize the variation and to increase the capacity of the service unit by improvements in method, the main task is to determine the optimum capacity of the service facility. How many bank tellers, waiters, market check-out stations, department store clerks, airport runways, freighter docks, repairmen, stockroom clerks, freight elevators, or shop cranes should be used in order to minimize total cost?

What constitutes the optimum service facility capacity in a given situation is not ordinarily immediately obvious. True, there are some instances in which the answer can be satisfactorily arrived at through good judgment, common sense, and experience. However, in many other cases it is necessary to depend on one or more of the following methods if reliable solutions are sought.

1. Experimentation.
   a. With the actual system, by systematically introducing changes in number of service channels and observing the effects on waiting. The obvious disadvantages of this method are the amount of time required and the cost of changing the number of channels. It is hardly economical to keep on altering the number of airport runways, shipping and receiving docks, number of cranes, and the like, in an effort to determine the optimum solution.
   b. With something other than the actual system but which behaves in an analogous manner, a procedure commonly described as simulation. Two methods of simulation will be described.
      (1) Simulation by means of a physical analogue, mechanical or electrical, for example.
      (2) Simulation by a digital technique described as the Monte Carlo method.
2. Analytical methods. By employing mathematics and statistics, special sets of equations have been derived for certain types of queuing situations. If such equations exist for the problem at hand, it is possible to obtain the information required to make a decision directly from these expressions or from tables or graphs based on them.

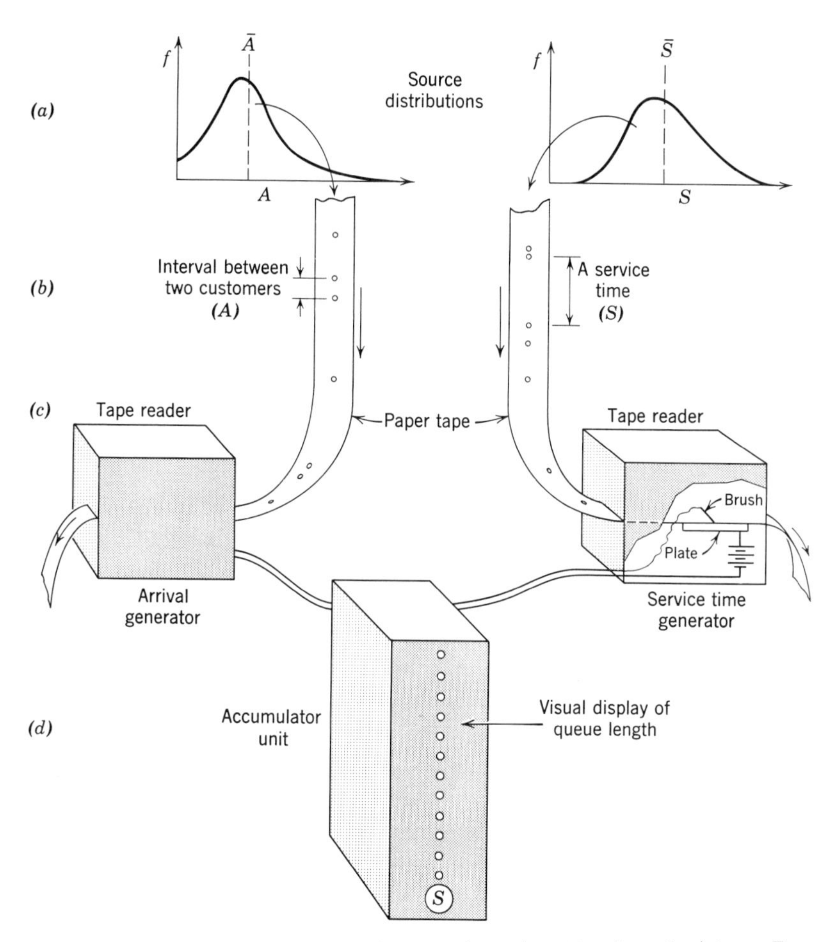

Figure 114. Main features of an electromechanical waiting-line simulator. Times between arrivals and service times are selected at random from distributions such as shown at (a) and punched into the paper tape (b) at distances proportional to the times selected. To perform the simulation, the tapes are fed to the tape readers (c) which sense the information and transmit it to the accumulator unit (d). This unit keeps track of and records all customer waiting time and service unit idleness, and displays by a row of lights the status of the system at any instant.

Solution of Queuing Problems by Analogue Simulation

Solution of queuing problems by a physical analogue is exemplified by the electromechanical device diagrammed in Figure 114. As indicated in part (b) of the figure, customer arrivals are simulated by holes punched in a moving paper tape. This tape moves at constant speed through a tape reader (*c*) which electrically "senses" the arrival of a "customer" and informs the accumulator unit (*d*) of the fact. This tape makes it possible to reconstruct the customer arrival pattern of the real life situation involved. There is a similar tape reader for generating service times. These two inputs are fed to the accumulator unit of the simulator, which consists of relays, timers, and other hardware. This unit keeps track of the queue length and records customer waiting time and idleness of servers. With this device it is possible to experiment with and collect data for one, two, three, or more service channels in order to determine the optimum number of servers for the problem at hand. The following must be known or assumed before a real life situation can be satisfactorily simulated by this (or any other) means.

1. The approximate nature of the frequency distribution of times (A) between arrivals. This includes the mean interval between arrivals ($\bar{A}$) and at least the general shape of the distribution. This information is obtained by observation of the actual arrival pattern that is to be simulated. A series of times between arrivals is obtained by stop-watch or other timing means, until a reasonably reliable sample is accumulated. This sample may indicate that the underlying population of times between arrivals may be satisfactorily approximated by one of the common, well-defined distributions, such as the normal or exponential. If the distribution of observed times between arrivals does not resemble any of the familiar basic types of distributions, which is quite possible, this does not hamper the use of simulation. The important matter is that the approximate nature of this distribution be known, whatever its form.
2. The approximate nature of the distribution of service times (S). This includes the mean service time ($\bar{S}$) and at least the general form of the distribution. This information, too, must ordinarily be obtained by timing of the actual service unit itself. An adequate sample of times must be collected so that satisfactory inferences may be made concerning the underlying population. This distribution of observed service times may resemble one of the familiar basic types of distributions, but as in 1, this is not necessary.
3. The queue discipline, which is a set of priority rules governing the order in which waiting customers will be served. The discipline might be first-come-first-served with a common waiting line or individual waiting lines for each service unit, or the selection might be on a random basis, or there might be a special priority system, and so on.

4. Interdependencies, such as the tendency for $\bar{S}$ to decrease as the queue length grows longer, which is likely when human beings control the rate of servicing. An interdependency between $\bar{A}$ and queue length is fairly common, for two reasons. If the customers are human beings, there is an increased tendency to refuse to wait as the length of the waiting line grows. Secondly, in many industrial waiting-line problems there is a relatively small population of potential customers, so that for each customer that joins the queue there is a significant reduction in the supply of customers and therefore an increase in $\bar{A}$. This interdependency usually exists in the crane, materials handler, repairman, and many other familiar queuing situations in the manufacturing plant. Sometimes the existence of an interdependency is obvious, whereas at other times it is determined only through observation of the actual system under investigation.

Once these characteristics of the situation have been learned or assumed, a series of times between arrivals is randomly selected from the distribution of same. As indicated at (*a*) in Figure **114**, holes are then punched in the tape at intervals proportional to the times selected. A similar procedure is followed in order to generate a service time tape. These tapes are then run through their respective tape readers, providing customers and service times, as the experiment is conducted with varying numbers of service channels.[2]

By repeating the experiment a sufficient number of times for each number of parallel service channels that seem reasonable for the situation at hand, the engineer can thereby generate the information necessary to determine the optimum number of channels. A device such as this may be used to experiment with changes in the system other than number of channels. For example, experiments may be conducted in a similar fashion using different queue disciplines.

[2] Two or more of the basic accumulator units may be connected in series to increase the waiting line capacity, to simulate queues and service units in series, or to simulate parallel waiting lines. If there are significant interrelationships such as mentioned in 4, physical simulation is less straightforward. For example, if $\bar{A}$ tends to increase as the waiting line increases, the nature of the relationship must be determined and "built into" the simulator accordingly. This particular relationship is provided for by automatically varying the speed of the customer tape drive according to the number of customers in queue.

Two simulators based on the principles just described are known to exist. One has been developed at the Office of Operations Research, The Johns Hopkins University, Chevy Chase, Maryland. The other is at Cornell University, Ithaca, New York, in the Department of Industrial and Engineering Administration, Sibley School of Mechanical Engineering.

Solution of Queuing Problems by the Monte Carlo Method

In its simplest form, the Monte Carlo technique involves doing "manually" and in digital form what the described waiting line simulator did electromechanically (and to be sure, much more rapidly). To do this the same basic information is required, namely, the mean time between arrivals and mean service time, the general pattern of variation about each, the queue discipline, and any significant interdependencies. As before, the general nature of the arrival time and service time distributions must be known, but these need not conform to any of the familiar distribution forms.

The simulation is achieved by making random selections from the distribution of times between arrivals and distribution of service times, and by keeping a running tabulation of the status of the system as illustrated in Table 16. For example, according to row 1, Table 16, the first time selected at random from the distribution of times between arrivals was 3 minutes. If a starting time of eight o'clock is assumed, this means that the first customer arrived at 8:03. A service time was then selected at random, the selection being 1 minute. This means then that the first customer arrived at 8:03 and departed at 8:04, that he did not have to wait for service, but that the service unit was idle from 8:00 to 8:03. This process is continued as illustrated in Table 16, until an adequate sample has been generated. At the conclusion of the run, total waiting time is accumulated by summing times in row 5, Table 16. This same procedure is repeated as the number of channels is varied, until sufficient information is generated to permit identification of the optimum number of channels.

The time intervals required for this simulation, rows 1 and 3, Table 16, may be generated by first writing times on slips of paper according to the relative frequency of various times in the distributions assumed for A and S, then randomly drawing times from this population of slips. The drawing should be with replacement. There are faster, easier, and more sophisticated methods of generating the A's and S's required than this "drawing from the hat" procedure. In many instances it is possible to generate the desired times through the use of a table of random numbers in conjunction with the cumulative form of the frequency distribution involved, or by means of a specially constructed roulette-type wheel, or by digital computer, and thereby reduce the time and labor involved.

In summary then, a series of customer arrivals and service times into which expected variation has been introduced are generated to

TABLE 16. Portion of a Monte Carlo Simulation; Single Channel, First-Come-First-Served, Unlimited Source of Customers

1	A (Selected at random from assumed distribution.)	3 minutes	5	1	2	3	2	5	6
2	Arrival time (start at 8 A.M.)	8:03 (8:00 + 3)	8:08 (8:03 + 5)	8:09 (8:08 + 1)	8:11	8:14	8:16	8:21	
3	S (Selected at random from assumed distribution.)	1 minute	3	3	2	4	1	2	
4	Departure time	8:04 (8:03 + 1)	8:11 (8:08 + 3)	8:14 (8:11 + 3)	8:16	8:20	8:21	8:23	
5	W (Customer wait time.)	0	0	2 (8:11 − 8:09)	3	2	4	0	
6	Idle service time	3 minutes (8:03 − 8:00)	4 (8:08 − 8:04)	0	0	0	0	0	

simulate the real life situation under investigation. What would happen to the actual system under these same circumstances is noted on paper so that total waiting time can eventually be accumulated. By this means behavior of the system under different experimental conditions can be observed and practical decisions made. Note that there is a sampling error involved in both of the simulation methods described.

Solution of Queuing Problems by Analytical Methods

If the distribution of times between arrivals and of service times may be satisfactorily approximated by certain quantitatively definable frequency distributions, equations are available from which it is possible to compute information needed to determine the optimum number of service channels. The simplest of these cases requires the following assumptions.

1. The times between arrivals are exponentially distributed. The basic assumption is that customer arrivals are random events in time, and thus that the probability of an arrival at any specific instant in time is independent of the time of the previous arrival. If this is so, the distribution of times between arrivals is an exponential one. For example, suppose that the times between arrivals of trucks at a company's receiving dock were recorded over a 2-week period, with the results shown in histogram form in Figure 115. The mean time between arrivals is approximately 30 minutes. An exponential distribution with a mean of 30 minutes—the shaded distribution in Figure 115—offers a good approximation of the observed distribution, so that the assumption of exponentially distributed times between arrivals is a safe one here.
2. The service times are likewise exponentially distributed. For the receiving dock example, the unloading times were observed for a 2-week period, indicating a mean service time of 22 minutes. The data observed indicates that an exponential approximation of the distribution of truck unloading times is satisfactory.
3. The queue discipline is first-come-first-served in a waiting line common to all service units, no limit on length of waiting line.
4. There are no significant interdependencies. This means that the source of customers is unlimited, which for practical purposes means 100 or more potential customers.

If these conditions are satisfied, the average amount of time a

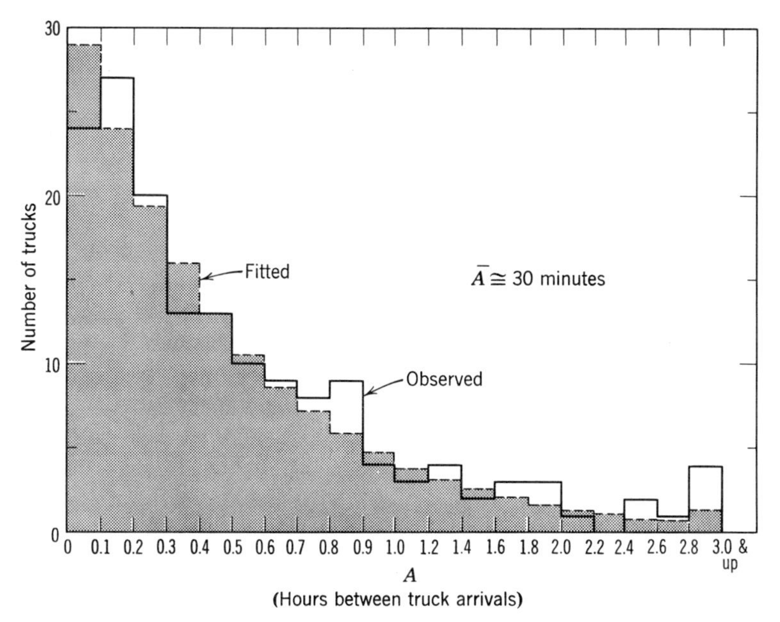

Figure 115. Observed distribution of times between truck arrivals at receiving dock, showing (shaded) exponential distribution fitted to observed data.

customer must wait until served in a single channel system is

$$\bar{W}_q = \frac{\bar{S}^2}{(\bar{A} - \bar{S})} \quad \text{when } \bar{A} > \bar{S} \tag{1}$$

where $\bar{A}$ is the average time between customer arrivals and $\bar{S}$ is average service time. The mean time spent by a customer in the system—the mean time required to wait and be serviced—is

$$\bar{W} = \bar{W}_q + \bar{S} \tag{2}$$

in a waiting-line situation. For the receiving dock example, if the dock has only one bay for unloading, according to equation 1 a truck must on the average wait

$$\bar{W}_q = \frac{\bar{S}^2}{(\bar{A} - \bar{S})} = \frac{(22)^2}{(30 - 22)} = 60.5 \text{ minutes}$$

According to equation 2 the average time spent by a truck waiting and being unloaded is 60.5 + 22 or 82.5 minutes.

For a multichannel system with the same assumptions stated, the equations are too complex for ordinary practical purposes. Fortunately for the practitioner, graphical solutions of these equations are available. Figure 116 presents a graphical solution of the equation for $\overline{W}_q$. The graph in Figure 116 is entered with the number of parallel channels (B) and a factor F, where $F = \bar{S}/B\bar{A}$. For the receiving dock problem, for two unloading bays (two channels) $F = 22/(2 \times 30) = 0.367$. Entering the graph with $B = 2$ and reading up to $F = 0.367$ yields a value of $\overline{W}_q/\bar{S}$ of approximately 0.16. $\overline{W}_q$ therefore is 0.16×22 or approximately 3.5 minutes. As before,

$$\overline{W} = \overline{W}_q + \bar{S} = 3.5 + 22 = 25.5 \text{ minutes}$$

For three bays at the receiving dock, $\overline{W}_q$ becomes approximately 0.4 minute according to this graph.

$\overline{W}_q$ can be converted to expected total waiting time for a given period of time by being multiplied by the expected number of customers over that period. For example, for a single unloading bay, $\overline{W}_q$ is 60.5 minutes. For a week, 80 trucks (40 hours $\times$ 2 arrivals per hour) are expected. Total time spent waiting for unloading over the week is expected to be 80×60.5 minutes or 80.6 hours.

Equivalent sets of equations are available for several other service time and time-between-arrival distributions; however, the expressions are rather complex. Graphical solutions exist in a majority of cases. In most of these instances it is possible to compute the probability that a customer will have to wait, the probability that the line length will exceed a given value, and other useful information.

Comparison of Methods of Solving Queuing Problems

Three important and useful methods of solving queuing problems have been briefly discussed, namely: analogue simulation, Monte Carlo simulation, and analytical means. There are some important differences in the applicability of these methods to different types of waiting line problems.

Simulation methods in general are comparatively high in cost but relatively free of stringent assumptions. Analogue simulation involves special-purpose hardware. Monte Carlo requires time-consuming manual computation or rather expensive time on a digital computer. A sampling error exists in both instances. In their favor, however, is the fact that it is not necessary to assume a particular distribution form for A or S. It is not necessary to be able to mathematically characterize these distributions. Regardless of how peculiar and unfamiliar the shape of these distributions, all that is necessary is an estimate of the general nature of each. Furthermore, a wide variety of special situations

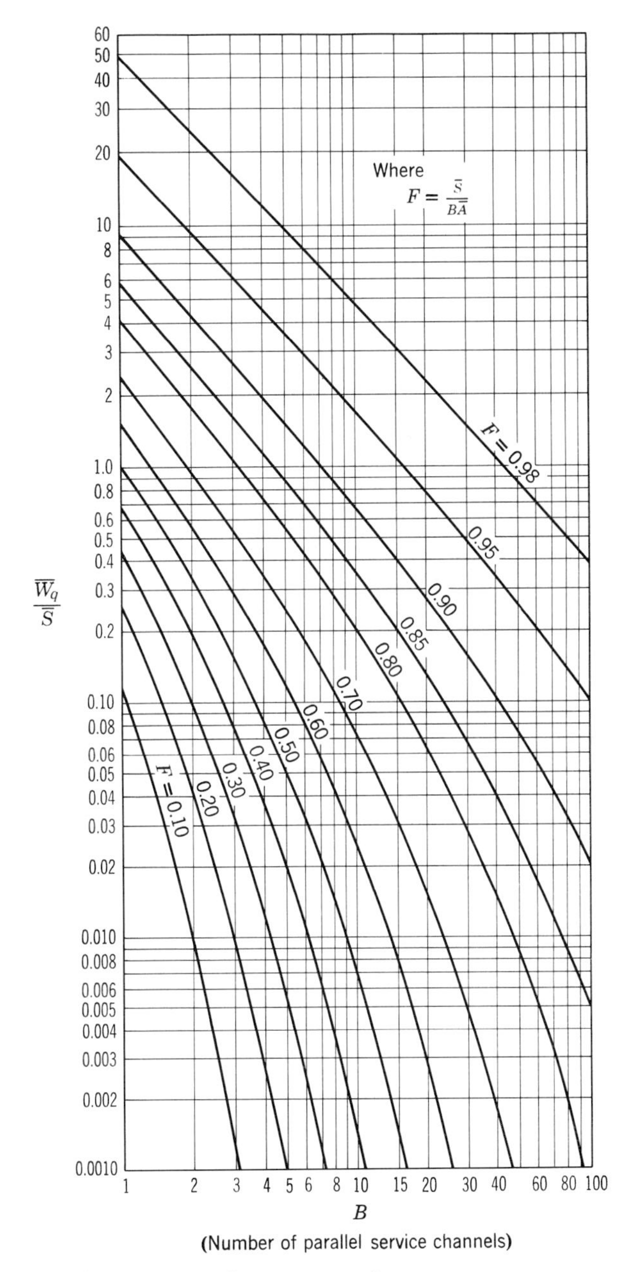

Figure 116. Graphical solution of an equation for average waiting time in a multi-channel system. Assumptions: A and S exponentially distributed; single waiting line common to all service units; first-come-first-served; unlimited source of customers; no interdependencies.

can be taken into account, such as a dependency of $\bar{A}$ or $\bar{S}$ on queue length, changes in $\bar{A}$ or $\bar{S}$ with time,[3] line quitters, and many other special circumstances. Thus, simulation, especially the Monte Carlo form, is a very versatile technique in contrast to the analytical methods currently available.

The latter excels, of course, in the matter of cost. If the equations apply satisfactorily, the desired waiting time information can ordinarily be obtained in a matter of minutes compared to the time-consuming simulation process. A major drawback to the analytical approach is the fact that there are only several distribution forms for which equations have been derived. If none of these distributions happens to offer a satisfactory approximation of the real life situation, the mathematical approach does not apply. Another important drawback is the fact that the analytical method offers a solution for what is generally referred to as a steady-state condition. The presumption is that the waiting line system has been in operation for a sufficiently long period to settle down to a stable condition. In the ordinary factory, with shift changes, lunch hours, official rest periods, changes in $\bar{A}$ or $\bar{S}$ through the day, and the like, there is often little opportunity for the system to settle down to a stable condition.

The Finite Source of Customers

In the preceding discussion an unlimited supply of customers has been assumed, which for practical purposes means the population of potential customers is 100 or more. In a majority of waiting-line situations in the typical industrial plant, the source of customers is very limited in number, considerably below 100. Thus, the entry of a customer into queue significantly reduces the probability of another customer arriving. This is true, for example, of the man who must tend a certain number of semiautomatic machines needing attention at varying intervals. One man might tend anywhere from one to as high as fifteen or twenty machines, depending on the characteristics of the machines. When a machine calls for service the man tends to it. If additional machines call for service while he is servicing the first, these machines go into queue for the man's services. As each additional machine goes into queue because the man is already occupied, the supply of machines still running is diminished and the probability of another call for service proportionately reduced. Under

[3] As for example a tendency for the rate of arrival of machinists to be higher at the tool crib in the early and latter stages of the work shift.

these circumstances the equations given earlier do not apply. A basically different approach and set of equations is necessary.

Under the topic of man-machine charts, the use of one operator to service more than one machine was discussed. At that time the assumption was made that the demand for service by a machine was cyclical, the machine running time always being the same so that a man could easily synchronize his attention to two or more such machines, as pictured graphically by the man-machine chart. A number of machines in the ordinary production plant are of this type so that synchronization is possible. However, there are some machines that require the attention of an operator at variable intervals, for such reasons as setup, resupplying with stock, repairing breaks and jams, etc. Variability also exists in the needs of equipment for a repairman, or the needs of operators for materials handlers, inspectors, and supervisory personnel, and so on. These and many other situations in the factory involve queues with a relatively small source population. Since the matter of assigning production machines having need for attention at varying intervals is characteristic of this type of queuing problem, and since most of the analytical work done so far has been devoted to this situation, the "machine assignment problem" will be used herein to illustrate the available approaches to a finite population queuing problem.

Simulation and analytical methods may be applied successfully to this type of problem. Since the analytical method is rapid and ordinarily low in cost, it might well be considered first. Available mathematical approaches to the machine assignment problem require the following major assumptions.

1. Calls for attention occur at random intervals. From this it follows that the frequency distribution of times between calls for service is of the exponential form.
2. Service times are either exponentially distributed or are constant. One or the other of these conditions must be assumed, whichever is more appropriate for the situation at hand.
3. The servicer will attend to the machines on a first-come-first-served basis.
4. There are no interdependencies other than that between $\bar{A}$ and queue length caused by the finite population of customers.

Sets of equations have been derived under the foregoing assumptions, one set for exponentially distributed service times and another set for constant service time.[4] Graphical solutions of these equations have been prepared, as illustrated in Figure 117 for random service calls and exponentially distributed service times. Figure 117 may be used to determine the optimum machine assignment (n)—the number of machines to assign to one operator for attention that will minimize the sum of the relevant costs (see Figure 113). To do this the graph must be entered with two ratios. One is the ratio $\bar{S}/\bar{R}$, where $\bar{S}$ is the average time required to service a machine and $\bar{R}$ is the average machine running time, the average amount of time a machine runs between calls for attention. The information for quantification of this ratio must ordinarily be determined by actual on-the-spot measurement of the machines in operation. The second ratio is

$$\frac{\text{Cost of waiting machine per unit of time}}{\text{Cost of servicer per unit of time}} = \frac{C_w}{C_l}$$

To illustrate the use of Figure 117, assume that the average time to service any one of a battery of machines to be tended is 2 minutes, and that the average time between calls for attention by a given machine in the group is 20 minutes. The penalty for downtime of a machine of this type is \$15 per hour, whereas the wage of the servicer is \$1.50 per hour. Under these conditions

$$\frac{\bar{S}}{\bar{R}} = \frac{2}{20} = 0.1 \quad \text{and} \quad \frac{C_w}{C_l} = \frac{15}{1.5} = 10$$

For these values, an optimum assignment of three machines per attendant is indicated by Figure 117. For a rather different set of conditions, namely $\bar{S}/\bar{R}$ of 0.08 and C_w/C_l of 2.5, the optimum assignment is six machines according to Figure 117.

This illustrates the analytical approach to the finite population type of queuing problem. For a more thorough description of the background and use of Figure 117 and discussion of other useful equations, tables, and graphs, see the references cited.[5]

[4] C. Palm, "Arbetskraftens fordelning vid betjaning av automatmaskiner," *Industritidningen Norden,* vol. 75, 1947; H. Ashcroft, "The Productivity of Several Machines under the Care of One Operator," *Journal of the Royal Statistical Society,* series B, vol. 12, no. 1, 1950.

[5] *Ibid.;* E. Bowman and R. Fetter, *Analyses of Industrial Operations;* Richard D. Irwin, Homewood, Ill., 1959; W. Feller, *An Introduction to Probability Theory and Its Applications,* vol. 1, John Wiley and Sons, New York, 1950; R. Fetter, "The Assignment of Operators to Service Automatic Machines," *Journal of Industrial Engineering,* vol. 6, no. 5, September-October 1955; T. F. O'Connor, "Productivity and Probability," *Mechanical World Monographs,* no. 65, Emmott and Company Limited, Manchester, England.

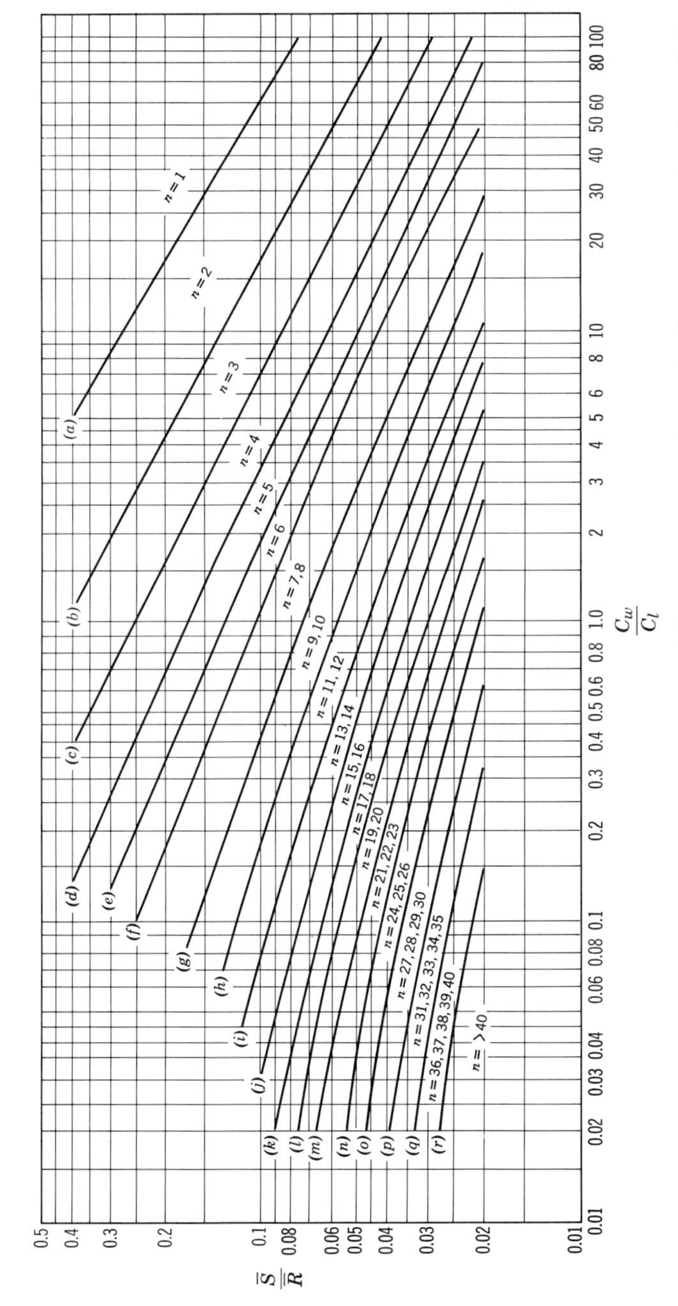

Figure 117. Curves for determining optimum number of machines (n) to assign to a worker, under the assumption of random calls for attention and exponentially distributed service times. (From E. H. Bowman and R. B. Fetter, *Analyses of Industrial Operations*, Richard D. Irwin, Homewood, Illinois, 1959. By permission of the publisher.)

The available analytical procedures for solution of this type of queuing problem offer assistance to the practitioner for certain types of situations. Much additional investigation and computation is needed, however. One very important difficulty in the use of a method such as illustrated in Figure 117 is that very seldom will the user be dealing with a total number of machines that is an even multiple of the optimum indicated. For example, a practitioner might have twelve machines to assign to X operators. He consults the curves and finds that for this type of machine and the costs involved, the optimum assignment is eight machines per worker. Obviously he must use less than the optimum assignment, but which assignment will cause the least increase in cost over the theoretical minimum? Should he assign all twelve machines to one worker, or eight to one worker and four to another, or six to each of two workers? Currently available literature is of negligible assistance to the practitioner in making this decision which arises in most machine assignment situations. Available mathematical methods suffer from some rather unrealistic assumptions, deviations from which are in most instances unexplored. For example, it seems rather unlikely that operators will attend to awaiting machines on a first-come-first-served basis, for a number of practical reasons. Also, $\bar{S}$ is almost certainly not independent of queue length. We would expect an operator to accelerate when the number of machines awaiting his attention increases. Also, it is necessary to assume that all machines being considered for grouping have identical $\bar{S}/\bar{R}$ and C_w/C_l ratios. The assumption of exponentially distributed service times seems unrealistic, as does the assumption of constant service times. However, investigators have shown that the conclusion as to optimum number of machines to assign is not strongly affected by the type of distribution assumed for S. For example, as long as there is a moderate amount of variation in service times, it appears safe to assume an exponential distribution, regardless of the exact shape of the actual distribution. All existing analytical methods assume random calls for service, and thus that times between calls for service are exponentially distributed. Little work seems to have been done to determine how satisfactorily this approximates the real life situation and the practical consequences of significant deviations from the assumption. There are, however, a variety of situations in which the calls for service are not random, yet running time is subject to variation. This assumption of randomness of calls seems especially weak in cases where the number of machines in the calling population is small, say less than ten. Fortunately, it is in such cases, and only so, that Monte Carlo simulation is feasible for this type of problem.

If the mathematical method of solution is to be used for such problems, the user is well advised to search the available literature and investigate some of the assumptions mentioned. The matter of concern is not whether or not a given assumption is violated; it is difficult to conceive of an analytical "model" in any field that perfectly fits reality. The relevant question in such cases is the following: how large a discrepancy between assumed and true state of affairs can exist and still not materially affect the conclusions that will be drawn? This question remains to be answered for a number of assumptions.

Application of the Monte Carlo Technique to Queuing Situations in Which the Source of Customers Is Limited

A Monte Carlo solution will be more expensive to obtain than an analytical solution if the equations are already available. However, a closer approximation of real life can ordinarily be obtained by the simulation method. To illustrate application of this technique to a finite population type of problem the machine assignment problem will again be used. Assume that a group of five machines, each of which requires intermittent attention of an operator at varying intervals, is to be assigned to X operators, hopefully in an optimal fashion.

A substantial sample of machine running times and machine service times has been obtained by stop-watch observation with the results summarized in Figure 118. From the frequency histograms shown, it is concluded that the machines are not homogeneous with respect to running time or service time characteristics. Also, calls for service are not random nor are service times exponentially distributed. The histograms shown were fitted and populations of "chips" were prepared on the basis of the relative frequency of running times and service times for the two types of machines. To start the simulation, a service time is selected for each machine from its respective chip distribution of S and entered on a tabulation sheet as illustrated in Table 17. A running time is also selected for each from the appropriate distribution. These selections are shown in Table 17, and the results illustrated graphically in Figure 119. The assumption is that the operator will attend to the nearest machine awaiting attention. (Figure 119 differs from the traditional man-machine chart in that variable rather than constant running times and service times are used, and the man has no fixed circuit to follow, as illustrated at the bottom of Table 17.) After a sufficient amount of data has been

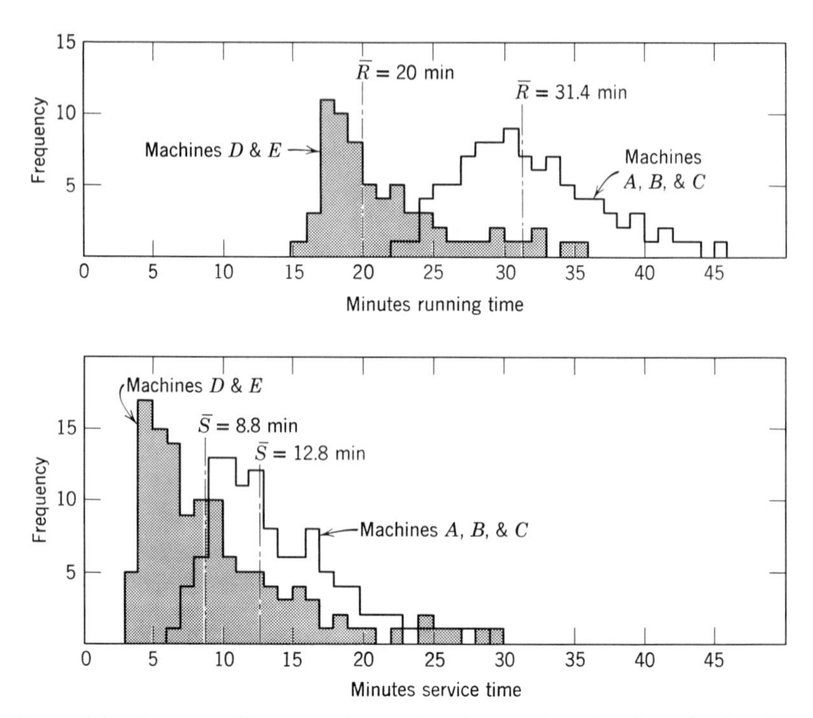

Figure 118. Frequency histograms for running times and service times for two types of machines being considered for a multiple machine arrangement.

generated, conclusions can be drawn concerning total waiting time using one operator on all five machines, two operators each assigned specific machines, or two operators collaborating in tending all five, or possibly other alternatives.

This is a laborious process to be sure, so that its use is ordinarily limited to the more important problems to which analytical methods do not apply satisfactorily. A variety of special conditions and assumptions can be introduced and investigated with this procedure as a result of the basic versatility of the Monte Carlo technique.

A Special Case of the General Machine Assignment Problem

Take the case of a group of machines requiring intermittent operator attention in which the running time is essentially constant but the service times are subject to variation. In this case it is feasible and in fact logical for the operator to follow a fixed circuit around the group

TABLE 17. Tabulation Sheet for Portion of a Monte Carlo Simulation of One Operator Tending Five Machines

Machine	A	B	C	D	E	A	B	C	D	E	A	B	C	D	E	F
Stop R	—	—	—	—	—	8:45	9:15	9:11	9:05	9:17	9:46	10:19	10:04	9:57	10:01	
Start S	8:00	8:12	8:30	8:43	8:47	8:58	9:40	9:14	9:29	9:36	9:52	10:27	10:11	10:01	10:05	
S*	12	18	13	4	11	16	12	15	7	4	9	7	16	4	6	
Start R	8:12	8:30	8:43	8:47	8:58	9:14	9:52	9:29	9:36	9:40	10:01	10:34	10:27	10:05	10:11	
R*	33	45	28	18	19	32	27	35	21	21	38	31	26	20	22	
W	0	12	30	43	47	13	25	3	24	19	6	8	7	4	4	
Order of servicing																

* Times in minutes are selected at random from distributions based upon the results shown in Figure 118.

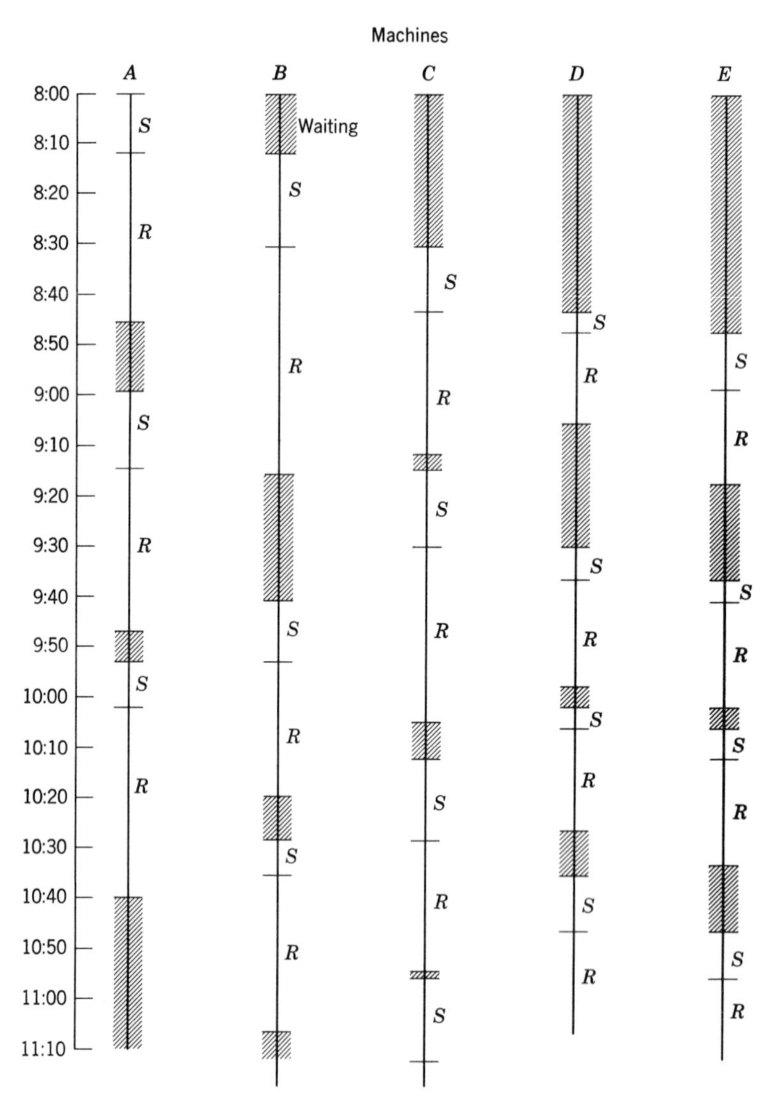

Figure 119. Graphical portrayal of the Monte Carlo simulation recorded in Table 17.

of machines assigned to him, as was suggested under the topic of man-machine charts. In fact, this is the same multiple machine situation described under that topic with one exception: the very realistic characteristic of varying service times has been introduced. Actually, since the human being is performing the service, variation in service times is inevitable, so that the assumption ordinarily made in using man-machine charts is not satisfied in real life. In some instances the variability in S has negligible effect on the cycle time, namely, when there is considerable idle man time in the cycle. However, if the man is busy servicing machines during most or all of the cycle, the variability in service times is likely to force a substantial increase in cycle time. Figure 120 is a man-machine chart for one man

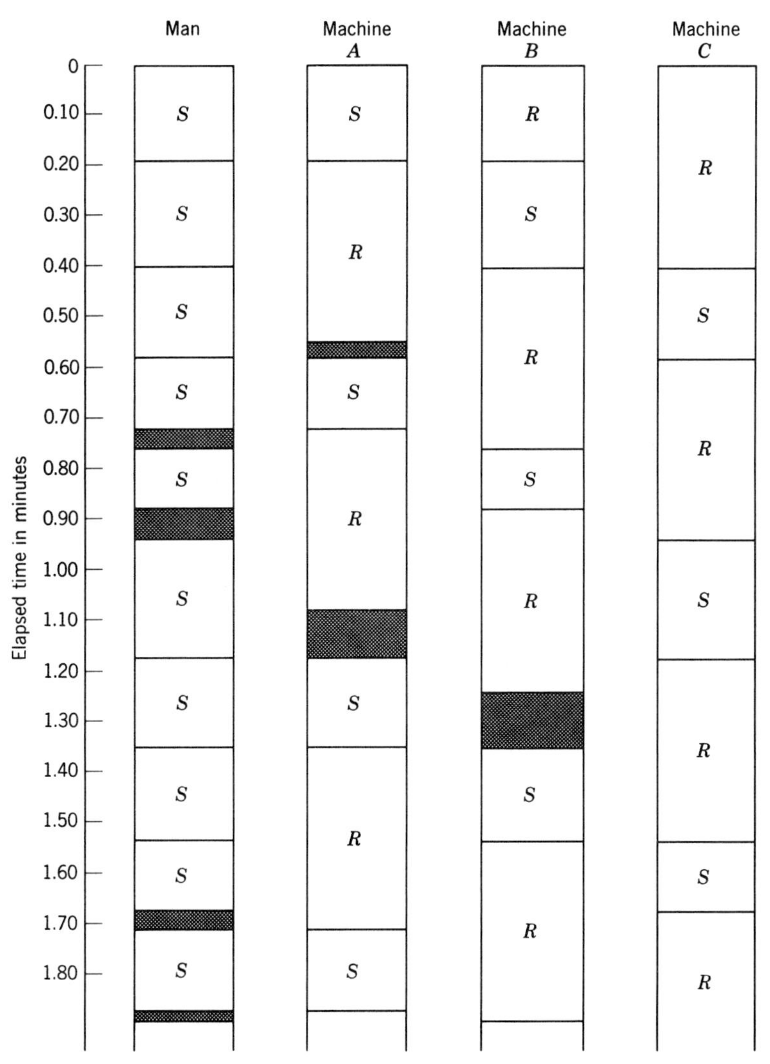

Figure 120. Man-machine chart for a series of cycles under the assumption of constant running time (R) and service times varying as indicated in Figure 121.

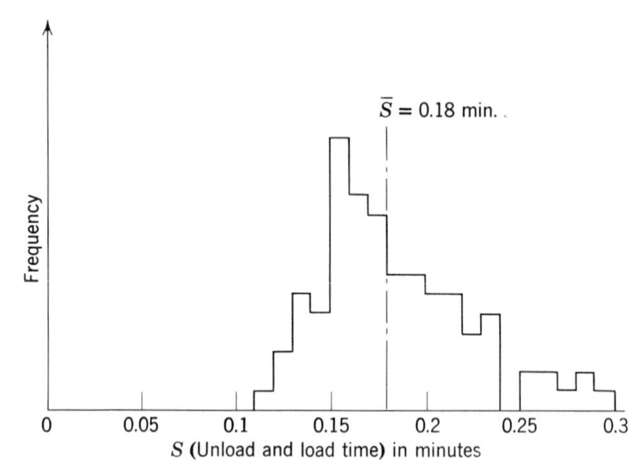

Figure 121. Sample of unload and load times for a production machine suitable for multiple machine operation. Although the mean service time is ordinarily used for construction of a man-machine chart, the service times actually vary about that mean approximately as shown. Times drawn from this distribution were used to construct the chart shown in Figure 120.

tending three machines, machine running time constant at 0.36 minute and service times selected at random from the distribution shown in Figure 121, which seems typical of the pattern of variation expected of unload and load times. A traditional man-machine chart would use a fixed service time of 0.18 minute and would show a cycle time of 0.54 minute with no idle man time or idle machine time. It would show only one cycle since all are assumed identical. However, as a consequence of variability in time to unload and load the machine, the man sometimes completes his cycle before the machines complete theirs, because of a chance series of shorter service times. On other occasions he finishes a cycle "late" and holds up the machines because of a chance series of longer service times. This situation works against him, for when the man finishes "late" with respect to the machines, he loses time and extends the cycle time beyond the expected 0.54 minute. If he should finish before the machines finish their cycle, he can do nothing but wait. For the period plotted in Figure 120 the man has "lost" an average of 0.03 minute per cycle or an increase of almost 6 per cent, solely as the result of variation in service time—mean service time constant.

When it is anticipated that variation will inflate performance time for this or similar reasons, Monte Carlo simulation can ordinarily be used to estimate the magnitude of the increase, as illustrated.

This concludes what is a brief and simplified introduction to the widespread effects of variation in the timing of events on the manufacturing system. The major methods of coping with the problems created by this variation have been described and illustrated.

EXERCISES

1. Times between calls for the services of a roving inspector, and inspection times, have been recorded over a prolonged period at the KATO Company. The results are shown in the frequency histograms on this and the following page. The total hourly cost of providing the inspection service is \$2.65 whereas the penalty associated with waiting for an inspector is estimated to be \$6.50. Inspectors are required to service operators on a first-come-first-served basis. What number of inspectors would you recommend?

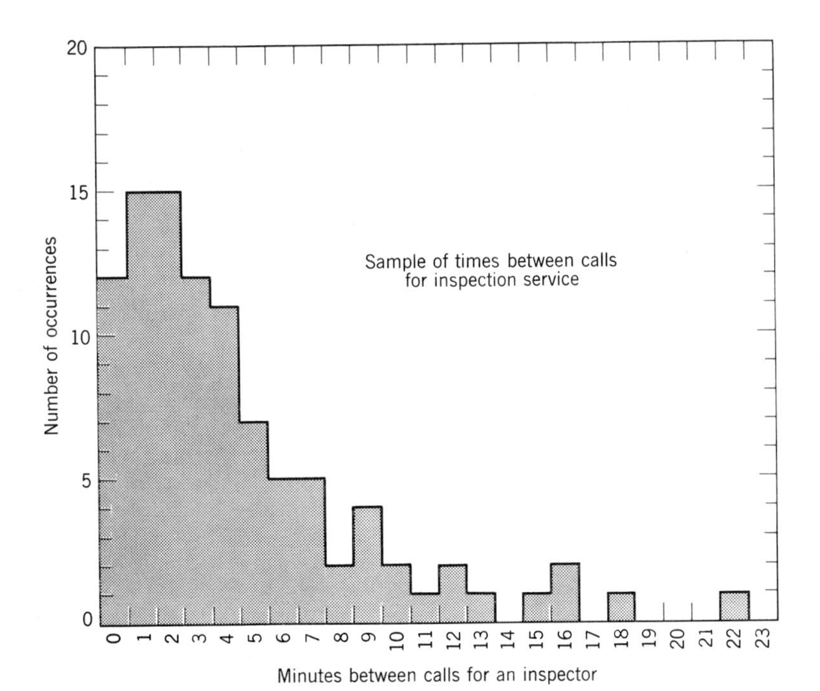

2. At the KATO Company, times between calls for the services of a repairman are approximately exponentially distributed, the average time between calls being 1.31 hours. The repair times are similarly distributed with an average repair time of 3.47 hours. If the total cost of a repairman is \$3.50 per hour and the average hourly penalty for lost production due to machine breakdown for the 130 machines involved is estimated to be \$7.00, what number of repairmen will minimize the total of these costs? Assume that the machines will be repaired on a first-come-first-served basis.

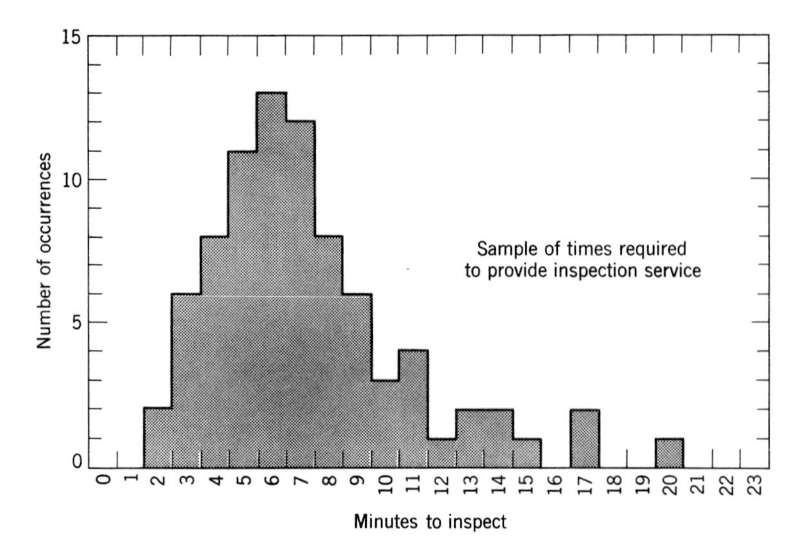

3. At a large printing firm, eighteen automatic type-casting machines are in use. A punched paper tape containing a program of type-setting instructions for the machine must be installed by an operator. An operator is required to periodically check and replenish each machine's lead supply, and to periodically inspect each machine's output. Studies have been made of running time and service time, indicating that both are approximately exponentially distributed with $\bar{R} = 6.4$ minutes and $\bar{S} = 1.9$ minutes. The cost of idle machine capacity is estimated to be \$3.50 per hour. The cost of the machine tender is \$2.50 per hour. Operators attempt to attend to machines on a first-come-first-served basis. How many machines do you suggest be assigned to each operator?

4. One operator is supposed to tend three semiautomatic machines. Service time and running time both vary, as indicated by the following samples of times. Estimate the idle operator time and machine waiting time that will result under the circumstances. Assume that the three machines are arranged side-by-side in a straight line and that the operator will service the nearest machine awaiting his attention.

Sample of Service Times		Sample of Running Times	
Interval (Minutes)	Number of Occurrences	Interval (Minutes)	Number of Occurrences
0–1	0	6–7	1
1–2	1	7–8	3
2–3	4	8–9	11
3–4	17	9–10	10
4–5	16	10–11	8
5–6	11	11–12	5
6–7	7	12–13	4

(Continued on following page)

7–8	4	13–14	4
8–9	3	14–15	3
9–10	3	15–16	3
10–11	2	16–17	2
11–12	2	17–18	2
12–13	2	18–19	2
13–14	2	19–20	1
14–15	1	20–21	0
15–16	1	21–22	1
16–17	1	22–23	1
17–18	1	23–24	1
18–19	1		

5. One man is operating two milling machines on a cyclical servicing pattern. R may be assumed to be constant at 0.24 minute. However, the unload-and-load time varies as indicated by the following sample of time study data. $\overline{S}$ is estimated to be 0.12 minute. How many parts per hour can realistically be expected from these two machines in view of this variation in service time?

Sample of Machine Service (Unload and Load) Times

Service Time (Minutes)	Number of Occurrences
0.05	2
0.06	2
0.07	9
0.08	14
0.09	11
0.10	10
0.11	11
0.12	9
0.13	6
0.14	3
0.15	5
0.16	4
0.17	4
0.18	2
0.19	1
0.20	3
0.21	0
0.22	2
0.23	0
0.24	1
0.25	2
0.26	1
0.27	1

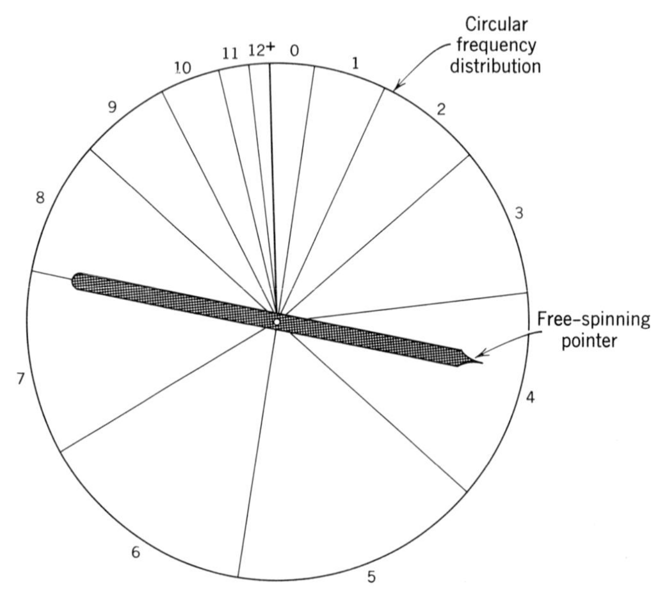

Time interval generator for
Monte Carlo simulation

6. The service times and times between arrivals for a Monte Carlo simulation may be generated by a roulette-type mechanism. This device in its simplest form consists of a free-moving pointer superimposed upon a specially constructed circular frequency histogram, as pictured above. The pointer should be unbiased (i.e., should exhibit an equal likelihood of stopping at any point on the circle). The circular frequency histogram is related to the conventional bar-type frequency histogram in that the arc occupied by each cell of the circular histogram is proportional to the height of the equivalent bar in the conventional form. Thus the arc for cell i of the circular histogram is $f_i \times 360°$, where f_i is the *relative* frequency of times in cell i and thus the relative height of cell i in the conventional type of histogram. With this device the desired time intervals can be generated by spins of the pointer.

Convert the frequency histogram for service times given in problem 4 to the circular form, and if possible, construct a service time generator of the type described.

REFERENCES

Fetter, R. N., "The Assignment of Operators to Service Automatic Machines," *Journal of Industrial Engineering,* vol. 6, no. 5, September-October 1955.

Knowlton, Stuart L., "A Simplified Approach to Waiting Lines," *Journal of Industrial Engineering,* vol. 10, no. 6, November-December 1959.

Malcolm, D. G., "Queuing Theory in Organization Design," *Journal of Industrial Engineering,* vol. 6, no. 6, November-December 1955.

Morris, W. T., "Queuing Theory," *Journal of Industrial Engineering,* vol. 6, no. 3, May-June 1955.

O'Connor, Thomas F., "Cyclic Servicing," *Journal of Industrial Engineering,* vol. 10, no. 4, July-August 1959.

Palm, Docent C., "The Assignment of Workers in Servicing Automatic Machines," *Journal of Industrial Engineering,* vol. 9, no. 1, January-February 1958.

Shelton, John R., "Solution Methods for Waiting Line Problems," *Journal of Industrial Engineering,* vol. 11, no. 4, July-August 1960.

24

Indirect Labor

Indirect labor is generally considered to be work in the production system not directly applied to the product. It includes the activities of the supervisor, the materials handler, the janitor, the toolmaker, the maintenance man, the repairman, the clerk, the secretary, the engineer, the manager, the salesman, and so on. Up until recently methods engineers have shied away from these activities. It has only been within the past decade or so that methods engineers have begun directing a significant portion of their effort to indirect labor, and even now this attention seems limited primarily to the larger, more progressive corporations.

Although it has not manifested itself to any great extent in actual practice, there has been a noticeable increase in interest in the application of methods engineering philosophy and techniques to indirect labor activity. The reasons for this increased interest and talk are several. First, indirect laborers have been putting the pressure on management to include these jobs under the incentive wage plan. Many indirect laborers must work near and service direct laborers that are on incentive, yet the former are on hourly rate and do not have an opportunity to earn incentive premiums. This situation has caused considerable dissatisfaction, higher turnover, and frequently voiced complaints on the part of indirect labor. This pressure directly affects methods engineering, for before putting these jobs on incentive it is important that they be subject to thorough methods design and work measurement procedures.

Second, as a consequence of the trend toward increased mechanization, indirect labor has for centuries been becoming a greater proportion of the work force. This growth trend has been greatly accentuated by the fact that for decades methods engineers have been hammering away at the direct labor operations in the factory whereas indirect labor has received negligible attention in comparison. Some firms report that indirect labor now constitutes more than half of their total work force, and that it continues to swell at an alarming rate. This somewhat dramatic increase in the proportion of indirect labor is finally attracting the attention of methods engineers and their managements.

A third factor generating interest in application of methods engineering to this type of activity has been the tremendous growth of that segment of the business world referred to as service industries, which encompasses wholesaling, warehousing, retailing, transportation, finance, and many others. And not to be ignored is the growth of government. These trends have contributed to a general increase in interest in the indirect labor payroll and the vast cost reduction potential therein. Another factor that is forcing more interest in indirect labor is the fact that many firms have grown so large that the efficiency of the organism has become heavily dependent on if not limited by the communication system, the upkeep of facilities, and the effectiveness of the flow of product between operations. These functions are performed by indirect labor. Therefore, it has become apparent to some that not only for direct cost reduction purposes, but also to improve the overall functioning of the business enterprise, is it highly desirable to direct more methods engineering effort to indirect labor activity.

Why Have Methods Engineers Avoided Indirect Labor Activity?

Inertia explains this condition in part. It seems that it has been customary from the inception of this field to focus attention on direct labor jobs. That is the way it has always been and for many practitioners it is difficult for them to see doing otherwise. In some instances managements have awakened to the fact that their methods engineers should be spending more time on indirect labor, but they find it difficult to sell the "old timers" on the idea.

The explanation lies in part, though, in the fact that in general there are more difficulties and higher costs involved in improving, standardizing, and measuring indirect labor activities than in direct labor operations. The increased difficulty is attributable mainly to the greater variability of conditions, method, and work content found in indirect labor activities. This is true of materials handling, maintenance, repair, janitorial work, tool making, and most indirect labor jobs. In addition, the decision or thought requirements of many indirect labor jobs makes standardization and measurement more difficult. This is true especially of supervisors, engineers, clerks, inspectors, and the like. The difficulty of measuring and standardizing quality is another troublesome factor. Of these factors, the primary deterrent seems to be greater variability inherent to many forms of indirect labor in contrast to the relatively repetitive and consistent work patterns found in direct labor.

Granted, the difficulties are greater, but so are the dollar savings achievable. In most firms, return on the investment attainable by directing a greater proportion of methods engineering's attention to indirect labor jobs is very high in contrast to that offered by the relatively thoroughly worked-over direct labor activity.

Design of the Indirect Labor Task

The remainder of the chapter will be devoted to three major topics. First, *design* of the indirect labor task. Second, *measurement* of indirect labor activities primarily for purposes of establishing time standards, and third, the application of *wage incentives* to this form of activity.

It is difficult to generalize concerning design of indirect labor jobs, one reason being the extreme differences in basic nature of different types of indirect labor. The procedures and techniques that might offer valuable assistance in designing maintenance activity might be totally useless for designing office procedures, and so on. Furthermore, many of the techniques that have been successfully applied to direct labor jobs are of little use to the engineer in attacking indirect labor tasks. There is a dire need for development of new descriptive and analytical techniques for application to this rather different type of activity. One safe generalization, however, is that the same basic design process that applies to a direct labor operation, as expounded in earlier chapters, applies equally well here. And, in fact, we usually find that ferreting out and challenging restrictions are often especially profitable in approaching indirect labor design problems. As a means of summarizing the analytical and descriptive techniques with which the design process may be supplemented in this case, the aids that a methods engineer might use in approaching two common forms of indirect labor will be outlined.

Design of maintenance and repair methods. Some previously introduced techniques that may well be useful in design of maintenance and repair methods are work sampling, techniques for solving queuing problems, the trip frequency chart (for layout of shops), balancing procedures (for crew activity), and time lapse photography. An information recording device that may prove of assistance is the so called activity journal, which is a log the worker keeps of his activities for a period of time. Application of the design process supplemented by such techniques as these and sufficient ingenuity should lead to significant improvements in maintenance and repair activity. Ordinarily, improvements are possible in work procedure, layout, and equipment, in procedures for planning and scheduling maintenance activities, in maintenance and repair "kits" and mobile units, in trouble-shooting strategy, and so on. It is obvious that the engineer must rely heavily on his creativeness in attacking indirect labor operations, for many of the improvements as well as the approaches are unique to the particular type of indirect labor involved.

Design of office procedures. Some previously introduced techniques that may well be useful in design of office (clerical) procedures are work sampling, the trip frequency chart (for office layout), the flow diagram and flow chart, and the activity journal. Two techniques to be illustrated are the "information requirements and source chart" and the "forms distribution diagram."

Most offices exist as the result of the need for a system to communicate and process information. Ordinarily, information comprises the input and the output of the system, and most clerical operations involve obtaining, handling, processing, storing, retrieving, or disbursing information. Thus the office is very much like the manufacturing phase of the business, with information as the product.

The design of an information processing and handling system should proceed in the following manner. First, the designer should determine what information is *required*. This step is crucial. In the typical manufacturing firm administrative and supervisory personnel are inclined to demand a great deal more information than they really need to successfully carry out their function. In many instances the need is imagined. In many other instances people are demanding information or files of same that they *might* need someday, never stopping to weigh the consequences of not having the information if it ever should be needed against the cost of collecting and storing that information perhaps for decades. Some of the major improvements in an information system, then, can be made during this stage, by eliminating the fictitious information needs. Once the real information needs have been identified, the designer should select the best source for each bit of information required. Then he should determine what must be done to convert the "raw information" to the final form desired; for example the processing that must take place, the storage that is necessary, and the like. When he has completed this, the designer has established the restrictions of the problem—the things that must take place. He can now proceed to specify the mechanics of how this is to be achieved in terms of the communication media to be used, the documents to be used and their specific form, how information is to be recorded, processed, filed, and retrieved, how computations are to be made, and so on. Thus, the general approach is to first establish the information flow pattern in skeletal form, then to build the communication and paperwork system around this.

A suggested aid to this process is the information requirements and source chart illustrated in Figure 122. The chart is the product of a thorough analysis of the information required of and by the system, and of the preferred sources of this information. On the basis of this the designer can decide on the paper flow and other communication media to be used, the design of documents, the method of performing the necessary operations, the files required, and the like.

A diagrammatic technique that is often useful in providing an overall, compact view of a paper flow system is called a forms distribution diagram. As illustrated in Figure 123, this diagram shows all participants in the system, all documents, the information added to and obtained from each document, the origin and distribution of each, and the final destination of each. This diagram is not intended to be a substitute for an information requirements and source chart; the former should should follow and be based on the latter.

The methods engineer can do a rather comprehensive job in design of an office system, starting with design of the basic system as described previously and continuing to detailed design of the documents. This includes specification of workplace layout, procedure, and equipment, and overall layout of the office. The application of methods engineering to the office has become somewhat of a specialty of its own.

Establishing Time Standards for Indirect Labor Tasks

Time standards for planning manpower needs, scheduling, balancing workloads, evaluation of performance, and incentive wage purposes, may be established for indirect labor jobs by any of the work measurement techniques previously discussed, although some are more suitable for this purpose than others. In general, work sampling

Information Requirements and Source Chart
for Order Handling and Billing Procedure of a Heating Equipment Supply Firm

	Information Required by						
Best basic information source ↓	Customer	Salesman	Credit dept.	Accts. rec. dept.	Pricing dept.	Inventory dept.	Warehouse
Customer		Customer ident. Items Quantity Delivery point					
Salesman	Availability of items ordered		Customer ident.	Customer ident. Items Quantity	Items	Items Quantity	Customer ident. Items Quantity Delivery point
Credit dept.				Discount Credit approval			
Accts. rec. dept.	Amount due						
Pricing dept.				Unit price			
Inventory dept.		Availability of items ordered					
Warehouse							

Figure 122. An effective means of summarizing the requirements and preferred sources of information in the course of designing an information flow system.

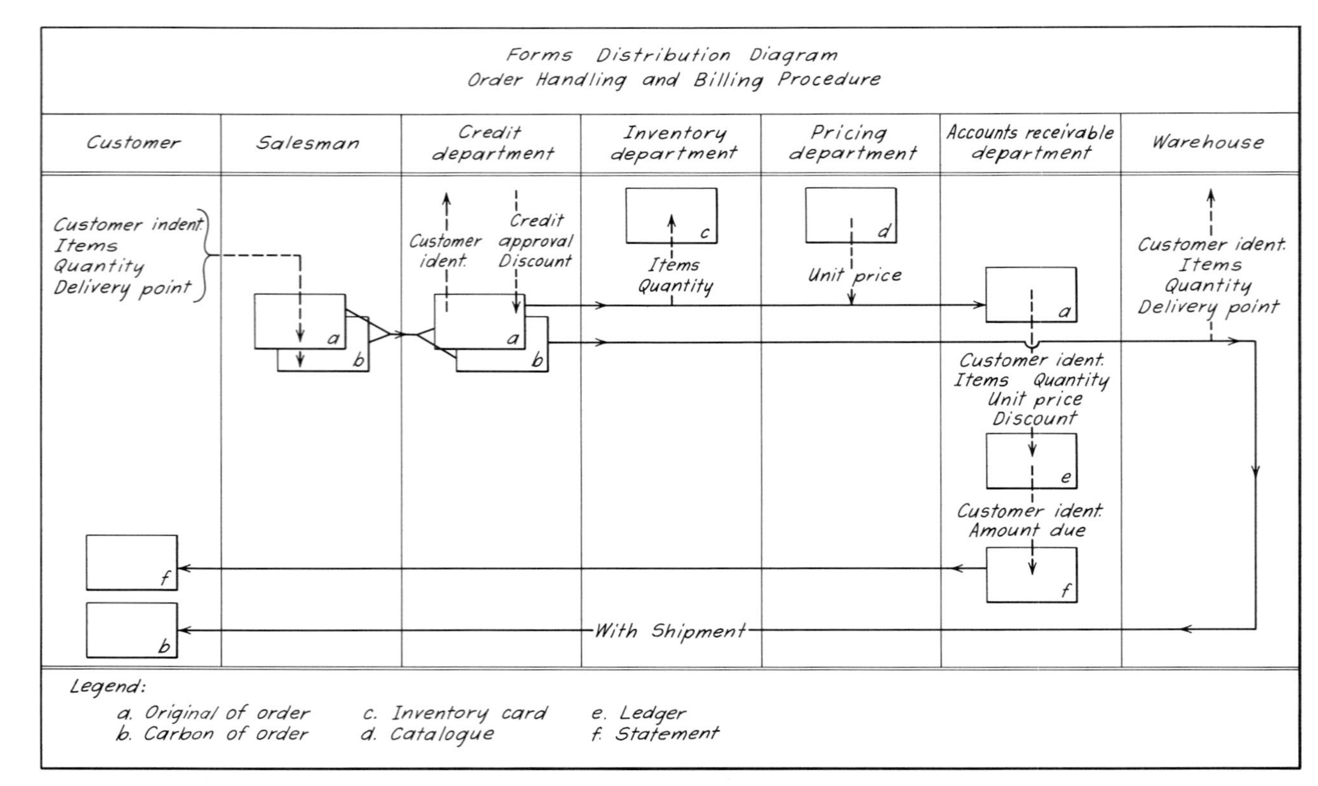

Figure 123. A forms distribution diagram for the same information flow system charted in Figure 122, showing origin and destination of each document in the system plus all information added to or removed from documents.

and macroscopic standard data are more suitable for setting standards on indirect labor than stop-watch time study or predetermined motion times. Because of the unusually high degree of variation in method, work content, and quality, stop-watch time study in many cases is too time consuming to be practical. The same is true perhaps to an even greater degree of the use of predetermined motion times. Work sampling is more suitable for indirect labor measurement because it provides a sample of many workers, jobs, and conditions, and yields a grosser standard or set of standards, which is what should be sought for highly variable indirect labor tasks.

Macroscopic standard data is often suitable for establishing indirect labor standards primarily because after a moderate amount of data has been collected it becomes possible to henceforth synthesize standards as work content varies. Take the case of the standard data developed for office cleaning described in Chapter 19. To set a time standard for cleaning each individual office by stop-watch time study would hardly be economical. However, with standard data, once standard times have been developed for the basic elements of the cleaning task, standards can be synthesized quickly for a wide variety of offices of differing contents. Or take the case of maintenance work. It is possible to develop standards for basic elements of maintenance activity, such as lubrication, replacement, inspection, etc., of various types of machine elements, then to synthesize standards for various jobs from this standard data. In setting standards for indirect labor it may prove worthwhile to develop standard data on the basis of the results of a work sampling study.

Since different types of indirect labor have so little in common, the problem of establishing time standards in each case requires a considerable amount of ingenuity, versatility, and alertness if the job is to be done economically and satisfactorily.

Application of the Incentive Wage Method to Indirect Labor Tasks

One method of putting indirect laborers "on incentive" is to pay then an incentive premium determined by the average incentive premium earned by direct laborers over the same pay period. The objection to this ratio method is that the indirect laborer's earnings are only very remotely dependent on the level of effort he exerts, so that the incentive offered by the plan is extremely weak.

Another major alternative is to put the indirect laborer on direct incentive by establishing a time standard for his task, measuring his actual output, and computing his incentive earnings accordingly. This is the procedure ordinarily applied to direct labor operations. However, it has already been implied that the cost of setting standards for indirect labor tasks is greater. Furthermore, the quality of the standards is not what can ordinarily be expected on the more straightforward, repetitive direct labor operations. Worse than this, however, is the problem of measuring performance of the indirect laborer, both with respect to quantity and to quality. In many instances the worker can adjust the quality of his work within rather wide limits, in order to earn the incentive earnings he wishes. This phenomenon is quite apparent in janitorial work, for example. An additional problem is posed by the fact that the amount of work for some types of indirect labor is limited. Suppose a janitor or maintenance man works at incentive pace and completes say 9 standard hours worth of work in 7 hours. He may well then sit around with nothing to do for the last hour because it is not possible or feasible to find another assignment for him for that period. The circumstances in this respect are different from those ordinarily true of direct labor production operations. Furthermore, a major reason for using the incentive wage plan is to obtain more efficient utilization of equipment. However, a number of different types of indirect labor involve no expensive equipment. What is gained by getting more efficient utilization of a broom and mop?

It is apparent that there is some question in the mind of this author about the economic feasibility of applying a direct wage incentive plan to some, if not a majority of types of indirect labor. Some companies have installed such plans on various types of indirect labor and claim rather dramatic reductions in man hours required for the jobs involved. There is no doubt that a significant reduction in indirect labor cost has been achieved in most of these instances. The doubts arise when the costs of setting up the incentive wage plan, maintaining the controls, keeping the necessary records, and computing earnings, are compared with the gross savings. In some instances the net savings have justified the investment involved, in others the economy of the undertaking appears doubtful.

Summary

There is no doubt about the high savings attainable by the methods engineer through improvement of indirect labor work methods. There is a good deal of "fat" to be trimmed off in this area in most businesses. The economy of setting time standards for such activity seems fairly certain in a majority of cases. However, it does not appear economical to attempt to obtain time standards for indirect labor activities of the same refinement sought for direct labor operations. The profitability of applying the incentive method of wage payment to this type of work seems open to question. When considering the desirability of establishing a wage incentive plan for a given type of indirect labor, it behooves the practitioner to make a thorough investigation of the expected cost of establishing the standards necessary and of administering the plan once it is in operation.

EXERCISE

1. You have been assigned the task of improving the paperwork system employed by a large university in its purchasing and requisitioning activities. In general terms, how would you proceed?

REFERENCES

"A Fair Day's Work in Maintenance," *Factory,* vol. 113, no. 1, January 1955.

"How a Pro Looks at Maintenance Cost Reduction," *Factory,* vol. 117, no. 6, June 1959.

"Measuring Maintenance Output," *Factory,* vol. 117, no. 2, February 1959.

"Slicing Maintenance Labor Costs," *Factory,* vol. 115, no. 5, May 1957.

"What Planned Maintenance Means to Me," *Factory,* vol. 115, no. 9, September 1957.

"Work Sampling in Maintenance," *Factory,* vol. 117, no. 12, December 1959.

"Work Simplification in Maintenance," *Factory,* vol. 115, no. 2, February 1957.

"16 Soft Spots to Hit in Maintenance Costs," *Factory,* vol. 116, no. 8, August 1958.

part VI

ADMINISTRATION OF THE METHODS ENGINEERING FUNCTION

25

Methods Engineering Administration: Determination of Function

The major problem areas confronting the administrator of an engineering department, whatever the field of engineering, are the following.

1. *Determination of function.* What function should this engineering specialty fulfill in the enterprise? What are its purposes?
2. *Procurement and maintenance of staff.* This responsibility includes employment and placement of personnel, initial training, long-run betterment of staff, motivation, and determination of salaries and promotions.
3. *Determination of methodology.* Methodology encompasses the procedures, techniques, practices, policies, organization, principles, and all other matters pertaining to the manner in which the department goes about fulfilling its function. This is the department's "modus operandi."
4. *Programming activities.* This responsibility involves allocation of the department's resources to potential projects as well as scheduling of department activities and assignment of personnel.
5. *Maintenance of favorable relations with other personnel in the organizations.* This is the administrator's "salesmanship" or "public relations" responsibility, the objective being to provide a favorable organizational atmosphere for conduct of the department's work.
6. *Evaluation of the function the department is supposed to fulfill in the enterprise and of the effectiveness with which it is fulfilling that function.* This is primarily for purposes of justifying the existence of the department and its current scale of operation, and of improving its performance.

Relatively little material pertaining to management of engineers has been published. However, there has been a noticeable increase in interest in the subject. The alarm that government and industry have displayed with respect to the growing magnitude of engineering cost and to the apparent undersupply of engineers is finally leading them to the rightful conclusion that the engineering function must be effectively managed in the same manner as any other phase of a business. It is gradually becoming clear that a partial remedy to the growing engineering cost and the talent shortage lies in better engineering management, directed especially to improved utilization of existing engineering staffs. Much of the material in Part VI is directed to this end.

Discussions of these major administrative responsibilities as they apply to methods engineering follow in this and succeeding chapters. In each chapter a general discussion of the particular administrative function is followed by a detailed treatment of that function as it relates to methods engineering. In this discussion it is assumed that methods design and work measurement have a common administrative head. In practice methods design and work measurement are sometimes performed by one department and under one administration, under a title such as Methods Engineering Department. Frequently though, they are separate departments and thus administratively separate, under such headings as Methods Department and Time-Study Department. Some of their administrative problems are in common, but a number of others are not because of the emphasis on design in one case and on measurement in the other. This makes for some confusion in a discussion of the administrative problems involved, and in practice makes it difficult to decide whether the two should be under one roof or separate ones. In the discussion that follows it is assumed that methods design and work measurement are performed by a single organizational unit under one administration.

METHODS ENGINEERING ADMINISTRATION

Determination of Department Function

Certainly the engineering administrator and his staff should have a clear understanding of the purposes they are to serve in the organization. But knowing what these purposes are and having a voice in determining them are different matters. That a department be the sole determiner of its function is a rather optimistic and impractical expectation, but that it be a major determiner is not. This prerogative is something that many administrators have still to strive for. Having the prerogative to determine or strongly influence its own objectives and limitations is a real expression of management's confidence in the department involved. In any event, continuing attention to this matter of function, regardless of the amount of say he has in the matter, is an important responsibility of the engineering administrator, especially in light of the ever changing nature of most enterprises.

Determination of the Function of Methods Engineering

Although the specific function differs from organization to organization, in general, methods engineering is responsible for design of the productive process insofar as the human being is involved. For the typical situation this might well include the following.

1. Determination of *what* role the human being will serve. If this is organizationally unacceptable or impossible, methods engineering should at least be the prime advisor in decisions of this nature.
2. Determination of *how* the human being is to carry out his function, both as a direct *and* as an indirect component in the system.
3. Serving as a motivator and catalyst in the improvement of methods of work by others in the organization.
4. Establishment of labor performance standards for planning and evaluation purposes.
5. Advising others in the organization on matters pertaining to this specialty, including management on matters of company policy, product engineering on design of the product for effective human use, and the labor relations department during collective bargaining deliberations.

Whatever the specific responsibilities and limitations might be, there should be a definite understanding with management as to what these are. So often this is left a vague matter subject to many different interpretations throughout the organization. These differences of opinion as to a department's function can cause much friction between methods engineering and other engineering departments and especially between methods engineering and the operating departments serviced by it.

Wage incentives. In general, the methods engineering department should not be responsible for installation and administration of an incentive wage plan. Combining the function of determining performance time standards and the function of pay administration under the same organizational roof could lead to havoc. It is ordinarily difficult enough as it is for the work force not to look on the time-study man as strictly a wage administrator, without fostering such impressions by mixing the two under the same administration. The closer labor associates time study with take-home-pay, the more tendency there is for bargaining to enter into the process of establishing time standards. Therefore every effort should be made to dispel the impression that the time study man is a wage determiner.

Even if, as suggested, the methods engineering department is not responsible for administration of the incentive wage plan, if the company employs such a plan the department cannot escape being strongly affected by it and vice versa. This being the case, the methods engineering administrator should maintain a keen interest in the manner in which the incentive wage plan is set up and administered, and in fact in its very worth to the company. Surely the administrator should be interested in and convinced of the economic justification for any scheme causing as much difficulty for his department's operations as an incentive wage plan does.

EXERCISE

1. Incentive wage plans are adopted primarily to reduce overhead unit cost by increase in worker productivity. This increase in productivity, however, is not obtained free-of-charge. There are a number of direct and indirect costs associated with the setup, maintenance, and especially the operation of this type of wage payment scheme. True, a number of these costs are virtually unquantifiable, but they remain as important factors to consider in deciding on the economic feasibility of this form of wage payment. Identify these costs.

REFERENCES

"Chief Industrial Engineers' Forum," *Journal of Industrial Engineering,* vol. 9, no. 5, September-October 1958.

Mellin, Warren R., "What Is An Organized Methods Improvement Program," *Journal of Industrial Engineering,* vol. 10, no. 6, November-December 1959.

26

Methods Engineering Administration: Procurement and Maintenance of Staff

The engineering administrator is responsible for procurement, placement, and initial training of his personnel. In fulfilling these responsibilities the administration should maintain a well-balanced program with respect to the development of these three characteristics so vital to success in any field of engineering.

1. Technical ability, which is a universally recognized prerequisite of superior performance.
2. Creativeness, the major importance of which is not so widely recognized.
3. The ability to effectively work with people at all levels in the organization. This characteristic is as crucial as technical ability and creativeness. An engineer's inability to effectively solicit the cooperation, respect, and confidence of others in the company, to gain acceptance and effective use of his ideas, to competently deal with the many "human relations problems" that are an inevitable part of his job, is likely to be his undoing. Do not underestimate the importance of this ability. This is partly a personality matter and partly a matter of developing a knack. The administrator can control the former in selection of personnel, the latter through proper training.

After initial indoctrination and training of the engineer, there should follow a continuing effort directed to long-run improvement. Without this an engineering department is destined for nothing better than mediocrity. Continuing betterment of staff ordinarily requires both encouragement and material assistance from the administration. Some of the features of a successful long-run betterment program are spelled out below.

1. Some "push" from the administration.

2. Some material assistance in the form of time off and financial aid for furthering education, attending conferences, and the like.

3. Knowledge of results with respect to quality of performance, strong points and weaknesses, mistakes, and so forth, upon which a man can base his improvement. It is an important responsibility of the administrator to keep each of his subordinates informed in these respects.

4. Training directed especially toward anticipated changes in methodology, with emphasis on providing the groundwork for understanding and acceptance of new techniques. The objective is to increase technical competence *and* to give the engineer confidence in his ability to master new developments. The latter is vital to gaining his eventual acceptance and use of new techniques and is one aspect of what might be called long-run conditioning for change.

With respect to motivation of engineers, another responsibility of the administration, it seems generally acknowledged that much could be gained through better utilization of time and higher quality performance if engineers were better motivated to these ends. But the question is how. What can the administration do other than "keep the boys happy" through what is generally considered to be good personnel policy and practice? The question is not an easy one to answer. One promising suggestion concerns "enlargement" of engineering jobs to make them more attractive. Ultra specialization that is true of many engineers' jobs in all likelihood is a strong stifler of interest and enthusiasm.

Some Special Personnel Problems Confronting the Methods Engineering Administrator

There is no engineering specialty that requires a man to work more closely with other personnel in the organization and over a wider span of authority than methods engineering. The ability to effectively work with people is especially important in this field not only because it is necessary to work so closely so much of the time with people in the shop, but because this specialty concerns a controversial matter. Furthermore, because of the antagonistic feeling customarily held by shop people for methods engineers in general, the man in this field has an uphill battle in this respect from the start. He must be especially skilled in dealing with human relations problems if he is to be successful in this job. Therefore, in selection and development of his personnel, the administrator of a methods engineering department should give special emphasis to this quality.

Although the number of technical developments in this field since its inception has not been impressive, new techniques have been introduced. Yet a surprising number of practitioners are inadequately informed and prepared even with respect to the relatively few innovations that have appeared. This state of affairs plus the fact that the rate of innovation in methodology of this field is increasing, means that many administrators have a major task before them in assisting and encouraging their staffs to overtake and then keep abreast of developments.

Remember that a majority of practitioners in this field have many years of experience in the application of a relatively stabilized methodology. Thus, being rather unaccustomed to change in technique, these men do not accept new developments overnight. Therefore many administrators have the task of technical enlightenment of their staffs *and* of conditioning their personnel for acceptance of change. A new development in methodology usually represents a real threat to job security in the eyes of the ordinary practitioner—a perfectly natural reaction. The practitioner is inclined to fear that the new technique cannot be mastered, that his value to the company will be lost, that the skill he has developed over years of experience will no longer be needed, that he has neither the time nor the ability to achieve the same level of proficiency with the proposed new procedure, and so on. These are very real anxieties to be expected whenever a significant change in methodology is in prospect. They give rise to a negative response that manifests itself in many forms, such as discrediting a new technique, avoiding it, rejecting it outright, and even attempting to undermine its effectiveness and value. If this fear and resistance occurs when predetermined motion times are introduced or a department begins to focus its attention on indirect labor activities, what sort of reaction can be expected when more rigorous techniques are introduced?

To minimize this negative reaction, it is necessary that the administration engage in a long-range program to strengthen ability and confidence in that ability to master new developments. This can only be accomplished by a continuing educational program in anticipation of the introduction of new techniques. To illustrate, take the case of the expected introduction of statistical techniques into time-study practice. In anticipation of this, a course in basic statistics might be given in order to provide general background and appreciation of the subject. This long-term preparational program should prevent the new statistical techniques from "suddenly" looming before the man when they are eventually introduced. This program also strengthens his ability and confidence in that ability to master these techniques. It also strengthens his confidence in the techniques themselves.

In the case of stop-watch time study, a question arises as to whether a college-trained man is necessary or even desirable. In this author's opinion, use of a college-trained man for routine time studies, for collecting observations for work sampling studies, and for similar tasks, is a waste of talent and money. Supervising, deriving standard data, setting up work sampling studies, and similar activities of a more demanding nature may well warrant the use of an engineer. There are some strong arguments for using a man with some "shop experience," with or without technical school background, for the more routine tasks. Such individuals usually have a superior knowledge of shop methods, are commonly believed to make better raters, and seem to be able to maintain more favorable relations with shop personnel, as compared to college-trained men.

27

Methods Engineering Administration: Determination of Methodology

It is the administrator's responsibility to select the procedures and techniques with which his department is to be equipped, and to decide where in general these are to be applied. Although he should make the decisions concerning methodology in general and at least in the case of major problems, many methodology decisions made in the day-to-day operation of the department must ordinarily be made by the department's operating personnel as the occasions arise. To illustrate, the administrator decides whether or not his department should be equipped with a predetermined motion-time system and if so which specific system to adopt. In addition he outlines in general terms the circumstances under which the technique should be used. However, in dealing with most specific day-to-day problems the designer himself will probably be the one to decide whether or not to employ the technique. Only when the problem is particularly difficult or important will the administrator make this decision. In order to make such decisions intelligently, the administrator should be well informed about current "enlightened" theory and practice and constantly aware of industrial and professional trends and developments.

Selection of Procedure and Technique in Methods Design

In the selection of methodology the objective should ordinarily be to employ those techniques and procedures that offer maximum return through results achieved, relative to the time and other resources expended. Such questions as these arise. Should a flow diagram be used? Motion pictures? A prototype setup? Brainstorming? And so on. If a prototype setup for the proposed solution to a methods problem is decided on, it is done so because it is anticipated that the cost will probably be economically justified by a superior method in the end, as well as by the opportunity for a thorough debugging before introduction of the method into the shop and by facilitation of introduction and training.

Because of the uncertainties involved, the difficulty of quantifying prospective benefits, and the limited amount of time that may be devoted to making these decisions, choice of methodology is primarily a judgment process. To illustrate, suppose that in the course of designing an important and expensive operation in the plant the question arises as to whether or not it would be worthwhile to have other members of the department join with the designer involved and have a brainstorming session to generate alternatives. It is impossible to satisfactorily evaluate in advance the benefits brainstorming will yield relative to what might result if it is not employed. The role of conjecture in making this decision is obvious. As a result of the degree of judgment required in selection of methodology, there is considerable difference of opinion among practitioners, great diversity in practice within and especially among companies.

Selection of Methodology in Work Measurement

Recall that there is a variety of methods of establishing time standards. Seldom is the administrator's problem that of deciding which of these alternatives his department should be equipped with, to the exclusion of the others, for the typical methods engineering department should have most if not all of the available techniques at its disposal. Instead, the decision confronting the administrator should be that of selecting the most suitable of the alternatives for each different set of circumstances in the plant. Thus he should be concerned with such questions as: "which standards-setting technique should be used to establish the time standards in our assembly department, on punchpress operations, inspection operations, packaging operations?", and so on, rather than, "which technique should we set time standards with at our company?"

There are three major criteria for selection of the most suitable technique to apply to a given situation.

Applicability. Some of the foregoing techniques do not apply to certain types of situations. For example, predetermined motion times do not apply to mechanically paced activity.

Acceptability to labor. In some situations the labor union will refuse to tolerate a certain standards-setting technique, and other times it will offer vigorous opposition. The latter, although not directly prohibitive, may well sufficiently inflate the cost of using a certain technique through friction, restrictive clauses in the company-union contract, etc., so as to render it uneconomical to use.

Total cost. The total of two relevant costs, the cost associated with setting time standards via a proposed procedure and the cost incurred as a result of the errors associated with that procedure, constitutes a major criterion in the selection of the technique to use in a given situation. This total cost criterion will be discussed at length.

Selection of the Preferred Measurement System on the Basis of Total Cost

With any practical measurement system certain costs are incurred as the result of errors in the measurements produced. For example, if a company makes errors in estimating time standards for its operations, it will consequently make errors in scheduling. This causes increased backlogging of production lots, increased time lost in tracing and trying to expedite orders, increased machine idleness, dissatisfied customers due to late shipments, and other costly situations. The same errors in time standards will cause difficulty in operation of the company's wage incentive plan. The measurement errors result in inconsistencies among standards on different operations, and these in turn contribute to considerable dissatisfaction among the employees, numerous grievances, excessive transferring between jobs and departments, restriction of output, and other costly consequences. The same errors cause erroneous decisions concerning whether or not to produce proposed products, whether or not to make or buy various components, concerning future needs for facilities and manpower, and so on. Therefore, errors in the company's time standards result in costly consequences when those standards are applied in control, wage payment, and decision making.

These costs ordinarily associated with measurement errors increase as the degree of error increases. In general, these costs appear to vary with the degree of measurement error as indicated by curve *A*, Figure 124. As a rule, a relatively small error is of little consequence, but as the error increases the resulting costs rapidly and sometimes rather abruptly accelerate and compound. The fact that some error, up to a point, is of relatively small consequence means that it seldom pays to attempt to eliminate that error, to strive for perfection.

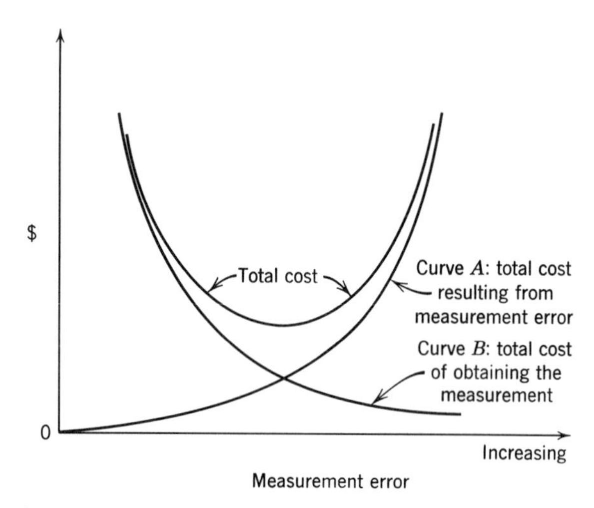

Figure 124. Expected magnitude of the costs relevant to selection of a measurement system for a given situation, for different degrees of measurement error.

Costs of obtaining measurements. It is unprofitable to reduce the error below a certain degree for another reason. Curve *B*, Figure 124, indicates the general relationship between cost of making measurements and the degree of error in these measurements. This indicates that to achieve incremental reductions in error necessitates increasingly larger measurement costs in the form of a larger original investment and higher costs of operating the system. For example, the cost of making a time study may be varied considerably, by varying the quality of the observer, the time spent observing, the number of observers, the elaborateness of the timing device, etc. But as more and more money is spent in the above ways to reduce the error, proportionately less error reduction is achieved, until additional expenditures for this purpose accomplish virtually nothing.

The minimum cost degree of error. The result of the two preceding situations, positively accelerating cost consequences as the error increases and positively accelerating costs of obtaining measurements as the error is decreased, is that there is a certain degree of measurement error for which *total* cost of obtaining *and* using the measurements is a minimum, as pictured in Figure 124. Thus, for a given measurement problem there is a system and expenditure in same that minimize total cost and that are optimum for the situation involved. This principle is applied at least in a gross sense in everyday life, for we do not ordinarily use a wrist watch for timing in ballistics tests, nor do we usually have a high speed electronic timer on our mantelpiece. In designing or selecting a measurement system we should attempt to attain a proper balance between error consequences and the cost of establishing and operating the system.

Application of this principle to selection of a work-measurement system. There are a number of methods available for estimating time standards and each of these methods can be operated at a rather wide range of costs, as was previously illustrated for the case of stop-watch time study. Which of these methods should be selected? How much should be invested in the work-measurement system in a given situation? The major criterion for selection of the appropriate work-measurement system is ordinarily total cost. This requires consideration of the *installation cost* (initial investment consisting of the costs associated with training, apprenticeships, equipment, professional consultants, and the like), the *operating cost* (the variable cost, which is primarily a function of the man-hours required to set a standard and the salary of the type of person required) of potential systems, *and the cost consequences of various degrees of error in the particular situation involved.* For example, the cost of obtaining time standards by judgment is certainly low, but the errors are high. Yet in some situations these errors may not be costly, because for example the production volumes are low, or the more demanding uses of time standards like wage incentives are not present. In this type of situation in which a more costly estimating technique is not justified, the judgment process may well be the minimum cost method. But in other situations the cost consequences of errors are great, because of high production volumes, because the standards are used in an incentive wage plan, because of an adverse labor-management relationship, and for other reasons. This type of situation justifies use of a relatively elaborate and expensive estimating technique, and use of anything less than this will increase the overall cost.

Quantification of the cost of setting a standard by any of the alternative methods is relatively straightforward, but not so with estimation of the errors involved or of the costs associated with these errors. Estimates of errors and their cost consequences must at present be made almost exclusively on the basis of judgment. As a consequence there are radical differences in opinions and practices in industry with respect to these matters. Although the difficulty of quantification exists, this concept of minimum total cost is a valuable one. The concept and objective of an economical balance between cost of measurement and cost of errors is an extremely important one in practical selection of a measurement technique.

There is no perfect system for establishing time standards; every known method has a sizeable inherent error. This point bears emphasis and repetition because some writers are prone to unrealistically advocate rejection of a particular technique (predetermined motion times for example) merely because it has *an* error associated with it. In this instance the given technique is being compared with an abstract and unattainable level of perfection. Clearly the problem is one of comparison and selection on the basis of total long-run cost, from a variety of imperfect estimating methods.

A General Comparison of Alternative Time-Study Techniques

There is no justification for offering a general ranking of the alternative procedures with respect to overall cost, quality, or preferability, for each is superior to the others under certain circumstances. Similarly, there is no basis for ordering them with respect to measurement cost, for this varies with the specific situation for which the standards are to be set, and too, each technique can actually be operated over a wide range of measurement costs. Similarly, they cannot be generally ranked with respect to error for each technique can be operated over a wide error range. Furthermore, the magnitude of the error of any technique depends on the particular situation involved as do the cost consequences of that error. And last but perhaps not least, the errors associated with current procedures have yet to be satisfactorily quantified. Thus, it is apparent that the specific nature of the situation for which a time study technique is being chosen is a major determinant of measurement cost, error, and cost of error, that the relative suitability of the techniques varies with the situation, and that it is impossible to generalize concerning preferability of the alternative techniques. Therefore, of necessity, the comparisons that follow are fractionated and encumbered with qualifications.

Extraction from past experience. Very low in measurement cost, high in error, this procedure is competitive when little accuracy and precision are required. As a method of establishing a *standard* this procedure is of questionable value under almost any circumstances. However, as a method of forecasting actual production rate rather than standard, extraction from past experience is competitive and worthy of more serious consideration.

Stop-watch time study. This, the general purpose tool among alternative time study techniques, has been the most extensively used method of establishing time standards since the early part of the century. It is the most flexible of the methods but also the most despised, troublesome, and ill-famed of those available. The obvious drawback is the highly subjective rating process, where the error of the technique is concentrated. The rating feature has proven objectionable and troublesome enough to be at least partly responsible for a continuing shift from stop-watch time study to standard data methods.

Stop-watch time study can be operated over a wide span of costs and errors. Such a study can be 5 minutes in duration, 5 days, 5 weeks, or longer. It can be operated with different numbers of observers, different levels of observer skill, with special equipment aids such as motion pictures or electronic timers. By such adjustments the measurement cost and error can be varied over a very wide range to suit a variety of demands.

Work sampling. Work sampling is ordinarily competitive with stop-watch time study and standard data methods as a means of setting standards only on very long cycle or nonrepetitive operations, as typified by indirect labor activities. It is especially worth considering if an aggregative or grosser standard is acceptable. However, up to the present time work sampling has not been widely used as a method of setting standards. Rather, it has been used and shows its greatest potential for studies of the distribution of man or machine time over different categories of activity, especially for the estimation of delay allowances.

Macroscopic standard data. The main virtue of this technique is its ability to establish a standard in what ordinarily amounts to only a matter of minutes. Whether or not it will be economically feasible to develop such data for a given situation depends primarily on the derivation cost and the volume of standards that can be set with it. Macroscopic standard data is likely to be economically worthwhile for the so-called "basic processes," such as press, drill, mill, and lathe operations, where there are a number of machines performing similar operations on a fairly large number of parts and where the likelihood of any major change in method of operation is small. One case in which this technique is difficult to compete with is the job shop situation, where standards must be set quickly, often, and at very low cost.

Predetermined motion times. In spite of the fact that this technique exhibits no general superiority with respect to initial cost, operating cost, or error, a significant number of companies have changed from stop-watch time study to predetermined motion times as their primary method of setting time standards on manual operations. There are several reasons for this not the least of which is the desire to avoid rating and use of the stop-watch insofar as possible. One situation in which microscopic standard data, like the macroscopic form, does have a clear-cut advantage is that in which reasonable standards for operations are needed in advance of actual production.

Summary. For overall flexibility stop-watch time study excels. For economy in setting standards for a type of operation on which a relatively large number of standards must be established, macroscopic standard data is often superior. The standard data methods are well adapted to setting standards prior to actual production. If there is a special aversion toward rating and use of the watch, the predetermined motion time technique is often a logical choice. For setting standards on long cycle or nonrepetitive activities, work sampling frequently excels.

The Responsibility to Remain Aware of Trends and New Developments

To keep the department's methodology effective and up-to-date the administrator must keep abreast of new developments in his profession. Perhaps not so obvious as this requirement is the necessity of maintaining an alertness for *trends in industry*, for example the continuing increase in the indirect labor content of the total labor force, *trends in the company's activity*, for example a gradual shift from mechanical to electronic components in a company's product, and *trends in the profession*, for example a trend toward the increased use of statistics. It pays to be cognizant of such trends, for adjustment to them is not ordinarily an overnight proposition. For example, a company suddenly and belatedly reacts to the fact that over a period of years indirect labor has increased to the point where it is 55 per cent of the labor force, whereas methods engineering has and still is devoting all of its attention to direct labor activity. Now it will be at least several years before the methods engineering staff can be properly trained and equipped to handle the different and more difficult type of problem indirect labor presents, and can make significant progress in reducing costs and setting standards in this area. Had the administration been alert and reacted to this trend many years ago as it should have, the staff would have been gradually prepared for and attention gradually shifted to indirect labor problems.

There are numerous trends on the industrial and professional scene with which the administrator should be familiar. Failure to react early and effectively, as in the preceding example, can be very costly and embarrassing. Typical of anticipated trends in and affecting the methods engineering field are the following.

1. Continued and probably accelerated rate of conversion to mechanization.
   a. Some consequences of this trend are:
      (1) indirect labor will continue to become an increasing percentage of the total labor force.
      (2) man will perform fewer and fewer direct production tasks, but the role he does play as a monitor, servicer, or emergency link in the system will become increasingly critical to speed and effectiveness of action.
      (3) the work of the human being will be less demanding physically.
      (4) the work of the human being will demand more technical skill and training because of man's enlarged role as a decision maker (repairman, adjuster, planner, etc.) and because of increased complexity of his task. Thus investment in the employee will be greater.
   b. These changes will affect methods engineering in a number of ways. For example:
      (1) since most indirect labor is much more variable than direct labor, both in content and in time, the need for analysis and measurement techniques different from those conventionally used on repetitive operations will be much more pressing.
      (2) motivation of the operator's attentiveness (vigilance) will become more of a problem, whereas motivation of physical exertion will become less important. The ordinary incentive wage plan does not seem applicable to the motivation of attentiveness.
      (3) as the machine assumes control of the rate of processing, prediction of performance time will become less of a problem.
      (4) the prediction and control of learning time will become of more interest as the investment in the employee in this respect increases.
      (5) there will be increased need for the methodology and body of knowledge presently characterized by the term "biomechanics" or "human engineering."
      (6) the process of troubleshooting will receive increasing attention, as breakdowns of equipment become more likely, more difficult to diagnose, and more expensive.
2. Increased interest in the human element of the productive process.
   a. Some consequences of this trend are:

(1) increased attention will be given to proper design of the task and equipment associated with it.
(2) there will be a greater interest in the currently unquantifiable costs of work, such as fatigue, monotony, effort, etc.
(3) there will be greater interest in nonfinancial incentives.

b. These changes will probably affect methods engineering in the following ways:
(1) there will be an increased demand for objective work design principles.
(2) there will be an increased demand for objective methods of measuring fatigue, monotony, effort, and other current "unquantifiables."
(3) wage incentive plans as we currently know them will become less popular.

3. Accelerated rate of development and application of mathematical and statistical procedures to management of the enterprise.
   a. Specific consequences are:
   (1) management planning and control techniques will become much more powerful.
   (2) a more objective attitude will permeate throughout management in general.

   b. How these changes will probably affect methods engineering:
   (1) there will be a demand for more accurate and precise time estimates to feed to these more powerful techniques.
   (2) there will be an increased demand for actual-time forecasts in addition to or perhaps even in preference to the traditional time standards.
   (3) there will be a demand for information on expected variation in performance times.
   (4) administrators of the methods engineering function will become more objective in their outlook, perhaps will do more looking at the forest and less at the individual trees, and will be more prone to self evaluation and correction.
   (5) there will be stronger pressures for more alert maintenance of performance time estimates.
4. Simulation will find increased use in management decision making. The consequences of this trend are similar to those of (3) above.
5. Increased pressure on the methods engineer to recognize inherent variability in behavior of man and machine and the queuing problems that arise therefrom, in his designs and standards.
6. A superior statistical background on the part of some personnel and the availability of computers. This will result in an increased interest in and ability to handle more rigorous procedures, and opens the way to vast improvements in work measurement methodology.
7. Increased popularity of human engineering. This will stimulate a general interest in proper design of the human's role in the system.
8. Increased interest in and utilization of "macro work measurement," the

establishment of labor-hour standards and forecasts of labor-hour content for aggregations of activity.

9. Continued pressure exerted on management by labor to yield certain prerogatives the company has previously held in the area of methods engineering and wage incentives.

EXERCISE

1. What method of setting time standards would probably prove the most economical in each of the following situations?

(*a*) Materials handling in a warehouse.

(*b*) A group of long cycle machining operations.

(*c*) A company at which there are a number of low-volume products involving many component parts processed on several types of basic metalworking operations.

(*d*) Job-order machine shop.

(*e*) Maintenance and repair activity.

(*f*) Furnace cleaning crews.

(*g*) Assembly operations for electrical appliances, where volumes are very high, the operators many, and the typical work cycle length relatively short.

28

Methods Engineering Administration: Programming Activities of the Department

The programming responsibilities of an engineering administrator involve two problems.

1. Allocation of the department's resources (funds, manpower, talents, etc.) in the most profitable manner. Especially important to the success of an administrator is his ability to profitably allocate his department's resources to potential projects and services. The hope is that the most profitable undertakings will be selected, but these are often difficult to isolate. Furthermore, there are frequently various pressures to get into relatively unprofitable endeavors.
2. Detailed scheduling of personnel and projects. The administrator must determine *who* should be assigned to given projects, *when* projects are to be undertaken, and approximately *how long* a time should be allotted for each.

Ordinarily the engineering administrator is confronted with the task of allocating limited engineering resources to a myriad of awaiting problems. This is a difficult task, yet one that is a primary determinant of the ultimate success of the department. On what basis is the administrator to select the problems most deserving of attention? Potential return on the investment is ordinarily the preferred basis, but it is difficult to predict reliably. As a consequence the order in which projects are undertaken is often far from optimum. The ordering of department undertakings is usually suboptimum for other reasons also, such as the interjection of the "pet projects" of persons within or outside of the department. If a vice president of the corporation gets an idea that he wants pursued immediately, it will be taken up regardless of how it compares with other problems awaiting attention.

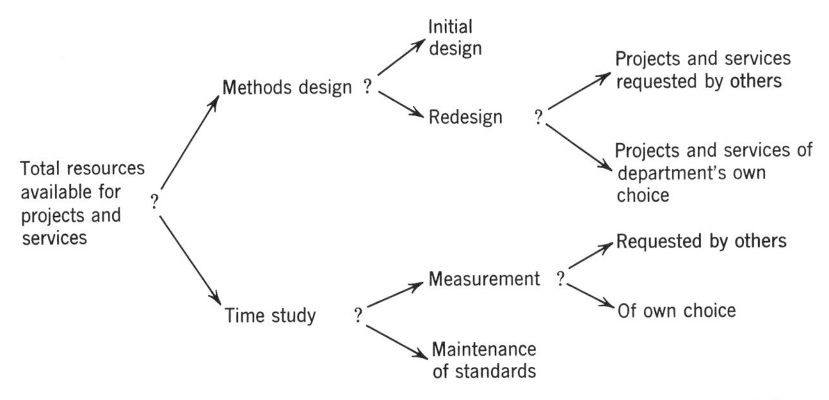

Figure 125. Distribution of the methods engineering department's "project dollar."

In the assignment of personnel, tasks should be properly allocated according to the technical knowledge required. It seems generally agreed that engineers devote too much of their time to tasks that clerks and technicians could perform. Therefore, it behooves the administrator to attempt to have a proper proportion of nonprofessional personnel available, and have them used whenever practical. Particularly profitable is the availability and use of a technician, a person who technically is somewhere between clerk and engineer, and who can be relied on to perform computations, to collect data, to construct models, etc.

Allocation of a Methods Engineering Department's Resources

The dollars available for department projects and services must ordinarily be distributed over a number of areas of activity, as summarized in Figure 125. This diagram indicates the major areas into which at least some effort should be channeled. The problem confronting the administration is that of determining what fraction of the funds available to allot to each area, then determining the most profitable assignment of these allotments to specific projects and services within each area.

Design vs. measurement. Note that the administrator must decide on the distribution of resources between methods design and work measurement. Like the similar decisions involved here, this would, we hope, be made on the basis of the relative profitability of the two types of activity. But especially in this case, relative profitability of the alternatives is difficult to estimate satisfactorily. There is a likelihood that the administrator will yield to external pressures and devote an unjustifiably greater proportion of effort to time study at the expense of methods design.

Initial design vs. redesign. Another difficult decision concerns the optimum distribution of funds between initial design (design of methods for new operations) and redesign (improvement) of existing methods. An argument advanced in favor of not investing heavily in initial design and thus starting the manufacture of a product with rather inferior methods, is that often during the early life of a product the sales volume is a very uncertain matter. The reasoning is that it is better to wait until long-run sales volume can be more satisfactorily predicted; then, if necessary, redesign methods to the extent justified by this volume. Another argument in favor of giving redesign the preference is that allowing inferior methods to be installed initially leaves a greater savings potential for later methods improvement efforts. This may not be good policy from the company's point of view, but from the administrator's personal point of view this is one way to inflate apparent return on investment in the department. In practice this is probably not a rare method of operation.

There is an argument against a disproportionately heavy expenditure on redesign at the expense of initial design. By later returning to redesign an operation for which a superior method could have been designed and installed in the first place, extra costs are incurred as a result of the added engineering time required, unnecessary duplication of installation costs, and savings foregone in the interim. This author's impression is that in general this field has overemphasized methods improvement at the expense of good initial design, that it has been preoccupied with improvement when better methods could and should have been installed initially.

Projects and services requested by others. It is inevitable that a number of the undertakings of methods engineering will be at the request of others in the organization, such as foremen, executives, and office managers. The advantages of this arrangement are that it relieves the methods engineering department of some of the responsibility in selection of projects, it brings to light worthwhile problems that the department would not otherwise be aware of, and it usually means receiving more cooperation from the persons making the requests. However, there is a distinct disadvantage to being obligated to undertake projects suggested by other departments, and that is the likelihood that an appreciable amount of methods engineering's time will be devoted to relatively unprofitable undertakings, especially trivial "pet" projects of various personnel in the company. The preferred policy seems to be to solicit suggestions but to reserve the right to decide which projects will be actively pursued.

Projects and services of the department's own choosing. It is preferable that the administrator have the prerogative to select the improvement projects to be undertaken by his department. He should be a superior judge in these matters, and under this arrangement he has much greater control over the fate of his department. On what basis should the methods engineering administrator select problems and services to be undertaken, when he has this prerogative? There are a few possibilities.

1. Potential return on the investment, considering the expected investment of the department's time devoted to the problem plus the expected installation cost of a solution, total expected savings in operating costs, and the uncertainties involved. Ordinarily return on the investment is the preferred criterion for project selection. The hope is that projects will be undertaken in order of overall profitability.
2. Observed "inefficiency" in the existing method, such as excess work-in-process, poor work procedure, inefficient layout, avoidable unbalance, excess handling, delays, backtracking, etc. *However, greatest return on the investment is not necessarily where the most obvious inefficiencies are.* One can be rather easily misled in this respect, such that better paying opportunities are forgone and too much time is devoted to relatively unimportant problems.
3. Special purposes "of the hour" that further return on the investment in the long run. For example, when a department is still very young, it is well to devote special attention to confidence or "character" building in the eyes of the organization. In this stage, it is desirable to select projects for which the probability of success, of satisfaction of the appropriate individuals, is a maximum. Such "sure-fire" projects may not be those offering the greatest potential return on the immediate investment, but they may well improve the position of the department in the long run.

In allocation of time to projects, the relatively large savings potential that indirect labor activity frequently offers through improvement in methods should not be overlooked. Considering the relatively unexploited nature of this problem area, an appreciable proportion of a methods engineering department's man hours might justifiably be devoted to indirect labor activity.

Another special factor to consider in allocation of effort is the inevitable unbalance that exists in production capacities of different components in the production system. One should be wary of devoting effort to the improvement of operations that already have excess production capability, thereby creating additional excess and perhaps unusable capacity.

Allocation of effort to time-study activities. Resources allocated to the work measurement phase of a department's activity is ordinarily shared between standards-setting activity and maintenance of standards. The tendency seems to be to devote only a fraction of the proportion justified to maintenance of standards. The costly consequences of inadequately maintained time standards are discussed at length in a later chapter. As in the case of design, it is desirable that the methods engineering department have the right to choose the activities it will measure, to determine the priority of jobs that will be undertaken, and to decide on the amount of attention to be devoted to each job. Otherwise, the department is likely to become bogged down in setting standards on jobs of trivial importance, whereas others, especially in the relatively untouched area of indirect labor, go unmeasured.

Detailed Scheduling—Use of Nonprofessional Personnel

In scheduling the activities of a methods engineering staff, the planner should give special attention to this matter of full utilization of semitechnical and nontechnical personnel. There is great potential in this field for use of the semitechnical man as an observer and as a data processor. Similarly, there is much routine paperwork and computation that can be satisfactorily performed by clerks. This point is stressed for more than one reason. One purpose is to get more efficient utilization of the higher paid personnel. Another is the fact that the respect held by shop people for engineers and the company is not enhanced by having these relatively highly paid individuals visibly performing tasks (such as taking work sampling observations) that obviously do not require a college education. The latter argument for conservation of technical manpower may well be stronger than the economic one.

REFERENCE

Thompson, James W., "The Economic Management of Methods Development and Industrial Engineering," *Journal of Industrial Engineering,* vol. 10, no. 2, March-April 1959.

29

Methods Engineering Administration: Maintaining Favorable Relations with Other Personnel in the Organization

Recall that engineering is ordinarily a staff function in the organization and therefore serves in a consulting capacity. It recommends rather than commands. This being the case, the success of the department in gaining acceptance of itself and the problem solutions it produces depends on the general attitude of operating personnel toward the department. If a superintendent has respect for and confidence in the industrial engineering department, he is likely to be generally receptive and cooperative. If he feels the opposite he can be a real obstacle.

Engineering departments must ordinarily rely heavily on cooperation of fellow engineering departments in the form of advice, data, alteration of specifications, and the like. Achieving and maintaining favorable relations with other engineering departments is extremely important. It is not uncommon for such departments to develop rather antagonistic attitudes toward one another over a period of time, resulting in their working at odds, in the rise of "empires," and in a large unnecessary expense to the enterprise.

Therefore, the engineering administrator should devote attention to achieving and preserving a favorable "organizational atmosphere" for the work of the department. He should make continuing efforts to "sell" himself, his staff, and the particular engineering specialty they represent.

Relationships to which the Methods Engineering Administrator Should Devote Careful Attention

The ultimate success of a methods engineering department depends heavily on the quality of its relationships with others in the organiza-

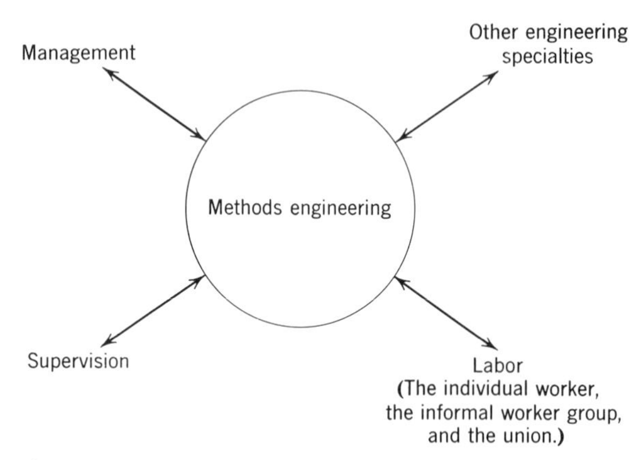

Figure 126. The "organizational environment" in which methods engineering must operate.

tion. The fact that methods engineering determines work methods and work loads, two vital matters to supervisor, worker, and union, makes it inevitable that there will be considerable contact between the methods engineer and shop personnel. The interest displayed by supervision and labor adds some unique and difficult problems to the ordinary organizational relations problems of an engineering administrator. The magnitude of the problem confronting a methods engineering administrator in this respect is indicated by Figure 126. This figure summarizes the various groups with which the department comes in contact daily. Failure to maintain a favorable relationship with any one of these groups is likely to result in a major headache for the administrator.

Relations with management. Obviously it is highly desirable to retain the confidence, cooperation, and support of key management personnel; they hold the purse strings, and in most instances they control the department's function in the organization. Management's satisfaction with the department's technical achievements is necessary but not sufficient to retain a favorable relationship. It is necessary also to maintain a harmonious and sympathetic personal attitude toward the department and its staff on the part of key management personnel. Some personal animosity between key individuals can go a long way toward nullifying what is otherwise a fine engineering performance.

Relations with other engineering specialties. Product, manufacturing, plant-layout, materials-handling, and plant engineers set an enormous number of specifications that become restrictions to the methods engineer. Methods engineers have occasion to question, challenge, and seek revocation of a number of these specifications. This is only natural and desirable, for not all of these specifications are optimum to be sure. However, if the relationship between methods engineering and the engineering specialties imposing these restrictions is unfavorable, characterized by friction, antagonism, and personal feuds, the methods engineer is likely to receive an arbitrary and uncooperative response when he attempts to challenge specifications. Furthermore, other engineering specialties make a number of decisions involving the product and manufacturing process in which methods engineering should be consulted for advice. And, too, methods engineers frequently require the technical assistance of other engineers. For these reasons it is vital that there be a cooperative relationship between these related engineering specialties, that there be sufficient liaison, consultation, and interchange of ideas and information.

Relations with production supervisors. The confidence, cooperation, support, and participation of supervisory personnel are extremely important to successful operation of a methods engineering department. Foremen serve to assist the methods engineer in numerous ways.

1. The foreman can direct the methods engineer's attention to a variety of profitable problem areas that might otherwise be overlooked.
2. The foreman can provide a wealth of information with respect to manufacturing practice, many "whys and wherefores" that the engineer might never learn otherwise.
3. The foreman can contribute many worthwhile ideas.
4. The foreman can usually do more than any other person to see that the new solution to a manufacturing problem is successfully implemented. In fact he can usually "make or break" it, depending on his attitude.
5. The foreman can usually do more than any other person in the organization to instill a favorable worker attitude toward the company and its representatives, such as methods engineers. The role of the foreman as a determiner of worker attitudes might well be underestimated. His influence is demonstrated by the fact that worker attitudes toward the company in general are seldom more favorable than the foreman's. Their attitudes might well be worse, but rarely better than that of their supervisor.
6. The foreman can, by the type and degree of cooperation he offers, "make or break" the standards-setting and wage incentive programs of the company. Without the foreman's cooperation it is very difficult

to establish acceptable time standards and even more difficult to maintain them. For example, when a standard is about to be set on an operation, the time-study observer must frequently rely on the judgment of the foreman as to whether the method observed is acceptable *and is the one that will be used after the time standard is established.* His advice is especially crucial on matters of equipment operating speeds and feeds. After the standard is established, the foreman must be relied on to maintain the method and conditions on which the standard is based, and to report significant changes in method and conditions that require resetting of the time standard. Take the case of machine speeds and feeds. If the operator is allowed to run his machine at one speed and feed during the time study, then is allowed by the foreman to increase these by 25, 50, or even 100 per cent after the standard is set and installed, the type of time standard structure that will result is obvious. The foreman is a key figure in such situations as this, and in fact, he is as crucial as anyone in the organization to ultimate quality of the time-standard structure.

These are things the foreman can do and things he can likewise fail to do if a favorable relationship with him is not maintained. Rest assured, a foreman who refuses to cooperate in the foregoing respects can be a major obstacle to the methods engineering department's progress.

One measure that has been found especially successful in soliciting the foreman's cooperation and understanding is to make a special effort, probably in the form of a series of training sessions, to familiarize him with the objectives and methodology of the department. If the foreman has an appreciation of the purposes of this engineering function and understands the need for and workings of the procedures used, he is likely to be much more cooperative than he can be expected to be if these matters are mysteries to him. Mystery of purpose and action breeds suspicion and contempt. In addition to a more favorable attitude, education of the foreman in these respects equips him for doing a much improved job of explanation and justification to those he supervises. Furthermore, this education usually makes it possible for the foreman to make a more substantial contribution to improvement in the company's manufacturing methods. He is more "methods conscious" insofar as his department is concerned and insofar as he assists the methods engineer with ideas and information. Some companies put special emphasis on building this "methods consciousness and savvy" through these training sessions, by teaching the foreman the means and principles of job improvement. This seems very worthwhile from the standpoint of the foreman improving methods in his department and of his working more closely with the methods engineer.

Relations with workers. A number of the points made with respect to importance of the foreman's cooperation and participation apply also to the worker on the job. He too *can* be very helpful as a source of information and ideas; he too plays a crucial role in implementation of new methods; he too can be equipped and encouraged to participate in the introduction of better work methods. Quite a number of companies have successfully engaged the worker's contribution in the form of better methods in his own job and others he is familiar with, usually as part of a special "work simplification program." Under this program, by education and considerable motivation and publicity an attempt is made to increase the worker's ability and interest in improving work methods.

There is a tendency to underestimate the worker's potential to contribute to, and his capacity to hinder if he is so inclined, the performance of the methods engineer. Every effort should be made to maintain favorable relations with him. Aside from fair treatment and due consideration of his point of view, the worker's attitude and response can usually be improved considerably if the trouble is taken to familiarize him with the objectives and practices of methods engineering. Misconception and ignorance of this field abound in the minds of workers, especially with respect to time-study purposes and practices. Such circumstances are hardly conducive to a favorable response. Probably the most effective means of minimizing the resulting suspicion and contempt is through frank and easy-to-understand explanations to all employees.

Relations with organized labor. This is another crucial relationship. Unions are not reputed for bending over backward to assist and cooperate with management, especially with methods engineering, because of its role in determining work methods and work loads, so this is not ordinarily expected. But the union can and frequently does bend over backward to resist, protest, and even control the activities of and results produced by methods engineering. This is to be avoided if at all possible, for a particularly unfavorable relationship with the union is a costly, severely restricting, and very troublesome matter to a methods engineering department. Avoiding an antagonistic labor relations climate is certainly worth considerable attention. Here are some measures that may well assist in maintaining a reasonable relationship with organized labor.

1. Attempt to familiarize union personnel with methods engineering's purposes and methodology.

2. Make a clear, explicit statement of policies in this area.

3. Demonstrate fairness and a willingness to be the first to make concessions.

4. Consult with union officials *in advance* of making or even announcing changes that may be of concern to them.

5. Attempt to gradually "condition" organized labor for certain changes in methods, practice, policy, etc., that the department wishes or will wish to make, but that probably will be vigorously resisted if introduced too abruptly.

6. Maintain an open communication channel through which the union may express its views and objections.

7. Attempt to anticipate labor's reaction in advance of deciding on, announcing, and installing changes in work methods, work loads, methods engineering practice, policy, and other matters of concern to the union. There are many cases on record to bear out the point that anticipation of labor's reaction is worth the administrator's careful attention. Time and time again methods engineers have installed some system, announced some policy, adopted some technique, or made some other type of change, only to have the situation blow up in their faces even to the point of precipitating a strike and the need for an arbitrator, when with a little forethought and good judgment this response could have been anticipated and avoided. Anticipation of labor's reaction is a matter worthy of additional discussion.

Anticipation of Organized Labor's Response to Activities of a Methods Engineering Department

Anticipation of labor's reaction to various decisions and actions is based partly on what appear to be certain generalities in labor policy and response. Typical of these apparent generalities are the following.

1. Policies announced by the "national" union seem to hold little weight with the local union organization when it is economically and politically more advantageous for the latter to do otherwise. Take the UAW's expressed policy against the use of wage incentives. In spite of this general policy, a substantial number of affiliated locals tolerate this method of payment; others even prefer it. Another example of deviation from national policy pertains to the matter of technological change. Although a number of nationals make policy statements to the contrary, many locals offer considerable resistance to a labor-saving innovation when it affects *them*.

2. Unions in general seek a veto power rather than the right to participate in making decisions, determining policies, setting standards, and the like. Although there are some notable exceptions, most unions avoid having any part in making specific management decisions, which includes the determination of time standards, apparently because once the union becomes a partner in making a decision it is identified with it and is obliged to support it. Unions prefer that management unilaterally set the time standard but insist on the right to protest that standard.

3. Organized labor seeks bargainability of the results of the matters pertaining to job evaluation, time study, and wage incentives. From the union's point of view, time standards produced by a time-study department provide a basis for collective bargaining. The union holds that management has only the right to initiate, not the absolute right to determine the time standard or wage rate. Thus almost all unions retain the right to bargain over time standards.

4. Organized labor seeks extension of the right to bargain and limit methods engineering techniques, practices, and procedures, as well as the results of same. For example, some company-union contracts contain provisions specifying how an operator is to be selected for time study, the number of watch readings that must be taken, the method of timing, and so on. Here are some actual provisions from various labor-management contracts to illustrate the point.[1]

Before the actual timing begins the selected operator shall first complete one-half dozen cycles and the actual time study shall then be made upon the next dozen and the average time consumed upon this dozen shall be the time set. To such time there shall be added an allowance of ten (10) per cent for unavoidable delay and fatigue.

All time studies are to be taken to assure a fair test under normal conditions. The time-study man shall remain on any job studied not less than one-half hour and the rates shall be set and the employees notified within 2 hours after a study is completed.

In event that an employee can change the method of operation through some timesaving device or method, thereby increasing his productive efficiency, such change shall not affect the bonus rate on that particular operation during the life of this agreement.

Permanent rates once established will not be increased or decreased unless such action is justified by change in materials; change in tools or equipment; change in methods; change in quality standards; work added to or removed from a job; mathematical error in setting the timing or the rate.

It is agreed that changes in method may take place gradually or may be of such a small nature as not to warrant a re-timing when such small change

[1] Selected from "Collective Bargaining Provisions: Incentive Wage Provisions; Time Studies and Standards of Production," Bulletin No. 908–3, U. S. Department of Labor, Bureau of Labor Statistics.

takes place. The effect of such gradual change or of minor changes may accumulate to the point where a re-timing of the job is required, at which time all previous gradual or minor changes will be taken into consideration. The union steward will be notified of each particular change as the result of which a new time study is not then being made.

The union is not a party to the time studies, but it shall have the right to bargain collectively concerning all matters pertaining to the time studies, including the basic formulas used, the choice of the operator to be timed, the defining of average conditions and the determining of the leveling factors and other time allowances.

All time studies shall be available to the employee through the officers of the union at any time.

Labor-management contracts also contain a myriad of provisions pertaining to operation of the incentive wage plan. These statements cover guarantees, amount of incentive coverage, downtime provisions, bases for computing earnings, and so on. Many of these provisions relating to time study and wage incentives are quite disadvantageous from the company's point of view, and certainly to be avoided. In some cases the company has allowed itself to be maneuvered into a position where they have very few prerogatives remaining in time study and wage incentive matters. This is forcefully illustrated by the fact that many companies no longer have the right to choose the form of wage payment. Some have an incentive wage plan and wish to change to a daywork plan; others are in the opposite position. Yet the union has gained the right to prevent them from making the desired change.

5. Unions in general, attempt to maintain a "one-way street" policy aimed at preventing the reversal of any change that is beneficial to labor. The outstanding illustration of this is the inability that most companies have to correct a standard if it is loosely set, but the obligation they have to correct that standard if it is tightly set. Another illustration is the occasionally encountered provision that whenever there is a change in method or time standard, the level of earnings achieved under the old method or standard cannot be reduced.

Not only are there differences between policies of the national unions and reactions of the locals; there is considerable variation between locals of the same union. Policies, reactions, contract provisions, and the like, depend much on the particular circumstances, and are matters that have evolved over a long period of time. Thus, a methods engineering department must rely heavily on the history of its own company-union relationship in attempting to anticipate labor's reaction to its activities.

Summary

There are methods engineering departments which on the basis of the caliber of personnel and of the techniques employed would be expected to produce gratifying results, but they fall far short of expectations because they have failed to achieve a favorable organizational atmosphere for their operations. One of the unfortunate aspects of being in this predicament is that achieving favorable organizational relationships is not an overnight matter; this is a long-term proposition requiring continuing forethought and attention. Note that failure to achieve a favorable relationship with any one of the four major organizational groups—management, other engineering specialties, supervision, or labor—can severely restrict department achievement.

EXERCISES

1. Assume that you have been appointed organizational relations director (official "character-builder") for a large industrial engineering department.

(*a*) Where would you direct your efforts?

(*b*) Suggest typical measures, policies, procedures, etc., you would resort to in achieving and sustaining a favorable organizational atmosphere for the department.

2. The manager has requested you to outline and direct a company-wide methods improvement campaign. The purpose is to encourage and solicit improvements in work methods from all personnel in the organization at all levels. Outline the manner in which you would set up and conduct this program.

3. Discuss the provisions quoted from labor-management contracts on pages 459 and 460, with respect to possible consequences and advisability from the company's point of view.

REFERENCES

Dylenski, E. P., and M. R. Wilson, *Union-Management Cooperation in Developing Standards: I, Management's Viewpoint, II, Labor's Viewpoint,* Production Series, No. 146, American Management Association, New York, 1943.

Goodwin, Herbert F., "They Live by the Clock," *Journal of Industrial Engineering,* vol. 11, no. 3, May-June 1960.

Guyatt, Cecil W., "The Art of Methods Engineering," *Journal of Industrial Engineering,* vol. 10, no. 5, September-October 1959.

Mellin, Warren R., "What Is an Organized Methods Improvement Program?" *Journal of Industrial Engineering,* vol. 10, no. 6, November-December 1959.

30

Methods Engineering Administration: Evaluation and Improvement of Performance in Methods Design

The results actually attained in a department's undertakings should be evaluated both for the aggregate of its activities over a period of time and for each project and service performed. The evaluation in aggregate is necessary for *justification* of the current level (scale) of operation of the department and in fact its very existence. Follow-up evaluation of the outcome of individual undertakings is essential if *improvement* in performance of the department is desired. Knowledge of results is a prerequisite to improvement of performance. If the administrator is to learn whether personnel and methodology are in need of corrective action, and if he is to know what personnel and methodology changes to make in order to improve results, he must know how effectively these are performing. Follow-up evaluation is a basic feature of the design cycle pictured in Figure 10, page 63. Through surveillance of the design in use and evaluation of its effectiveness, the engineers involved can profit from experience and perform more effectively in similar situations in the future. The importance of this feedback to improvement of departmental performance cannot be overemphasized.

What to Evaluate

Basically, the administrator should attempt to evaluate the *function* his department is intended to serve in the enterprise and the *effectiveness* with which that function is being fulfilled. The former is primarily a matter of appraising the overall objectives of and boundaries imposed on the department, and determining if these are in the best interests of the enterprise. Appraisal of effectiveness of performance is to be discussed at length.

Some of the activities of an engineering department yield results that can be appraised satisfactorily in terms of dollar savings and return on the investment. How relatively easy the administrator's evaluation would be if all activities could be appraised in this manner! The relative profitability of individual projects and classes of projects could be conveniently determined, and from these it would be possible to compile an aggregate return on the investment figure for the whole of the department's activities for a given period of time. This is the figure most engineering administrators and company executives would like to have for appraisal of the department's overall performance and of its current scale of operation. Unfortunately, only a portion of a department's activity, primarily the redesign or improvement projects, can be satisfactorily evaluated in this direct fashion. In original design, say the design of a new product, or providing advice to others in the organization, or providing numerous other services, it is impossible or prohibitively difficult to quantitatively estimate the benefits in terms of dollars. Since it is likely to be a major portion of a department's man-hours that yields unquantifiable benefits, an evaluation of this phase of a department's activity should be attempted by some means. This may be accomplished by one or preferably a combination of the following indirect means.

1. Opinions of the administrator and others in the organization as to the quality of job the department is doing.
2. The extent to which the services of the department are sought by others in the organization provides a rough indication of the kind of job being done.
3. Comparison of the manner in which the department is equipped and the method with which it operates against what is considered in the profession to be the most likely to yield good results. If the personnel are competent, time is effectively utilized, the techniques, organization, practices, etc., are sound, the allocation of effort is effective, and relations with remainder of the organization are good, then in all likelihood the performance of the department in the activities under question is yielding substantial benefits.

Utilization of Knowledge of Results

Knowledge of the results being achieved should be used periodically as a basis for deciding if the activities of the department should be expanded, remain stable, or be reduced. Perhaps potential return on the investment in unexploited problems and services is such that expansion of the department is highly desirable. Perhaps also, the size of the department is such that projects offering little or no return must be undertaken to keep the current manpower occupied, thus diluting the overall benefits to be gained. Under these circumstances a reduction in scale of operation would probably improve the return on investment. To make such a decision intelligently requires a reasonable estimate of the results achieved by the present size of the department. This same information is often valuable to the administrator in his efforts to retain or expand his budget for the coming year. Ordinarily, an increase in a department's budget is not achieved unless management is "sold" on the idea by the administrator. To do this he must usually have facts and figures to back his case for expansion.

Mistakes, failures, miscalculations, oversights, errors in judgment, and the like, as well as the particularly favorable results of various undertakings, learned of through follow-up evaluation, should be put to practical use through corrective action. Examples of what might be learned and remedied through effective follow-up evaluation are that engineer *A* is in need of certain special training, that engineer *B* does not perform satisfactorily in the problem area to which he is currently assigned, that in a certain problem area the department's efforts are not paying off, that the administrator is overestimating the engineering hours required in scheduling projects, that the department is not maintaining adequate liaison with other engineering specialties, and so on. There is always opportunity for improvement in the performance of an engineering department, much is to be learned; the opportunity is there for the seeking. The completeness and effectiveness with which an administrator obtains *and benefits* from knowledge of the results of his department's activities is an excellent index of his administrative ability.

Summary of Evaluation Responsibilities

In summary then, evaluation of an engineering department should ordinarily involve the following.

1. *Evaluation of the function* the department is intended to fulfill in the organization.
2. *Evaluation of the effectiveness* with which the department performs this function. This is accomplished primarily by:
   a. estimating the return on investment achieved through savings in operating costs, for those department undertakings in which savings *can* be satisfactorily measured in dollars.

b. estimating the results achieved by those activities that do not yield benefits measurable directly in dollars. An attempt should be made to evaluate these benefits by a combination of the following means.
   (1) Opinions as to results obtained.
   (2) Demand for the department's services.
   (3) Evaluation of the department's facilities and mode of operation in relation to what experience leads us to believe produces the best results.

3. *Utilization* of knowledge of results,
   a. to appraise the department's scale of activity.
   b. to improve performance through correction of weaknesses in personnel, methodology, programming of activities, organizational relations, and even in methods of evaluation.

The specific manner in which the administrator of a methods engineering department should fulfill these responsibilities will be discussed.

EVALUATION AND IMPROVEMENT OF A METHODS ENGINEERING DEPARTMENT'S DESIGN ACTIVITIES

Evaluation of the Function

The design function of methods engineering deserves frequent reappraisal in view of the rapid developments that have been taking place in industrial management and industrial engineering, and of the fact that methods engineering still has considerable opportunity for growth and expansion in many companies. In the course of evaluating function, consideration should be given to such matters as scope of responsibility, purposes, limitations, areas of emphasis, and services provided. It is highly desirable that the administrator occasionally take the time to reflect on such matters.

Quantitative Evaluation of Effectiveness in the Design of Work Methods

The most direct and meaningful means of evaluating a department's methods design performance is through estimation of return on the investment, in the form of the ratio of operating dollar savings to total cost of design and installation of new work methods. Yet, since this can only be estimated for methods improvement projects and not for initial design of work methods and other department services, this return on the investment figure is not the complete picture. Furthermore, as noted earlier, the poorer the job a department does in the initial design of shop methods, the more savings potential that remains for improvement projects. This makes satisfactory interpretation of a return on the investment figure rather difficult. One factor that should be recognized in this situation is the department's distribution of engineering hours between initial design and redesign projects. If a high percentage of the hours devoted to design work is spent on redesign projects and the department's return on its investment in this type of endeavor is low, this is an unfavorable sign. However, if a high percentage of hours is devoted to initial design, a return on the investment figure tells little about the department's design effectiveness.

Qualitative Evaluation of Effectiveness in the Design of Work Methods

To obtain a satisfactory evaluation of the unquantifiable benefits to be credited to a methods engineering department, it will usually be necessary to rely partly on the judgment of the administrator and other key personnel in the organization. The opinions of management personnel, shop supervisors, and other engineering administrators should be helpful in this respect. An important basis for these opinions is a comparison of existing shop methods with what are commonly acknowledged to be good work methods and with the methods found in what are known to be well-managed plants.

Another index of the quality of job the staff is doing is the extent to which the services of the methods engineering department are sought by others in the organization. If there is a substantial waiting list of projects suggested by management and supervision, this is a healthy sign. The backlog of project requests is an indirect expression of the organization's opinion of the department that is perhaps more reliable than direct expressions of opinion.

A third indirect source of information on effectiveness of performance is an appraisal of the methods engineering department's staff, techniques, and facilities. The presumption is that a department that is well equipped and well operated is probably making a worthwhile contribution. The weakness in this means of appraising overall effectiveness is obvious. Also obvious is the value of this evaluation of the modus operandi in leading to improvements. To illustrate the type of appraisal to which a methods engineering department's facilities and manner of operation should be periodically subjected, the following checklist of questions is presented. To this sample list the experienced administrator could add numerous other illuminating questions.

1. What percentage of the workday is actually devoted to profitable endeavor?
2. With reference to personnel:
   a. what about the capability of each staff member with respect to technical ability, creativeness, and ability to effectively work with others in the organization?
   b. is each adequately trained and indoctrinated?
   c. what about the enthusiasm of each?
   d. what is the estimated return on the company's investment in each staff member?
3. With respect to methodology:
   a. is the department's general methodology in accord with current "enlightened" theory and practice?
   b. is the methodology of the department being developed to meet anticipated trends in the business, in industry, and in the profession?
   c. what is the typical manner of approach to a design problem? Is an intelligent problem-solving approach used? Is sufficient time devoted to problem formulation and analysis? Is too much time spent instead in analyzing and scrutinizing existing methods?
   d. is adequate use being made of the knowledge relatively recently made available by the field of experimental psychology, under the popular heading of human engineering data?
   e. is adequate use being made of the full range of techniques available to the methods engineer, especially the more recent ones such as predetermined motion times, work sampling, time-lapse photography, tape recordings for on-the-job instruction, and a number of others?
   f. is the department doing an effective job of implementing (selling and installing) its proposals?
4. With respect to programming:
   a. are the projects selected in accord with overall objectives?
   b. does the distribution of resources between methods design and work measurement appear optimum? Between initial design and redesign?
   c. is an adequate percentage of the department's efforts being directed to indirect labor?
   d. is the amount of time devoted by professional persons to semitechnical and nontechnical tasks kept to a minimum?
5. With respect to relations with remainder of the organization:
   a. has the department been successful in retaining management's confidence and support?
   b. is there a harmonious relationship between the department and other engineering specialties? Or have opposing "empires" arisen, manifestations of certain personal ambitions of various administrators?

c. has the department been able to maintain a satisfactory relationship with shop supervisors? With workers? With organized labor? What is the attitude of each toward the department and the individuals in it? How much cooperation is received from them?

d. does the department hold a position of positive leadership in the organization with respect to the matter of work methods? Have its educational and promotional efforts been successful in arousing enthusiasm and active participation of the organization in the improvement of work methods?

6. With respect to evaluation itself:

a. is a sincere attempt made to follow up on each undertaking of the department, in an effort to learn, to improve future performance?

b. is a satisfactory effort made to periodically reappraise the function of this specialty?

c. is a periodic attempt made to evaluate the department's scale of operation, to determine if expansion or contraction is desirable?

d. is an attempt ever made to classify estimated benefits according to type of department endeavor, in an effort to determine the most profitable allocation of the department's resources? For example, what is the relative "payoff" on efforts devoted to methods design as opposed to work measurement? On attention devoted to direct labor as opposed to indirect labor? On efforts devoted to improving methods in production department *A* as opposed to production department *B*?

As stated earlier, the fact that a given administrator can justifiably answer favorably to such questions as these does not guarantee, but makes it a reasonably safe assumption, that the department is performing favorably in the activities that cannot be quantitatively appraised.

Knowledge of results of the individual project, of the results attained in different types of undertakings, and the results of the aggregate of the department's activities, should be put to effective use in correction and improvement of future performance. This applies to department function, scale of activity, personnel, methodology, programming, organizational relations, and evaluation and administration themselves.

EXERCISE

1. As a consultant you have been employed by a large corporation to evaluate the methods design activity of the company's methods engineering department, and ultimately to make recommendations for improvement. How would you proceed to evaluate this phase of the department's activity and arrive at the requested recommendations?

REFERENCE

Kelly, Thomas E., "How to Measure and Appraise Your Industrial Engineering Department Performance," *Journal of Industrial Engineering,* vol. 10, no. 2, March-April 1959.

31

Methods Engineering Administration: Evaluation and Improvement of Performance in Work Measurement

The general pattern of evaluation outlined at the beginning of the previous chapter applies to work measurement also. However, a number of the specific procedures required and difficulties encountered in follow-up evaluation of work measurement methodology and results are basically different from those described for methods design.

Evaluation of Function

Rapid change in manufacturing and managerial technology and the ever powerful yet often deceptive force of precedence make it worthwhile to frequently reappraise the basic purposes of work measurement activity in the organization. The general purpose of this phase of methods engineering activity in a majority of situations appears to be to provide time standards for the company's operations. However, there are a number of basic questions with which this rather limited purpose might profitably be challenged.

1. Why does management want this time standard?
2. If the answer to the first question is "for evaluation purposes," then we might ask, "Why spend all this money and effort on evaluation?" Perhaps the organization might be further ahead in the long run if it used the bulk of the resources it ordinarily spends on evaluating how well it is doing, on the very act of doing better, by obtaining better equipment, methods, workers, and the like. The typical manufacturing firm spends an enormous sum in the course of "studying itself in the mirror."
3. Why use the incentive wage method of payment? The main

pressure for a time standard of the type ordinarily established by time study comes from the incentive wage plan. Furthermore, many of the difficulties encountered in time-study work exist, or are as severe as they are, because of the incentive wage plan. Finally, the long-run profitability of this method of wage payment is uncertain.

4. Why use this time standard in those instances in which a straightforward, actual-time forecast is clearly appropriate, as in the case of scheduling?

5. Why not provide an estimate of expected variation about the time standard and the actual-time forecast?

6. Why not show an active interest in measurement of performance characteristics other than time? If this is truly *work* measurement, why operate as if it is solely *time* measurement?

7. Why limit measurement to direct labor only? Why not extend activity to indirect labor, including engineering, sales, and clerical help?

Evaluation and Improvement of Work Measurement Effectiveness

A basic requirement in improvement of any measurement system is knowledge of results. Recall the blindfolded dart thrower. In work measurement, this feedback is achieved through a measurement cycle, an equivalent to the design cycle. This cycle, pictured in Figure 127, is fundamental to the evaluation concepts and procedures to be described. The point of interest at present is the follow-up feature of this cycle. The surveillance and evaluation of a time standard while it is in use are important for several reasons.

1. So that the system may profit from experience, learn, and eventually improve, with respect to error and cost. It is impossible to make intelligent decisions concerning alteration, supplementation, or replacement of current measurement procedures if the administrator has little or no notion of how satisfactorily each is performing.

2. Sizeable errors are often made in establishing time standards. Efforts should be made to learn of these, not only that personnel may profit from experience but so that they may correct (if permitted) the erratic standards.

3. Methods and conditions on which time standards are based naturally have occasion to change over a period of time. There is a tendency however, for the same standards to remain in effect long after they have been rendered inapplicable by these changes. There-

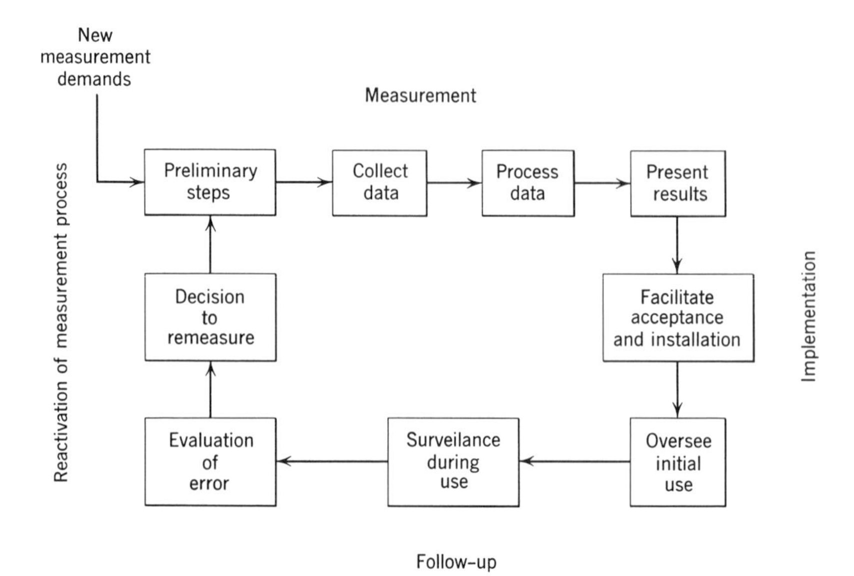

Figure 127. The measurement cycle.

fore, there is need for surveillance of time standards in use, to keep them abreast of changes in shop methods and conditions.

Failure to perform this follow-up function after the time standard is set virtually guarantees a preponderance of erratic and obsolete standards with the attendant costs, and of course severely retards improvement of the measurement system.

Quantitative Evaluation of Work Measurement Results

Unfortunately, quantitative appraisal of the results attained by a work measurement procedure or department is a difficult and rather unreliable matter. The cost of setting up and operating a particular work measurement procedure is a relatively simple matter to estimate. This is not true, however, of an attempt to estimate the errors in the standards produced.

Turning to the problem of quantitatively appraising the error in the standards established by a particular work measurement system, the following choices are available.

1. A work measurement system known to have less error than the

one under evaluation could be used to check standards established by the latter.

2. The standards under appraisal could be compared with the performance actually attained from operators on the jobs involved, in an effort to isolate the erratic standards. Under this proposal, if the operator(s) cannot moderately surpass a standard, it would be deemed tight (an underestimate). If the performance of the operator(s) far surpasses the standard for the job, that standard would be deemed loose (an overestimate).

For 1 this might be accomplished by using very elaborate and expensive stop-watch time studies to remeasure a sample of jobs for which time standards have been established by the system under evaluation. The presumption is that the special check studies are considerably less erratic than the system under appraisal. An obvious drawback to this procedure is the large and often prohibitive expense and consequently the very limited sample of jobs to which such elaborate check studies could be applied. And, too, when the comparisons are made, there is uncertainty as to how much of the disparity is attributable to the evaluated system and how much to the evaluating system.

There are several objections to 2. For one, it seems somewhat illogical and impractical to go to the expense of establishing a standard for purposes of evaluating actual performance, only to later reverse the process and use actual performance as a basis for evaluating the standard. The procedure described in 2 is unreliable as well as illogical, because of the tendency for operators on the job to adjust their output rate to agree rather closely with the standard output rate. A physical analogy to this situation is the object that appropriately expands or contracts in order to make a previously made measurement of that object appear more representative than it really is. The existence of this phenomenon in work measurement will be better appreciated after it becomes clear that it was common practice in the past for the company to arbitrarily tighten up a time standard if it was discovered from experience to have been set too leniently. The giveaway that the standard was loose was ordinarily the high earnings made by the operators under the incentive wage plan. Although this practice of outright "rate cutting" is no longer common, management is still frequently suspected and accused of it. Regardless of whether or not their suspicions are justified, most workers under an incentive wage plan *do* believe to this day that if the time standard established for their job turns out to be loose and the company discovers this fact, that standard will sooner or later be tightened by one means or another. Yet a loose time standard under an incentive wage plan can be a profitable proposition for the worker; a situation worth preserving if possible. Most workers seem to feel that it is more profitable to them in the long run to protect and prolong the life of a lenient standard than it is to take full advantage of such a standard and risk having it noticed and tightened. Therefore, under a loose time standard, operators ordinarily restrict their production to a level that they believe is not likely to attract management's attention to the looseness. Although not every worker may feel this same way, pressure from the work group usually causes him to restrict his output as the others do. As a consequence, loose time standards are made to *appear* as if they were accurately set even though they may be grossly in error.

As a result of this tendency for workers to make standards appear more accurate and precise than they actually are through judicious conformance to those standards, there is ordinarily a very wide range of time values, any one of which could be assigned as the time standard for a given operation, and all of which could be forced by the workers to *appear* to be quite satisfactory *on the basis of the ratio between actual and standard output rates.* That is, for any value in this range, the operators can and probably will produce at a level that causes the standard to look reasonable on the basis of the criterion under consideration.

Therefore, a company's time standard structure can *appear* on the surface to be quite satisfactory in that there are few grievances over tight standards and few or no cases of exorbitantly high incentive earnings, when in actuality a substantial portion of the standards are loose but concealed by restriction of output. Appraisal of a standard or group of standards by comparison to actual operator performance is not a reliable procedure.

Another situation that adds to this unreliability is the ordinarily small number of operators performing a given operation. As is so often the case, suppose there is only one person working under a given standard and that his performance ratio is 1.60.[1] Is this standard loose? Or is the standard accurately set and this exceptionally high performance ratio actually the result of a high level of skill and effort? Or is the 1.60 a result of both? Suppose the performance ratio were 1.10 for the one operator on that job. Is that standard tight? Or is the operator lacking in skill or effort and the standard okay? Or is it

[1] The average ratio of standard hours to actual hours experienced over a significant period of time under the standard in question will be referred to as the performance ratio for that standard.

a combination of these? Now suppose the performance ratio is at the expected 1.30 level. Does this necessarily mean that the standard is satisfactorily set? Actually, we *might* be able to obtain reasonable answers to these questions if we had satisfactory evaluations of the levels of skill and effort the worker has been exerting to obtain the performance ratio in question, but these are difficult matters to ascertain.

In light of these circumstances, only the following generalizations are justified. If the performance ratio is consistently very high, and the underlying levels of skill and effort obviously not commensurate, the standard is probably loose. An equivalent statement can be made with respect to probable tightness. Recall, however, that the absence of such extreme performance ratios by no means assures that a given standard is not in error!

Although the level of output under a standard is an unreliable indicator of its satisfactoriness, there is another characteristic of the operator's performance pattern that seems useful in the isolation of a loose standard concealed by restriction of output. This characteristic is the variability of operator output. In brief, detection of loose standards under restricted output conditions involves alertness for variation in output which is suspiciously less than would ordinarily be expected. Two sources of variation are of interest. One is variation of the performance ratios among different operators on the job. The other is variation in the individual's performance ratios over a period of time, say his daily performance ratios. The assumption is that when there is very little variation in the performance ratios of different operators or in the day-to-day performance ratios of a given operator, the probable cause is restriction of output caused by the desire to protect a loose time standard. A number of specific procedures for detecting a loose time standard by this means are proposed in Appendix C.[2]

From this discussion it should be obvious that evaluating the quality of a time standard or group of them is not a straightforward matter, and that it is easy to be misled. In spite of these inadequacies, it is highly desirable that attempts be made to isolate the measurement errors made and to obtain a cumulative error picture so that, among other things, effective corrective action can be taken.

[2] Because of the degree of speculation involved and the fact that the procedures described in Appendix C have not been adequately tested in practice, they are not presented as an integral portion of the text.

Qualitative Evaluation of Work Measurement Effectiveness

The opinions of the administrator, his staff, and other personnel in the organization as to the effectiveness of the work measurement phase of the department's activity must usually be given some weight in the evaluation process. A number of responsible persons in the organization might be asked for their opinions as to the quality of the time standards the department is producing. But whom can we rely on for an informed, unbiased expression of opinion in this case? In many instances, as implied earlier, management personnel may well be deceived as to the true quality of the time-standard structure. Shop supervisors are probably much better informed as to the errors the time-study staff has been making. However, they have some rather strong reasons for not expressing the true situation. A foreman is aware of the fact that he can make his men very unhappy if he should reveal the loose standards in his department. Unless the foreman is extremely "management oriented," this can not ordinarily be expected of him. The personnel probably the most qualified to pass judgment on the errors in time standards are the operators who work under those standards, but an unbiased expression of opinion is out of the question in this case and understandably so.

The extent to which time standards are requested by others in the organization is not a reliable indicator of the quality of standards being established. We can conceive of the situation in which, because a fairly high percentage of standards set by a department are loose, there is considerable and continuing pressure to have time standards established on the unmeasured jobs remaining. Or, because of a preponderance of tightly set standards, there is considerable demand in the organization for restudies.

Again, by comparing the manner in which the department is equipped for the task and the manner in which it operates with what is believed in the profession to yield favorable results, it is possible to obtain a reasonably reliable appraisal of this phase of methods engineering activity. The following sample checklist of questions illustrates how this appraisal of facilities and methodology should proceed.

1. What percentage of the workday is devoted to profitable endeavor?
2. With respect to personnel:
   a. what about the technical capability of each staff member? What about the ability of each to get along satisfactorily with other personnel in the organization, especially shop personnel?

b. is each adequately trained and indoctrinated? Equipped with up-to-date "know-how"?
c. what is the general quality of standards produced by each observer?
d. what is the "productivity" of each observer with respect to standards set?

3. With respect to methodology:
   a. is the department's general approach in accord with current "enlightened" theory and practice?
   b. is the methodology of the department being developed to meet anticipated trends in the business, industry, and profession?
   c. is adequate use being made of the full range of techniques available to the time-study practitioner?
   d. is the department doing an effective job of implementing the standards it sets?
   e. is the department doing an effective job in following up on the standards it sets?
4. With respect to programming:
   a. is an adequate percentage of the department's work measurement capacity being devoted to maintenance of standards? To indirect labor?
   b. is the amount of time devoted by skilled time-study observers to routine clerical tasks kept to a minimum?
5. With respect to relations with the remainder of the organization:
   a. has the department been successful in retaining management's support?
   b. has the department been successful in retaining the prerogative to set standards in the manner it sees fit?
   c. has the department been able to maintain a satisfactory relationship with shop supervision? How cooperative are foremen with the department?
6. With respect to evaluation itself:
   a. is a sincere attempt made to follow up on, evaluate, and improve performance?
   b. is an effort made to periodically reappraise the functions of this activity and the measurements produced?
   c. is a periodic attempt made to evaluate the scale of the company's standards-setting activity, to determine if expansion or contraction is desirable?
   d. is a conscientious and effective attempt made to obtain *and* benefit from error feedback? Are rating films used periodically? Are erratic standards sought out, analyzed, and benefited from? Are results analyzed to detect differences in the level of standards in different departments, in different labor grades, on different types of operations, or set by different time-study men?
   e. is a systematic attempt made to evaluate the quality of standards established by each technique used, and to compare effectiveness of these techniques?

Utilization of this Knowledge of Results

As discussed and illustrated in Chapter 30, knowledge of results should be put to good use in the improvement of all phases of the department's activity. Knowledge of source, type, and relative frequency of errors permits the administrator to make intelligent decisions concerning such matters as correction and replacement of individual staff members, alteration, augmentation, or replacement of various techniques, reallocation of work measurement effort, and the like. For example, on the basis of feedback obtained:

1. various observers can be informed of their tendencies to rate higher or lower on certain types of operations, or in certain departments;
2. a certain observer might be prevented from setting standards on some types of operations because of errors made;
3. a certain observer might be transferred to another type of work because of inability to set acceptable standards;
4. another observer might be given some special training or tutoring;
5. the administrator might decide that a certain set of standard data can no longer be relied on and should receive a major overhaul;
6. the administrator might decide that delay allowances are too far out of line, and should be re-estimated by work-sampling studies.

In reference to efforts to isolate loosely set time standards, the following question often arises: "Why bother attempting to detect a loose standard if company-union contract provisions prevent the company from correcting such a standard if it is brought to light?" In spite of this predicament there are several sound reasons for wanting to learn what standards are loose. One, as emphasized earlier, is to permit current standards-setting procedures and personnel to be intelligently evaluated and effectively improved. Another is that if a standard is loose and output is being restricted, the operators should ordinarily be encouraged to increase their production and take advantage of the liberal standard. Holding productivity back under the circumstances is going to increase, not decrease, unit cost. Of course, if workers are to be encouraged to do this, management must locate the cases in which looseness exists. And finally, many loose standards are such because of changes in method that have taken place since these standards were established. In this instance management does have the prerogative to reset the standard and should do so. Of course, before a loose standard can be revised it must be found.

EXERCISES

1. A certain drillpress operation is run intermittently, in batches of 2000 units. Over the past year the performance ratios for runs of this job have been as follows:

Date of Run	Performance Ratio (Ratio of Standard Hours to Actual Hours)	Operator
6/15	1.48	*C*
7/24	1.48	*C*
9/7	1.48	*C*
9/29	1.47	*C*
1/21	1.48	*C*
2/8	1.48	*A*
3/4	1.48	*C*
3/27	1.47	*C*
4/17	1.48	*C*
5/19	1.48	*A*
5/30	1.47	*C*
6/12	1.48	*C*

What do you suspect is causing this remarkable consistency in incentive efficiencies? What would you conclude if a similar situation prevailed for most standards in this drillpress department? Explain.

2. A representative of the management of a firm tells you that their time standards are accurate since they rarely have an employee complain about a tight standard and very seldom find an operator earning "excessively" high incentive pay. How do you view this statement? Why?

3. On a certain operation the performance ratio over a prolonged period has been approximately 160 per cent. What can you conclude about the time standard? What could you conclude if the long-term performance ratio were 105 per cent? What could you conclude if it were 135 per cent? Explain.

4. How would you answer problem 3 if in each case only one operator performed that particular job? How would you answer it if there were ten operators performing the job in each case?

5. As a consultant, you have been retained by a large corporation to evaluate the work measurement activity of the company's methods engineering department. Your appraisal is to be accompanied by proposals for improvement. How would you proceed to make your appraisal and arrive at suggested improvements?

32

Methods Engineering Administration: Maintenance of Time Standards

Recall that one of the primary purposes of the follow up on time standards is to assist in keeping the standard on each operation representative of current method and conditions. The term "maintenance of time standards" will be used to describe the process involved in attempting to keep the standards on operations abreast of changes in methods and conditions. A time standard structure, like machinery, will lose its effectiveness at a rapid rate unless it is properly maintained. The obsolescence of time standards means unsatisfactory results in manufacturing planning and control and from the incentive wage plan. To minimize the decay of a time standard structure it is essential that the methods engineering department take steps to minimize the chances of obsolescence, to keep informed of the status of its standards, and to make adjustments in time standards as dictated by changes in plant methods and conditions.

The Need for Maintenance of Time Standards

A methods engineering department is naive if it operates under the assumption that a job and its environment will remain "fixed" after the time standard has been established. Yet this is exactly the tacit assumption under which many departments appear to be operating. A time standard becomes obsolete just as a piece of equipment does; in fact, the time standard is considerably more vulnerable to obsolescence than a piece of machinery is. In addition, the obsolete time standard is usually less obvious than an obsolete machine. If methods engineering is to prevent inconsistencies from creeping into the time standard structure, it must keep abreast of changes in methods and conditions in the shop.

There is no logic in striving to *set* more consistent time standards and then making little or no effort to prevent these same standards from drifting out of line *after* they are set. Many managements are making elaborate efforts to improve the consistency of their time standards. In most instances, however, they are concentrating these efforts on setting a more consistent standard, and paying little or no attention to preserving that consistency after the standard is set and installed. It does not take long for a time standard structure to develop numerous inequities if this maintenance function is neglected. The sensible approach is to submit time standards to an audit procedure that will maintain, as far as possible, their original consistency.

Administrators responsible for failing to adopt a maintenance procedure sometimes assume this position because they believe that the time standard structure is functioning satisfactorily without such control. Perhaps these managers do not realize that their time standards may *appear* to be consistent when the true situation is quite the reverse. As mentioned earlier, restriction of output can be surprisingly effective in concealing loose time standards and thus the true quality of the time standard structure. The following situation is probably more common than the preceding. The company is concerned over the inferior quality of its time standard structure but does not realize that inadequate maintenance is an important, if not the most important factor causing the dilemma. In this case the administration fails to understand the true nature of its problem. In the author's opinion, many companies, either because the problem is diagnosed incorrectly or because it is not even recognized, are neglecting this maintenance function when it would be much to their advantage to do otherwise.

The reasons that some managements express for wanting more precise methods of establishing time standards are the very same reasons why they should be attempting to keep their existing standards as much "in line" as possible. Most of these reasons concern inconsistencies among standards and their effect on costs. When changes in method and conditions take place and appropriate adjustments are not made in the corresponding standard times, tight and loose standards result. These inconsistent standards mean extra costs to management by way of the following.

1. Errors in scheduling, budgeting, estimating, and other management planning and control techniques.
2. Employee dissatisfaction with the inequities in incentive earnings

arising from inconsistent time standards. This dissatisfaction increases costs through grievances, employee turnover, intraplant transferring, antagonism, and indifference. (Some readers ask, "Why not let the operators who devise better methods be rewarded for them through the increased earnings they can make under unchanged time standards?" Under this arrangement the earnings differentials would soon be tremendous, eventually leading to a complete breakdown of the incentive system. To cite just a few reasons: workers' earnings would depend considerably on chance since there is a large chance element involved in the "discovery" of methods improvements; earnings would depend considerably on the potential that the operator's particular job offered for improvements, and operations vary greatly in this respect.)

3. When an improvement in method increases an operator's earning capacity (under the unchanged standard) beyond a certain point, he will usually resort to restriction of output to protect his loose standard. Under these circumstances, the full production potential of the improvement is not realized. Consequently, management loses money through restricted output. Unfortunately the monetary loss associated with such situations is not obvious to the ordinary management. It is difficult to "see," for example, the money that is lost through poor employee morale or through restriction of output.

It is well known that some incentive wage installations have developed into costly headaches, causing the managements to discard or wish they could discard the entire program. There is reason to believe that inadequate maintenance of time standards has been a major factor contributing to the failure or malfunctioning of these plans. This is a contention worth serious consideration, particularly since the desirability of incentive wage plans is becoming increasingly uncertain.

It is important that the time standard be adjusted at the time a change in method takes place. If there is a long delay between the time a method improvement is made and the time standard is reset, the worker is prone to view the adjustment as an arbitrary "rate cut," which results in violent objection. If the standard is changed at the time the change in method occurs, the connection between the two changes is apparent, and provides less opportunity for such objection and dissatisfaction. It is important then, not only that the standard be reset but that it be done at the time the method change occurs. Therefore, the company should not rely on chance or on excess earning reports to locate obsolete standards, for these procedures are undependable and likely to result in undesirable delays. Rather, it should adopt a maintenance procedure that will offer reasonable assurance that improvements will be known before or at the time they are made.

In summary, if the development of serious inequities in the time standard structure is to be avoided, if time standards are to be kept representative of current methods and conditions, the methods engineering department must keep abreast of change. This is the main objective of the maintenance procedure. The department must keep informed about changes in method and conditions and take appropriate action, otherwise the time standard structure will gradually decay, bringing dissatisfaction to both labor and management.

To satisfactorily maintain the quality of a time standard structure the following procedure is necessary.

Prevention—which consists of certain measures the methods engineering department should take in order to minimize the chances that a standard will become "out of line" after it has been established.

Detection—which involves certain techniques and procedures that methods engineering should utilize to isolate tight and loose standards.

Remedy—the appropriate corrective action that should follow when a tight or loose standard (resulting from change in method or conditions) is noted.

Preventing Standards from Becoming Tight or Loose

Substantial dollar savings may be achieved through proper preventive maintenance of equipment. The same is true of preventive maintenance of time standards, for it is more economical to take steps that will minimize the chances of standards becoming tight or loose than to face the higher costs of grievances, loose standards, low morale, restriction of output, malfunctioning planning procedures, and the like, that result if such steps are ignored.

Following is an outline of the measures in this preventive maintenance procedure. These recommendations are directed at the period that begins immediately after the standard has been established. They are precautions designed to minimize the chances of a standard becoming tight or loose *after* it has been set. Note that some of these measures have advantages other than the one cited here. These other advantages are usually ample justification in themselves for adhering to these recommendations.

1. Make an attempt as extensive as economy will allow, to *improve the method before* the standard is set. Of the benefits received from this practice, the one of particular interest here is that there will probably be fewer improvements made *after* the standard is set. Every company should be following this practice if only for the dollar savings that are to be gained through less costly methods of operation. This alone is ample reason. But in addition, here is one of the most effective steps that management can take to prevent standards from becoming loose.
2. Improve the conditions surrounding the job *before* the standard is set. Reducing the number and duration of delays to a minimum is particularly important.
3. Make a detailed record of the method and conditions on which the time standard is based.
4. Obtain maximum benefit from the standard method description by making it the detailed record of method and conditions *and* by having it used as a basis for periodic checks of the method and conditions to determine if changes have taken place. This record should include the standard time for the operation, to encourage personnel to associate that time with the method shown and vice versa.
5. Educate supervisory and staff personnel, particularly foremen, as to their responsibilities in maintaining the time standard structure. This program should stress:
   a. the importance of keeping standard the method and conditions of the operation. Of course the supervisor or staff member should attempt to restore the method and conditions to standard only when the change has been an adverse one, for example worn machine bearings. When the change is a desirable one, such as an improvement in method, or when it is an adverse change that cannot be rectified for the time being, the fact should be reported so that the standard can be checked and adjusted if necessary. The second point to be stressed then is:
   b. the importance of reporting changes in method and conditions if these are desirable changes or adverse changes that cannot be rectified, so that standards can be promptly revised. It is particularly important that supervisory, engineering, methods, quality control, plant layout, maintenance, and tool design personnel be urged to cooperate in this respect.
   c. the importance of the foreman's cooperation to the success of a standards maintenance program and in turn the importance of such maintenance to the quality of the time standard structure. Without his support, several of these preventive steps will be to no avail, others will be seriously hampered. To help the foreman understand and carry out his part in maintenance of time standards, it is recommended that he be well informed as to the functions of and procedures used in time study. A series of training sessions seem well worth the investment for this and other reasons.

In summary, one of the most vital of the preventive maintenance steps is assuring that supervisory and staff personnel know their responsibilities in the upkeep of the time standard structure.

6. Establish a successful suggestion plan: In addition to the advantages usually associated with such an arrangement, it also plays a part in the prevention of loose standards. The operator should at all times be encouraged to develop an improved method of performing his job. He should *also* be encouraged to submit his idea so that he may be rewarded directly for it. We do not want the originator to use his improvement under an unchanged time standard. A suggestion plan with sufficient rewards, reasonable policies, and adequate publicity should aid in achieving these objectives.
7. Establish a routine paper procedure that notifies the methods engineering department of changes in method or conditions that may affect performance time. A 3-by-5-inch slip of paper will suffice. Supervisory and staff personnel should be required to use these forms to report changes that may affect time standards. Here are some of the intended users of these forms along with a sample of the type of information that the forms are designed to convey from each.
   a. Foreman. Necessity of using a less efficient machine than the one on which the time standard was originally based.
   b. Quality control department. Unavoidable decrease in the quality of raw stock.
   c. Product engineering department. Redesign of a part.
   d. Maintenance department. Installation of a new machine.
8. Establish responsibility for maintenance of time standards. Fixing of this responsibility is essential if the maintenance procedure is to succeed.

Detection of Standards that have Become Tight or Loose

A procedure should be established to call attention to those standards that have drifted out of line as a result of change, for in spite of the best preventive efforts a certain percentage of time standards in use will become obsolete and in need of detection and revision.

In brief, such standards may be detected via their cause or via operator output patterns under those standards. In the former method, detection is achieved by locating the change in method or conditions causing the tightness or looseness. In the latter method it is achieved by noting from production records certain characteristics in output patterns that are indicative of tightness or looseness, as described in Chapter 31. Both methods are of limited effectiveness.

Many of the standards that are set tight or later become that way are brought to management's attention by operators on the job. Unfortunately, however, operators are not always aware of the situation when their standard is tight. In addition, many operators complain about a "fair" standard in an attempt to have it loosened. Therefore, the problem for tight standards is one of detection in some instances, of verification (of a claim) in others. One method of detecting or verifying a tight standard is to find the change in method or conditions that has caused the tightness. For example, burrs on parts cause an increase in performance time for a certain assembly operation. Investigation of a protested standard for that operation reveals that the percentage of parts with burrs has increased since the standard was set, so it is concluded that the standard is tight. If the situation cannot be remedied, a restudy of the operation should be made to determine how much of an adjustment in time standard is warranted.

It is also possible to detect a loose standard via the cause. As previously recommended, the time study department should establish a formal communication channel with supervisory and staff personnel, to assist it in keeping informed of changes. Furthermore, operations should be periodically checked against the standard method description to determine if changes have taken place. A periodic review of this type should be particularly effective in diminishing the effects of "creeping changes."

In summary, detection of an obsolete time standard "via the cause" involves finding a deviation in method or conditions from those on which the standard was originally based.

Adjustment of Tight and Loose Standards Caused by Job Changes

Some method changes affect performance time to such a small degree that no adjustment in the standard is justified. Therefore this question arises: to what degree must a method change affect performance time to justify an adjustment in the standard? This is determined by that degree of change at which the costs involved in retaining the obsolete standard become larger than the costs of making the adjustment. This is a very difficult matter to quantify even in a general way, so that management must use its judgment in selection of this minimal effect. Regardless of the point selected, whenever a change occurs that apparently does not justify an adjustment, it is essential that the nature of the change and its estimated effect on performance time is recorded. Then when the cumulative effect of such relatively small changes becomes sufficiently large, this will be known and the standard can and should be adjusted accordingly.

appendix A

The Methods-Time Measurement System for Estimation of Manual Performance Time

A Caution

The purpose of the following material is to familiarize the reader with a popular and promising method of estimating manual performance times, namely, predetermined motion times. This is best accomplished by using one of a number of specific predetermined motion-time systems as an illustration of the general technique. The Methods-Time Measurement (MTM) system has been selected to serve as this example. MTM was chosen because it is the most popular system in industry and because its originators have been very liberal in revealing the source and details of their system to the public. Selection of this system was not based on any apparent superiority of MTM over other available systems.

It is important that the reader realize that the training in application of the MTM procedure offered herein is the *minimum amount* required to enable the student to make a *sample application* of the system under guidance. This exposure to the system is at the appreciation level only and is *no substitute for the comprehensive course considered necessary to qualify a person as an MTM practitioner.* The time that can justifiably be devoted to a specific system in a college level course does not permit this.[1] Furthermore, a substantial number of the system's details have been omitted from this write-up; others have been simplified. As long as this material is used as the basis of practice applications only, the omission and simplification of certain details will be of no practical consequence.

[1] Approximately 105 hours of class time are considered minimal to prepare the industrial practitioner for application.

METHODS–TIME MEASUREMENT (MTM)— A SPECIFIC PREDETERMINED MOTION–TIME SYSTEM

To develop the MTM system, the originators took a number of motion pictures of industrial manual operations. A careful study of these films indicated that most industrial motion patterns could be synthesized from eight basic motions: reach, move, turn, grasp, position, release, disengage, and apply pressure. From the films the originators obtained a quantity of time values for these eight motions and proceeded to determine what task variables affect the expected performance time for each motion. Their investigations indicated for example that the time to move an object is influenced by the distance the hand moves, the amount of control that must be exercised, and the weight of the object. As an outcome of their investigations, the originators concluded that to yield acceptable estimates of task performance times, a system of this type must provide a set of time values for each motion to recognize variables like the ones preceding. The results of their investigations are shown in Tables A-1 through A-7 (see end of Appendix A), which include MTM time values for finger, hand, and arm motions.[2] In order to be able to properly identify and classify these motions, it is necessary to become thoroughly familiar with a considerable number of details concerning each.

The MTM Motion Reach

Reach is the motion employed when the predominant purpose is to move the hand or finger to a destination. Note that it may be a movement of the finger, hand, or forearm. Reach begins the instant movement starts and ends when the body limb is approximately one inch from the destination.

One variable recognized is the length of the movement. Results of the investigations conducted to develop this system indicated that reach time varies with length of movement as indicated in Figure A-1. Therefore, in the MTM table of time values, reach times are provided for motion lengths ranging from 1 to 30 inches. Note that this distance is that which the limb actually traverses, and not the straight line distance between the point of origination and point of termination. It is suggested that measurement be made of the path of some reference

[2] One motion, disengage, has been omitted in this presentation because of the infrequency with which it is encountered.

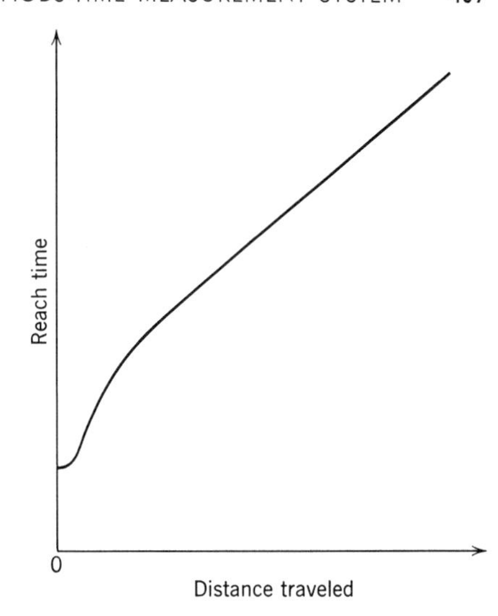

Figure A-1. General shape of the time-distance curve for the MTM motion reach.

point like the large knuckle of the index finger. Also, this length is the *net* distance traversed by the hand or finger. For example any contribution made by movement of the body trunk should be deducted from the gross distance traveled by the hand.

The second important variable accounted for is the amount of control required to successfully execute a reach. Control here is recognized as being a function of the size, shape, and location of the object to which the hand is reaching. For example, more time is logically allotted for reach if the object is jumbled with other objects than if it is lying on a surface by itself. Note that these variables, size, shape, and location, that logically affect the grasp time of an object are treated as factors also influencing the reach time that precedes that grasp. This influence has been verified by a number of independent investigations. Control is recognized by providing four "cases" of reach, each case representing a certain degree of control and assigned a separate set of time values. Thus, to select a time for a particular reach in an operation, it is necessary to determine two things about that reach.

1. The length of the motion;
2. The appropriate case of reach.

To assist the beginner in making these determinations, Table A-8 provides definitions, explanations, examples, and other details.

To facilitate the recording of various MTM motions, a system of symbols is used. For reach, the letter *R* is followed by the number of inches traveled and then by a capital letter designating the proper case. Thus a case *B* reach of 10 inches is abbreviated as *R*10*B*, or a case *C* reach of 8 inches as *R*8*C*.

The purpose of tables A-8 through A-15. Tables A-8 through A-15, appearing in the latter portion of Appendix A, summarize the definitions and rules required to identify, classify, and record motions in synthesizing a motion pattern by means of the MTM procedure. These tables were especially prepared to eliminate the need to memorize a vast number of details, or to continually thumb through pages of text, in order for the novice to apply MTM. When applying this system, the novice should have Tables A-8 through A-15 available for ready reference and rely on them in the selection, classification, and recording of motions required. After the motions have been recorded, Tables A-1 through A-7 may be consulted to obtain the time values for the motions involved.

Tables A-9 through A-13 will serve to provide the necessary familiarity with the motions move, grasp, turn, release, and apply pressure. Since there are more variables and complications involved in the classification of the motion position, some discussion will supplement Table A-14.

The Motion Position

There are three (often overlapping) phases to the element position.

1. Alignment, bringing the axis of the plug parallel and close enough to the axis of the hole to permit engagement. This is indicated by the dashed arrows in Figure A-2.
2. Orientation, rotation of the plug about its axis of travel to bring the mating shapes near enough to coincision to permit engagement. The nature of this movement is indicated by the solid arrows in Figure A-2.
3. Engagement, placing the plug into the hole.

The major variable affecting position time is the degree of control required, which is recognized in the MTM system as being a function of three factors: the amount of clearance between plug and hole, the amount of orientation required, and the degree of difficulty of handling.

To account for the first of these factors—amount of clearance—three "classes of fit" are allowed: loose, close, and exact. Thus in selecting a position time for a particular instance, the appropriate

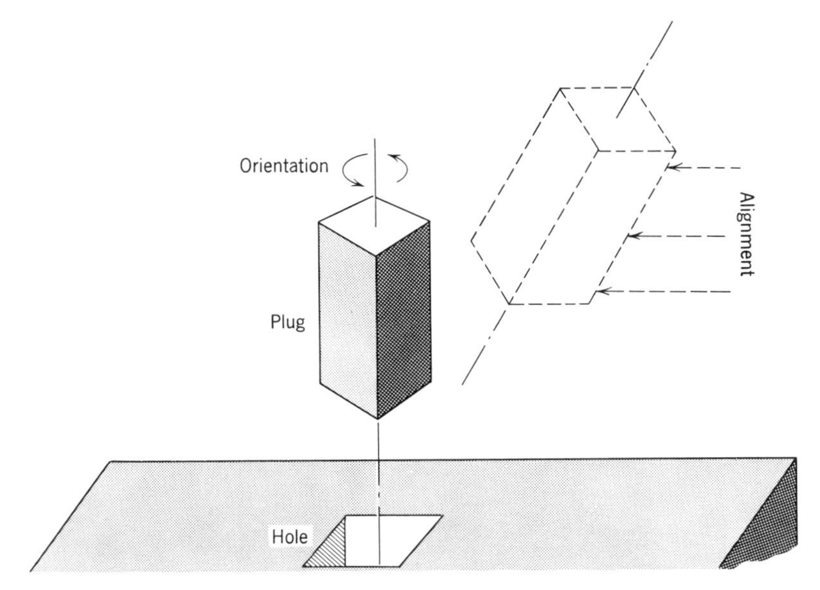

Figure A-2. Illustration of the alignment and orientation phases of the MTM element position.

"class of fit" must be assigned. These classes are described in Table A-14. To account for the second factor—degree of orientation required—three conditions of "symmetry" are allowed: symmetrical, semisymmetrical, and nonsymmetrical. To facilitate selection of the appropriate condition of symmetry for a particular instance, detailed descriptions are presented in Table A-14. The final factor considered is the difficulty offered by the parts in handling during positioning. Two degrees of difficulty are provided for, as described in Table A-14.

As clearance between the plug and hole increases, the time required to position decreases until no time allowance over and above the move time is justified. The clearance at which this occurs in MTM is $\frac{1}{2}$ inch. Thus, when the hole is at least $\frac{1}{2}$ inch larger than the plug, no position time is allowed.

Synthesizing a Motion Pattern with MTM

Being able to identify and classify the various MTM motions is not the most difficult aspect of applying MTM. More critical and more demanding of the analyst's judgment is the problem of determining in what manner these various individual motions can or should be put together to form an integrated motion pattern. The rules specified in the MTM system for building up a motion pattern are to be introduced.

The simplest task to synthesize is that involving a series of motions performed in succession by one hand only, the other hand being idle. Determination of the time for this task is the straightforward summing of the times for the various motions involved.

Still assuming a one-handed operation only, it is possible that two motions can and should be performed concurrently. For example, the arm may perform a turn as it reaches to an object, or the hand may perform a regrasp as the arm moves an object. Two or more motions performed concurrently by the same main body member (arm or leg) as in the foregoing cases are referred to in the MTM system as combined motions. When combined motions are encountered in a task, for example an *R5C* and a *T*45° performed concurrently as the hand reaches to a pile of parts, the time to allow for the combination is determined on the basis of the following rule.

Find the appropriate time for each motion in Tables A-1 through A-7. Allow the longer of the two times for performing the combination.

The motion with the longer time is described as limiting. The other is said to be limited out. In the example cited, Table A-1 indicates the time for an *R5C* is 0.0056 minute, Table A-6 indicates the time for a *T*45° is 0.0021 minute. Therefore the *R5C* is the limiting motion, and the time allowed to perform the combination is 0.0056 minute. Thus, in synthesizing the total time for the task of which these motions are a part, the limiting motion is treated as if it were performed by itself, and the motion limited out has no bearing on the total time. The foregoing rule may be extended to three motions performed in combination such as a move, turn, and regrasp.

Synthesizing the time for the case where both hands are working simultaneously is not so straightforward. When two main body members perform motions concurrently, the motions are referred to in the MTM system as simultaneous motions. A common example of simultaneous motions is reaching concurrently with left and right hands to bins of parts.

The key to the synthesis of a two-handed (bimanual) motion pattern under the MTM system is the following assumption. Given a left-hand and a right-hand motion, *either these motions can* be performed simultaneously *or they cannot* (they must be performed successively), depending on the particular motions involved and on the opportunity for practice afforded by the job. The former alternative is illustrated

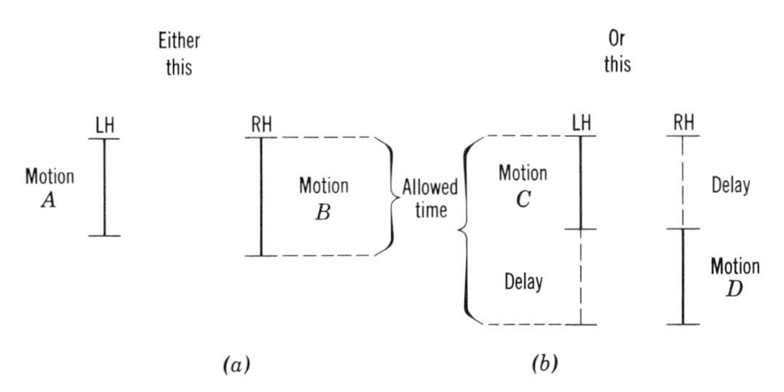

Figure A-3. The MTM system allows one or the other of these situations for bimanual motion combinations: simultaneous performance (*a*), or successive performance (*b*).

graphically in Figure A-3*a*, the latter in Figure A-3*b*. Actually, most left- and right-hand motion combinations can be overlapped to some degree, depending on the particular motions involved. As a consequence of the discreteness of the system, the actual degree of overlap is approximated by one or the other of the extremes allowed.

The assumption that performance of a pair of bimanual motions will be either concurrent or successive underlies Table A-15, the Table of Simultaneous Motions. The table, an integral part of the MTM data, indicates to the MTM practitioner whether or not he can expect simultaneous performance of any given pair of motions. The practitioner proceeds as follows.

1. If Table A-15 indicates that the two motions can be performed simultaneously under given conditions of practice, he specifies concurrent performance of the two motions as illustrated in Figure A-3*a*. Here the time allowed for the motion combination is the *longer* of the two times as given in Tables A-1 through A-7. The motion requiring the longer time is limiting, the other is limited out.

2. If Table A-15 indicates that the two motions cannot be performed simultaneously under the given practice conditions, the practitioner specifies successive performance of the two motions, as illustrated in Figure A-3*b*. Here the time allowed for the motion combination is the *sum* of the times for the two motions as given in the tables of motion times.

For a specific application of this principle, assume that in a certain operation it is desirable to reach to and grasp two parts (on the table by themselves) with simultaneous left- and right-hand motions. The questions that must be answered with the aid of Table A-15 are: (1) can a reach case *B* be performed simultaneously with a reach case *B*? and (2) can a *G1A* be performed simultaneously with a *G1A*? The answer is yes to both questions, for Table A-15 says both pairs are "easy to perform simultaneously." Therefore an MTM synthesis of this bimanual motion sequence assuming a ten inch reach would be as follows.

LH	Time in 0.001 minute	RH
R10B	6.9	*R10B*
G1A	1.0	*G1A*
	7.9	

For a second sample application of this principle, assume that in a certain operation it is desirable to reach to and grasp two parts jumbled with other objects in a bin, with simultaneous left- and right-hand motions. Table A-15 must now be consulted to answer these questions: (1) can a reach case *C* be performed simultaneously with a reach case *C*? and (2) can a *G4* be performed simultaneously with a *G4*? According to Table A-15 these reaches and in fact all reach-move combinations can be performed simultaneously, so the answer to question 1 is yes. But the answer to question 2 is no, for Table A-15 says *G4*'s cannot be performed simultaneously even when within 4 inches of one another. The MTM synthesis for this motion sequence, assuming that the reach distance is 10 inches, is as follows.

LH	Time in 0.001 minute	RH
R10C	7.7	*R10C*
Delay	5.9	*G4*
G4	5.9	Delay
	19.5	

Notice the successive performance of the *G4*'s; one hand hesitating while the other hand performs the grasp. Either hand could have performed its *G4* first.

For a third sample application of the simultaneous motions principle, assume that in an operation it is desirable to move two different parts 6 inches to an assembly fixture with simultaneous left-hand right-hand motions, one part requiring a *P1SE* and the other a *P1SD*. Of course the moves can be performed simultaneously, but what about the positions? According to Table A-15 these positions can be performed concurrently with practice. Assuming that this is a low-volume job, so that there is little opportunity for practice at performing these motions, the MTM synthesis for this sequence is as follows.

LH	Time in 0.001 minute	RH
M6C	6.2	*M6C*
P1SE	3.4	Delay
Delay	6.7	*P1SD*
	16.3	

If this were a high-volume job, the MTM synthesis for this motion sequence would be:

LH	Time in 0.001 minute	RH
M6C	6.2	*M6C*
(*P1SE**)	6.7	*P1SD*
	12.9	

* It is conventional in this system to circle a motion that is limited out, indicating that this motion has no bearing on the task time allowed.

By following the patterns illustrated in the preceding three examples, it will be possible to handle a majority of situations that arise in bimanual motion patterns. These typical patterns are summarized in Table A-16.

Procedure to Follow in Making an MTM Study

1. Break down the operation into intermediate sized elements composed of not more than about twelve motions each. A "get and place" breakdown is well suited for this purpose. Examples of such elements are: get part, place part in fixture, close fixture, engage feed, etc. This step is important to successful visualization of the motions required, especially to the beginner.

2. Identify the MTM motions required to execute each of these intermediate sized elements. Concentrate on one element at a time, using Tables A-8 through A-15 to aid in classifying the motions.

3. Record these elements and motions following the conventions shown in Figure A-4. Note in this figure the method of designating combined motions (second last line of right-hand motions). An *M5C*

Synthesis Sheet

Operation: *Remove pen from pocket and prepare to write.*

Left-Hand Description	Symbols LH	Limiting Time in 0.001 Min.	Symbols RH	Right-Hand Description
1 Remove pen from pocket	Idle	7.7	R12B	Pen
	Idle	1.0	G1A	Pen
	Idle	4.1	M4B	Out of pocket
To right hand	(R3A)	8.6	M14A	To left hand
2. Loosen cap with fingers and remove	G3	3.4	G3	Pass to L.H.
	Hold	1.5	R1B	To cap
	↓	1.0	G1A	Cap
		1.7	M1B	Cap
		1.0	RL1	Cap
		1.5	R1A	Cap
		1.0	G1A	Cap
		1.7	M1B	Cap
		3.4	M3B	Remove cap
3. Turn pen, put cap on other end				
Turn pen end for end	T120°	4.1	Idle	
To cap	(M1C)	3.1	M2C	To pen
Into cap	P1SE	3.4	P1SE	On pen
	AP2	6.4	AP2	Force cap on pen
4. Shift fingers, prepare to write				
Pen	Hold	1.0	RL1	Release cap
	Hold	2.4	R2A	To pen
To paper	G3	3.4	G3	Transfer to R.H.
	(R4B)	5.5	M5C and (G2)	To paper and shift fingers
Paper	G5	3.4	P1SE	To paper
Total		70.3		

Figure A-4. Synthesis of the performance time for a simple task using the MTM system.

and a $G2$ are performed as combined motions with the $G2$ limited out.

4. After all motions have been identified and recorded, consult the tables of time values (Tables A-1 through A-7) and enter the times on the synthesis sheet.

5. Total the limiting times to obtain the expected time for the task.

6. Record the workstation layout and describe the equipment used. When this is done the method on which the resulting standard is based is adequately specified.

METHODS–TIME MEASUREMENT TABLES OF MOTION TIMES[3] (Time in Thousandths of a Minute)

TABLE A-1. Reach Times

Case	Description	Length of Reach in Inches	Time Values for Case *A*	Case *B*	Cases *C* and *D*
A	Reach to object in fixed location or to object in other hand or on which other hand rests.	1	1.5	1.5	2.2
		2	2.4	2.4	3.5
		3	3.2	3.2	4.4
		4	3.7	3.8	5.0
B	Reach to a single, easily grasped object in location which *may* vary slightly from cycle to cycle.	5	3.9	4.7	5.6
		6	4.2	5.2	6.1
		7	4.4	5.6	6.5
		8	4.7	6.1	6.9
		9	5.0	6.5	7.3
C	Reach to object jumbled with other objects.	10	5.2	6.9	7.7
		12	5.8	7.7	8.5
D	Reach to a single, very small object or object requiring an accurate grasp, in a location which *may* vary slightly from cycle to cycle.	14	6.3	8.6	9.4
		16	6.8	9.5	10.2
		18	7.4	10.3	11.0
		20	7.9	11.2	11.9
		22	8.4	12.1	12.7
		24	8.9	12.9	13.5
		26	9.5	13.7	14.3
		28	10.0	14.6	15.2
		30	10.5	15.5	16.0

[3] Reproduced with the permission of the MTM Association for Standards and Research.

TABLE A-2. Grasp Times

Case	Description	Time Values
G1A	*Simple grasp of single object.* Small, medium or large object by itself, easily grasped.	1.0
G1B	*Hook or pinch grasp of single object.* Very small object or flat object lying close against flat surface.	2.1
G2	*Regrasp.*	3.4
G3	*Transfer grasp.*	3.4
G4	*Object jumbled* with other objects.	5.9
G5	*Contact grasp.*	0.0

TABLE A-3. Move Times

Case	Description
A	Move object against stop or to other hand.
B	Move object to approximate or indefinite location.
C	Move object to exact location. (Always precedes position)

Weight allowance

Weight range in pounds	Multiply table time by the following factor	And add this value
0–2.5	1.00	0
2.6–7.5	1.06	1.3
7.6–12.5	1.11	2.3
12.6–17.5	1.17	3.4
17.6–22.5	1.22	4.4
22.6–27.5	1.28	5.5
27.6–32.5	1.33	6.5
32.6–37.5	1.39	7.5
37.6–42.5	1.44	8.6
42.6–47.5	1.50	9.6

Example: Time for an *M6B10* is $(5.3 \times 1.11) + 2.3 = 8.2$

Length of Move in Inches	Time Values for Case *A*	Case *B*	Case *C*
1	1.5	1.7	2.0
2	2.2	2.8	3.1
3	2.9	3.4	4.0
4	3.7	4.1	4.8
5	4.4	4.8	5.5
6	4.9	5.3	6.2
7	5.3	5.8	6.7
8	5.8	6.4	7.1
9	6.3	6.9	7.6
10	6.8	7.3	8.1
12	7.7	8.0	9.1
14	8.6	8.8	10.1
16	9.6	9.5	11.2
18	10.6	10.2	12.2
20	11.5	10.9	13.3
22	12.5	11.6	14.3
24	13.4	12.4	15.3
26	14.4	13.1	16.4
28	15.3	13.9	17.4
30	16.3	14.6	18.4

TABLE A-4. Apply Pressure Times

Case	Description	Time Values
*AP*1	With a regrasp	9.8
*AP*2	Simple	6.4

TABLE A-5. Position Times

Class of Fit	Easy to Handle			Difficult to Handle		
	Symmetrical	Semisymmetrical	Nonsymmetrical	Symmetrical	Semisymmetrical	Nonsymmetrical
*P*1 Loose No pressure required.	3.4	5.5	6.2	6.7	8.8	9.6
*P*2 Close Light pressure *or* careful alignment and orientation required.	9.7	11.8	12.6	13.1	15.2	16.0
*P*3 Exact Heavy pressure required.	25.8	27.9	28.7	29.2	31.3	32.0

TABLE A-6. Turn Times

Turn	Time Values	Turn	Time Values
*T*30°	1.7	*T*120°	4.1
*T*45°	2.1	*T*135°	4.4
*T*60°	2.5	*T*150°	4.9
*T*75°	2.9	*T*165°	5.2
*T*90°	3.2	*T*180°	5.6
*T*105°	3.7		

Degrees of turn determined by rotation of forearm about long axis.

TABLE A-7. Release Times

Case	Description	Time Values
*RL*1	Normal release performed by opening fingers as independent motion.	1.0
*RL*2	Contact release.	0

TABLE A-8. Selection of Reach Time

Definition	Motion employed when the predominant purpose is to move the hand or finger to a destination.
Begins and Ends	Begins when movement starts. Ends when the limb is approximately 1 inch from target.
Must Determine	(1) Length of motion. (2) Case of reach.
Notation	*R*4*A*—4-inch reach case *A*; *R*10*C*—10-inch reach case *C*.

Selection of Case of Reach

Case	Explanation	Examples
A. Reach to object in fixed location or to object in other hand or on which other hand rests.	Requires practice as well as a fixed location.	Gearshift lever in auto. Spindle lever on drill press.
B. Reach to a single easily grasped object in location which *may* vary slightly from cycle to cycle.	Location of object is approximate. Allow case *D* if object is very small.	Eraser by itself on desk. Pencil in pocket.
C. Reach to object jumbled with other objects.	Applies to dissimilar as well as similar objects.	Parts in tote pan. Pile of paper clips on desk.
D. Reach to single, very small object or object requiring an accurate grasp, in a location that *may* vary slightly from cycle to cycle.	Allow for objects $\frac{1}{8}'' \times \frac{1}{8}''$ in cross section, or flat objects close to flat surface, or where danger is involved, etc.	Straight pin, wooden match, coin, wrench handle, or piece of paper, each on a flat surface by itself. Part with sharp burrs.

TABLE A-9. Selection of Move Time

Definition	Motion employed when the predominant purpose is to transport an object to a destination.
Begins and Ends	Begins when movement starts. Ends when object is approximately 1 inch from target.
Must Determine	(1) Length of motion. (2) Case. (3) Weight of object.
Notation	*M6B*—6-inch move case *B*, under 2.5 pounds. *M8C*10—8-inch move case *C* of 10-pound object.

Selection of Case of Move

Case	Explanation	Examples
A. Move object against stop or to other hand.		Close door. Slide jig against stop.
B. Move object to approximate or indefinite location.	Only requirement is that object be placed in general location.	Move object to tote pan. Raise hammer to strike blow. Toss object.
C. Move object to exact location.	Always precedes the MTM motion position.	Move part to jig. Move thread to eye of needle. Move cap to pen.

TABLE A-10. Selection of Grasp Time

Definition	Motion employed when the predominant purpose is to secure sufficient control of an object with the fingers to permit the performance of the next motion.
Begins and Ends	Begins when reach ends. Ends when control is secured.
Must Determine	Case of grasp.
Notation	$G1A$—Grasp, case $1A$; $G4$—Grasp, case 4.

Selection of Case of Grasp

Case	Explanation	Examples
$G1A$. Simple grasp of single object.	Includes small, medium, or large object, grasped by a simple closure of fingers.	Eraser, fork, door knob, faucet handle, fountain pen in pocket.
$G1B$. Hook or pinch grasp of single object.	Very small object (less than $\frac{1}{8}'' \times \frac{1}{8}''$ in cross section), or thin object, both against flat surface.	Eraser shield, piece of paper, page of book, coin (all hook grasps). Straight pin, string (pinch grasps).
$G2$. Regrasp.	A shifting of position of the fingers.	Shifting of pencil in fingers preparatory to writing.
$G3$. Transfer grasp.	Transfer of object from one hand to the other.	
$G4$. Object jumbled with other objects.	Applies to similar and dissimilar objects. Does not cover time consumed by separation of interlocked parts.	Parts in tote pan.
$G5$. Contact grasp.	Control is attained by simply touching object.	Touch book to slide across desk. Touch light switch preparatory to flicking.

TABLE A-11. Selection of Turn Time

Definition	Motion employed to turn the hand either empty or loaded by a movement that rotates the hand, wrist and forearm about the long axis of the forearm.
Begins and Ends	Begins when movement starts and ends when movement ceases.
Must Determine	Number of degrees forearm is rotated about its long axis.
Notation	*T*45°—Turn empty hand 45°.

Note: If there is no rotation of the forearm about its long axis there is no turn involved. Rotation of an object such as a screwdriver is not necessarily a turn, rather it is usually a series of short finger reaches and moves, i.e., *R*1*A*, *G*1*A*, *M*1*B*, *RL*1.

TABLE A-12. Selection of Release Time

Definition	Motion employed to relinquish control of an object by the fingers.
Begins and Ends	Begins when finger motion starts. Ends when fingers are far enough away from object to permit start of next motion.
Must Determine	Case of release.
Notation	*RL*1—Release case one; *RL*2—Release case two.

Selection of Case of Release

Case	Explanation	Examples
*RL*1. Normal release performed by opening of fingers.		Remove fingers from pen after placing in pocket.
*RL*2. Contact release.	Occurs at termination of sliding or pushing motion.	Remove fingers after sliding book across desk.

TABLE A-13. Selection of Apply Pressure Time

Definition	The action employed to exert the additional force necessary to overcome resistance too great to be overcome by the momentum of a move or turn.
Must Determine	Case of apply pressure.
Notation	*AP*1—Apply pressure case one.

Selection of Case of Apply Pressure

Case	Explanation	Examples
*AP*1. Apply pressure with regrasp.	A regrasping or adjusting of muscles followed by a noticeable application of force.	Tighten large valve. Tighten nut on auto wheel.
*AP*2. Simply apply pressure.	No regrasping or adjusting of muscles is necessary. Takes place immediately after completion of previous motion.	Push door bell button (takes place immediately after contact grasp). Tighten small bottle cap.

Note: Usually there is little or no noticeable movement involved in an apply pressure.

TABLE A-14. Selection of Position Time

Definition	Position is the basic element employed to align, orient, and engage one object with another.
Begins and Ends	Begins when object is approximately 1 inch from target. Ends when parts are engaged up to 1 inch of depth.
Must Determine	(1) Class of fit. (2) Symmetry. (3) Difficulty of handling.
Notation	*P*1*SE*—Position, loose fit, symmetrical, easy to handle. *P*2*SSE*—Position, close fit, semisymmetrical, easy to handle. *P*3*NSD*—Position, tight fit, nonsymmetrical, difficult to handle.

Selection of Class of Fit

Class of Fit	Explanation	Examples
No position.	When the difference in size of plug and hole is greater than ½-inch, no position motion is required.	Drop part anywhere in tote pan.
*P*1. Loose fit, no pressure required.	Difference in size of plug and hole is ½-inch or less yet no pressure is required during engagement.	Place match in ½-inch diameter hole.
*P*2. Close fit, light pressure *or* careful alignment and orientation.	Allow if careful alignment and orientation are required even if no pressure is required. If in doubt about class of fit required, allow *P*2, it is most common.	Start nut on bolt. Plug in wall outlet. Key in ignition.
*P*3. Exact fit, heavy pressure required.	Considerable friction during engagement. Rarely encountered.	Cork in wine bottle.

TABLE A-14 (continued)

Selection of Case of Symmetry

Case	Explanation	Examples
S. Symmetrical.	Plug can be positioned into hole in an infinite number of ways *about the axis of travel.* No rotation is necessary.	Round plug in round hole.
SS. Semi-symmetrical.	Plug can be positioned into hole in a limited (but more than one) number of ways about the axis of travel. Some rotation is necessary.	Square plug in square hole, octagonal plug in octagonal hole, etc. Plug in wall outlet.
NS. Non-symmetrical.	Plug can be positioned into hole in only one way about the axis of travel.	Key in keyhole. Keyed plug into keyed hole.

Selection of Difficulty of Handling

Condition	Explanation	Examples
E. Easy to handle.	Includes most positions.	Cap on pen.
D. Difficult to handle.	Very large or heavy parts, flexible parts, or parts that must be grasped more than three inches from point of initial engagement of the mating parts.	Thread needle.

Note: Position time allows for engagement motion of up to one inch. If further engagement is necessary allow additional motions.

TABLE A-15. Table of Simultaneous Motions

Enter Along This Axis With One Motion.

Enter Along This Axis With Other Motion.

Reach			Move			Grasp					Position							
A	*B*	*C, D*	*A*	*B*	*C*	*G1A* *G2* *G5*	*G1B* *G1C*		*G4*		*P1S*		*P1SS* *P2S*		*P1NS* *P2SS* *P2NS*		Case	Motion
							*W**	*O*	*W**	*O*	*E*†	*D*	*E*†	*D*	*E*†	*D*		
All combinations EASY to perform simultaneously.						*E*	*E*	*E*	*E*	*E*	*E*	*E*	*E*	*P*	*P*	*P*	*A*	Reach
						E	*E*	*P*	*P*	*D*	*P*	*P*	*P*	*D*	*D*	*D*	*B*	
						E	*P*	*D*	*D*	*D*	*D*	*D*	*D*	*D*	*D*	*D*	*C, D*	
						E	*E*	*E*	*E*	*E*	*E*	*E*	*E*	*P*	*P*	*P*	*A*	Move
						E	*E*	*P*	*P*	*D*	*P*	*P*	*P*	*D*	*D*	*D*	*B*	
						E	*P*	*D*	*D*	*D*	*D*	*D*	*D*	*D*	*D*	*D*	*C*	
						E	*E*	*E*	*E*	*E*	*E*	*E*	*E*	*D*	*D*	*D*	*G1A, G2, G5*	Grasp
							D	*D*	*P*	*D*	*D*	*D*	*D*	*D*	*D*	*D*	*G1B*	
									D	*D*	*D*	*D*	*D*	*D*	*D*	*D*	*G4*	
											P	*D*	*D*	*D*	*D*	*D*	*P1SE* *P1SD*	Position
													D	*D*	*D*	*D*	*P1SSE, P2SE* *P1SSD, P2SD*	
															D	*D*	*P1NSE, P2SSE, P2NSE* *P1NSD, P2SSD, P2NSD*	

E—Easy to perform simultaneously. Allow the longer time.

P—Can be performed simultaneously with practice.

D—Difficult to perform simultaneously even after long practice. Allow both times.

Motions not Included in Above Table:
Turn—Normally easy with all motions.
Apply Pressure—May be easy, practice, or difficult. Each case must be analyzed.
Position—Class 3—Difficult to perform with all motions.
Release—Easy to perform with all motions.

**W*—Within the area of normal vision. This area is defined as a circle 4 inches in diameter at a distance of 16 inches from the eyes.
O—Outside the area of normal vision.
†*E*—Easy to handle.
D—Difficult to handle.

TABLE A-16. Representative MTM Motion Patterns for Bimanual Activity

	Identical Motions			Dissimilar Motions		
	LH		RH	LH		RH
All of the given motions *can* be performed simultaneously.	*R6B*	5.2	*R6B*	*R9B*	6.5	(*R7A*)
	G1A	1.0	*G1A*	*G1B*	2.1	(*G1A*)
		6.2			8.6	
	LH		RH	LH		RH
Some of the given motions *cannot* be performed simultaneously.	*M10C*	8.1	*M10C*	*M8C*	7.1	(*R8C*)
	P2SE	9.7	Delay	*P2SD*	13.1	Delay
	Delay	9.7	*P2SE*	Delay	5.5	*G4*
		27.5			25.7	

appendix B

Common Causes and Means of Minimizing Resistance to Change

Two Types of Problems

In a discussion of resistance to change it is desirable to distinguish between major technological changes and everyday shop changes. The former, typified by introduction of mechanization in the coal mining industry, is an extensive technological innovation that affects whole populations and frequently causes major economic and social problems. However important these may be when they do occur, these major technological changes are rare relative to the day-to-day changes in product, process, and procedure, that the engineer is constantly involved with. Therefore in the following discussion attention is devoted to changes other than the major technological innovations.

The intent of the following outline is to provide a practical checklist for preventive and diagnostic purposes. For such purposes it is useful to distinguish between the person who has authority to accept or reject an idea, for example, an executive or supervisor, and the person who has no voice in acceptance or rejection of the proposal but is affected by it, and whose cooperation is important to successful implementation of that proposal. The latter category is comprised predominantly of shop and office workers.

Specific Common Cause of Resistance to Change on the Part of Persons Having a Veto Power Over the Proposal

A person who has the authority to accept or reject a proposal may reject it for any one or more of the following reasons.

1. Inertia, an innate desire to retain the status quo, even where the present situation is obviously inferior. It is the tendency to want to do things in the accustomed manner. For example, a person ordinarily wants to sit in the same seat he sat in during the first class of the term, even though that seat was originally a random choice. (In fact, many a student has become disturbed to find someone in "his" seat.) This tendency is analogous to resistance offered by the gyroscope to change in position. Resistance is aroused because a new method is *a* change and has no relation to the specific proposal. Thus a supervisor, for example, opposes the new method *merely because it is different* from what he is accustomed to doing.

2. Uncertainty. Regardless of how inferior the existing method may be, at least it is *known* how well it functions. However, how well a proposed change will function is a matter of prediction that is sometimes grossly in error. Any deviation from the current procedure involves a risk; there is no guarantee that the new method will bring better results after the cost and trouble of installation has been incurred. The feeling that many have is: "Why should we create a potential source of trouble by introducing a new system when the existing one does work?" In this case the person is unwilling to trade inferiority of which he is certain for superiority of which he is uncertain.

3. Failure on the part of the proposee to see the need for the proposed change.

4. The proposal may not be understood by the rejector. Even though this man is not directly affected by the proposal, a failure to understand the nature and functioning of the new system may well arouse overcautiousness and a feeling of inferiority and resentment.

5. The fear of obsolescence. A man who has invested years of experience to build up a high level of skill, knowledge, and judgment in administering a certain system, stands to have these fruits of experience obsoleted by adoption of a new procedure. When a new system is proposed, a fear of the inability to become equally proficient under the new system may well cause any man to be apprehensive as to his future value and security in that job. Under such circumstances resistance to the proposal is certainly likely.

6. Loss of job content. A change may reduce the skill required, scope, importance, or responsibility offered by a person's job. For example, a proposed change may reduce the size of the work force supervised by the objector. Such effects are commonly referred to as job dilution. This usually reduces the prestige value of the job and

the value of the job holder to the organization, and therefore may well arouse resistance.

7. Desire to retain favor of the work group. To avoid making life more difficult, a foreman might well be expected to act in the interest of his men and against the interests of management in passing judgment upon a proposal. Thus, if a change is unpopular with his subordinates, a supervisor is likely to resist the proposal.

8. A personality conflict between the proposer and proposee.

9. Resentment of outside help. For example, when an engineer is assigned to cope with a supervisor's problem, it is quite likely that the supervisor will fear a loss of prestige in the eyes of his subordinates. The implication as he sees it is that he cannot handle his own problems.

10. Resentment of criticism. This might well be so if the person to whom you are trying to sell your proposal has originated the present method. Noncritical statements are often *construed* as criticism; so beware not only of critical statements but of ones that might be taken as such.

11. Lack of participation in formulation of the proposed change. Embarrassment at not having conceived of an idea which on hindsight appears obvious, probably will cause resentment. There might well be a feeling of this kind if some or all of the ideas are somebody else's, giving rise to some sort of face-saving reaction.

12. Tactless approach on the part of the proposer. *Sometimes a few words can make the difference.* Remember those famous last words: "That's no way to do it. . . ."

13. Lack of confidence in the ability of the person proposing the idea. This situation is commonly encountered by inexperienced engineers.

14. Inopportune timing. Rejection may have been received only because the proposal was made when the rejector was emotionally or physically upset, or when he was unusually busy, or when business was in a temporary slump, or when there was an unusually heavy demand on investment funds, or when labor relations were strained.

Specific Common Causes of Resistance to Change on the Part of Persons Directly Affected by a Proposed Change But Having No Veto Power

Some common causes of resistance to change on the part of persons having no direct voice in the matter of acceptance or rejection of the proposal, but directly affected by it, are as follows.

1. Inertia, especially when the change is sudden or radical.

2. Uncertainty as to just what a change will bring. Even though the current situation may not be satisfactory to the worker, he may not care to risk the possibility of a poorer situation, such as lower pay, poorer working conditions, less desirable working mates or a more difficult job.

3. Ignorance of the need for or purpose of the change. Only too often changes are made with little or no explanation to the workers, sometimes with the feeling that this is none of their business!

4. Failure to understand the new method or policy may arouse suspicion or an insecure feeling.

5. A loss of job content. A change that means a reduction in skill required, or of importance, or of responsibility, may readily arouse resistance. Aside from the important fact that job dilution usually reduces the prestige value of the job and the value of the job to the company, a task is likely to be more monotonous as it becomes simplified and shortened.

6. Pressure of the work group. Each member of the work group often reacts so as not to offend the others, even though as an individual he does not feel as strongly as his actions would indicate. Every work group has certain ingrained policies, some expressed and others implied, that constitute a "code of behavior" and that help to govern the actions and reactions of its members. A person's reaction to a change is usually influenced by what he knows or anticipates that the group wants, even to the point of sacrificing personal gain for continued approval of his co-workers. And this is important: older members, who are the most frequent and stubborn resistors, are often the most influential persons in the work group.

7. Fear of economic insecurity. A change may result in displacement of the employee or a reduction in his earnings. The latter may arise in any one or more of the following ways:

a. a reduction in job classification.

b. a tightening of the time standard as a result of the change in method. The company should time study a job again if a change in method is made, then adjust the time standard if justified. If there has been a loose time standard on the operation, this will probably be remedied when a standard is set for the new method. Thus, a change in method may well be affecting the worker's economic status in an indirect manner. Partly as a result of past practices, a majority of workers suspect that whenever the method is changed on an operation that has a loose standard, the change was introduced for the sole purpose of providing the company with an opportunity to reset the standard. Under such circumstances resistance to the change is certainly to be expected.

c. inability to master the new method, or at least to reach the level of proficiency that the person had attained under the replaced method. This is particularly true of older operators.

8. Alteration of social relationships, or fear of same, such as breakup of a closely knit work group.

9. An antagonistic attitude toward the person introducing the change or

what he represents. It may be a personal antagonism, or it may be an antagonism toward his function, or toward management in general. The latter, an attitude that is often hostile, causes individuals and groups to resist almost any change, in fact almost everything but the pay check, that comes from the direction of management. This might be called resistance on general principles. The attitude underlying it is common.

10. Origination or introduction by an "outsider." Executives, higher supervisors, and engineers are usually considered as external to the worker economic and social group; in fact they are frequently quite unpopular. Resentment and resistance are quite likely if such persons introduce or are known to have originated a change.

11. No participation in formulation of the new method or policy. It appears to workers that changes are often adopted without their best interests in mind. They want an opportunity to express and protect these interests, to have a part in deciding what they must do and how.

12. Tactless approach on the part of the person introducing the change.

13. Inopportune timing. Resistance may have been received only because the change was introduced when feeling was running high between the work group and management, or because the change was made with little or no advance notice.

Suggested Methods of Minimizing Resistance to Change

The following recommendations should be considered in planning the introduction of an idea *and* in modifying the idea to make it more saleable and palatable.

1. Convincingly *explain the need* for the change. Don't overlook the worker in this respect.

2. Thoroughly *explain the nature* of the change. Use straightforward, clear, well-organized language to assure that persons understand the method or policy. Do not overlook the importance of this understanding. Tailor your written and oral reports to suit the particular reader. For example, executives in general should be given a condensed description of the proposal, emphasizing the overall picture and making liberal use of charts, graphs, and other visual aids. Reports to persons who must administer the new procedure should include a thorough and easily understandable description of how the procedure is to function.

3. Facilitate participation or at least the *feeling of participation* in formulation of the proposed method. In general, people are concerned about making their own ideas and recommendations succeed, whereas in general they have a neutral or negative attitude toward the ideas of others. The feeling of participation may be imparted in several ways.

a. *Consult* operators, inspectors, supervisors, tool makers, maintenance men, setup men, managers, *et al.;* ask for information, opinions, suggestions. Show a real interest in what these people have to say. Seek advice even if you think you do not need it, and you may benefit in more ways than one. The mere opportunity to express himself, the mere request to contribute may well give a person a feeling of participation in formulation of a new method, even though none of his ideas have actually been included in the new setup.

b. Of course, whenever possible, *include the worthwhile* suggestions of others in your final proposal and give credit to the appropriate individuals.

c. In some instances it may be advisable or even necessary to incorporate a person's idea that is inferior to that which you would otherwise use, in order to get him into the act. This "sugaring up" or compromising of proposals in order to get them accepted and to minimize adverse reactions will probably make the difference between success and failure on many occasions.

4. Use a *tactful* approach in introducing your proposal. Watch your wording and mannerisms, and above all, avoid criticism or anything that may be construed as such.

5. Watch your timing.

a. In attempting to gain adoption of an idea, avoid making your proposal when the recipient is upset, busy, etc. Allow sufficient time for him to think it over, don't rush the matter. Who knows, sometime later he may concede or may even propose the ideas as his.

b. In introducing a new method or policy to employees, provide ample advance notice. Avoid introducing certain changes when labor relations are abnormally strained.

6. In the case of major changes, if possible introduce the change in stages. The mere magnitude of some proposals sometimes frightens people and arouses objection.

7. In attempting to gain acceptance, capitalize on the features that provide the most personal benefit to the person(s) you are trying to sell.

8. If possible, by appropriate questioning maneuver a prospective rejector into "thinking" of (your) idea himself. This procedure of "planting" the idea in his mind is usually effective if it can be worked. The difficulty is that this is not easily accomplished, nor are we generally so noble that we make a habit of letting others get the credit for our ideas.

9. Show a personal interest in the welfare of the person directly affected by the change by:

a. paying particular attention to older persons, close social relationships, individual differences, etc.

b. finding equivalent jobs for displaced employees, where they may make maximum use of their previously developed skills.

c. providing, if possible, guarantees as to security of employment and earnings.

d. providing thorough training in the new procedure.

e. attempting to avoid drastic reductions in job content. If the skill, performance time, or responsibility have been reduced, perhaps this may be counteracted by allowing the worker to do his own setup, inspection, planning of jobs, etc., or by combining several short-cycle jobs into one.

10. Whenever possible, have changes announced and introduced by the immediate supervisors of those affected.

The foregoing measures, concerning the minimization of resistance to a *specific* change, are no substitute for a long-term "conditioning" for change. These measures should be supplemented by a long-term effort to prepare personnel of the organization for, and harden them to, change in general. This conditioning process, involving both a technical and a "psychological" preparation, should provide the following.

1. Adequate technical training, so that the personnel affected will feel and be capable of mastering new manufacturing methods and managerial techniques (such as those requiring a moderate knowledge of mathematics, statistics or computers). This is the provision of a grounding in preparation for technological change, not instruction in a specific new technique.
2. "Psychological" conditioning of personnel for change, by:
   a. education as to the importance of change to the economic welfare of the enterprise. The inevitability of change, the consequence of stagnation, the role of competition, and so forth, should be emphasized.
   b. keeping appropriate personnel informed of trends and expected developments in practice, policy, and technology, so that they can anticipate changes and be psychologically and technically prepared.
   c. maintaining a policy of, *and reputation for*, fair treatment of employees affected by change, on matters of replacement, retraining, job content, etc., as per the suggestions made earlier.

Not to be undersold as an effective long-term measure in minimizing resistance to change is the very awareness of the phenomenon itself. If a person is aware of the causes, manifestations, and frequency of this reaction, he will be less inclined to resist change. Thus, a worthwhile countermeasure is to instill this awareness in the minds of all personnel, especially supervisors and executives.

appendix C

Proposed Methods of Detecting Loose Time Standards Protected by Restriction of Output

Detection of loose time standards under restricted output conditions is aided by the fact that two types of variation are expected in human performance. The first concerns differences in ability among human beings. Individuals do not have equal capacities for performing a task; in fact, ability ranges of greater than three to one are occasionally found. The nature of the typical distribution of productive abilities has not been definitely established; however, it is generally pictured by speculators as being approximately normal. Regardless of the exact nature of this distribution, the important fact is that abilities of the working population do form *a* distribution, with a range commonly believed to be at least two to one.

This brings us to an important point. Differences in operator ability are known to exist. If the differences in *actual production rates* of operators on the job are small or nil, the situation may be an unusual instance in which operators with very similar productive aptitudes were selected for the job. Frequently, however, it is the result of restriction of output arising from the desire to conceal a loose time standard. This situation originates in the following manner.

1. A standard on an operation is set loose or becomes loose due to change.
2. Then, for fear of having the standard tightened, the better operators limit their output, agreeing either explicitly or implicitly not to produce above a certain level.
3. The poorer operators need not and most of them will not restrict their output. In their case they are not capable of making the high earnings that would attract the attention of the management.
4. The result then is that the *actual* frequency distribution of worker

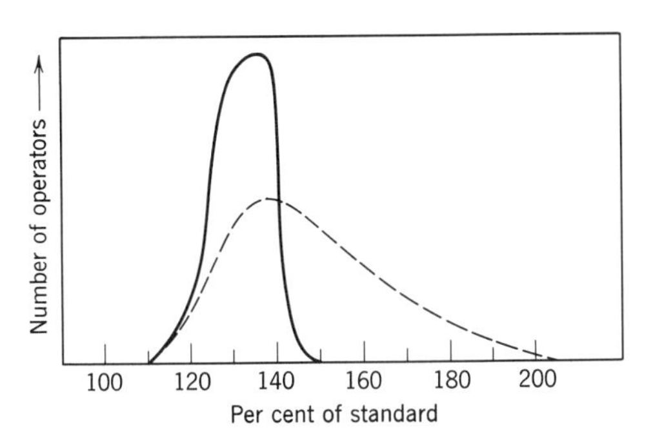

Figure C-1. Dashed curve represents the distribution of operators' outputs expected on the basis of their varied abilities. Solid curve indicates the distribution of operators' outputs actually experienced when output is being restricted by the group.

outputs is considerably different from that expected on the basis of operator output abilities. In Figure C-1 the dashed curve represents the expected distribution of operator abilities for performing an operation. This distribution is represented as positively skewed because of the expectation of selection in employment and the transfer and release of poor producers. Assume that the standard, represented by 100 per cent, is a loose one. The solid, negatively skewed curve in this illustration approximates the distribution that may be expected when output is restricted, to avoid drawing management attention to the loose standard.

To summarize then, under a loose time standard workers tend to limit their output to what they think is a safe maximum. This restriction may be expected to yield an output distribution of small variance and negative skewness.

This suggests that whereas the mean of the distribution of operator output rates is an unreliable indicator of restriction and looseness, the shape, that is, the variance and skewness, of that distribution is a very promising indicator. It is proposed therefore that if there is a relatively large group of persons performing an operation, a representative production average may be obtained for each operator for a period of weeks, months, or lots, and these plotted in the form of a frequency histogram, as shown in Figures C-2 and C-3. This plot will be referred to as an operation histogram. If there is restriction of output rates is an unreliable indicator of restriction and looseness, skewness will result; however, a histogram displaying these characteristics is not necessarily the result of restriction of output. These characteristics may result because the abilities of the operators may, by chance, actually be very similar. Furthermore, although the cause is probably restriction, that restriction may result from something

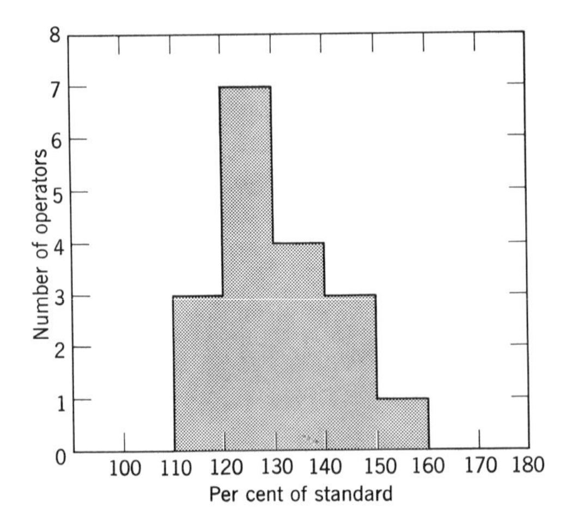

Figure C-2. Operation histogram for a time standard, demonstrating the spread of operator performances expected when restriction of output is negligible.

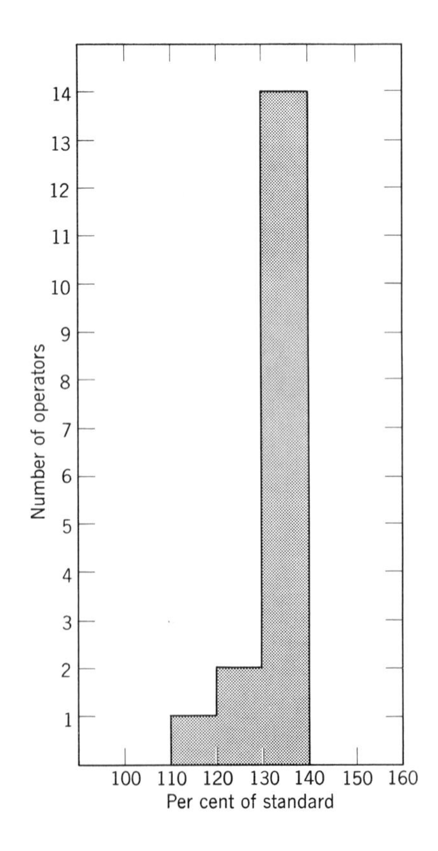

Figure C-3. Operation histogram for a time standard, demonstrating the spread of operator performances expected when output is restricted.

other than a loose time standard. In spite of these other possibilities, there is justification for investigating an operation for causes of looseness when the operation histogram exhibits these characteristics of small variance, peakedness, and negative skewness.

It would be advantageous to have quantitative information with respect to the range and variance to guide interpretation of the operation histogram. Unfortunately there is no reliable information available as to what variation in aptitudes, abilities, or actual output should be expected on different types of operations. Because of this situation, and because of the relatively small sample sizes ordinarily involved, interpretation of the variability of an operation histogram is of necessity rather dependent on judgment. Another factor limiting its usefulness is that the number of operators performing a given operation is rarely large enough to permit the plotting of an operation histogram. However it is an approach that can be used to test en masse the standards in a given department, the standards on a given type of operation, or the standards set by a particular time-study man, for general restriction and looseness.

A second source of variability that proves useful in detection of loose standards is expected variation in performance of the individual operator. The performance of a human being, unless artificially controlled, will exhibit variation. Ordinarily an individual's production is expected to vary from minute to minute, from hour to hour, and from day to day. The explanation for this is simple. A person's productive capacity varies considerably due to fatigue, monotony, emotional status, fluctuations in health, and the like. In addition, his motivation varies with the passage of time. Finally, all operations are subject to unavoidable delays, retardations and other chance factors which also introduce variability into the production rate. This expected variation in output rate may be represented by a frequency distribution, the general nature of which is unknown. We do know, however, that *a* distribution is to be expected, that a person's output *should* exhibit variation. Therefore, if a person's daily output is devoid or nearly so of variation, unless it is a process-paced operation, we may suspect that an unnatural situation exists.

When a person restricts his output, he is producing below the rate at which he could be working and still not be exerting undue effort nor incurring an undue amount of fatigue. He may readily increase his output wherever necessary or desirable with no inconvenience. Consequently his daily production will be relatively constant, for whenever he experiences delays or retardations he is able to compensate for the "lost production" and to meet his daily limit by temporarily working at or near his normal capacity. He is able to remove output fluctuations that would ordinarily be expected to appear. This results in relatively uniform daily outputs, represented by a frequency distribution having a small range, small variance, and a peaked and skewed shape. Thus, when we find an individual whose output is very similar from day to day or from lot to lot, we may suspect that he is placing a ceiling on his production (except where the output rate is process controlled). Lack of variation, therefore, should arouse suspicion of a loose standard. Two effective means of graphically revealing the variation in an individual's output are the operator histogram and the performance chart, to be described in detail.

The Operator Histogram

This is a frequency histogram representing one person's performance, usually his daily outputs, over an extended period of time, as shown in Figure C-4. To facilitate the comparison of such plots and to

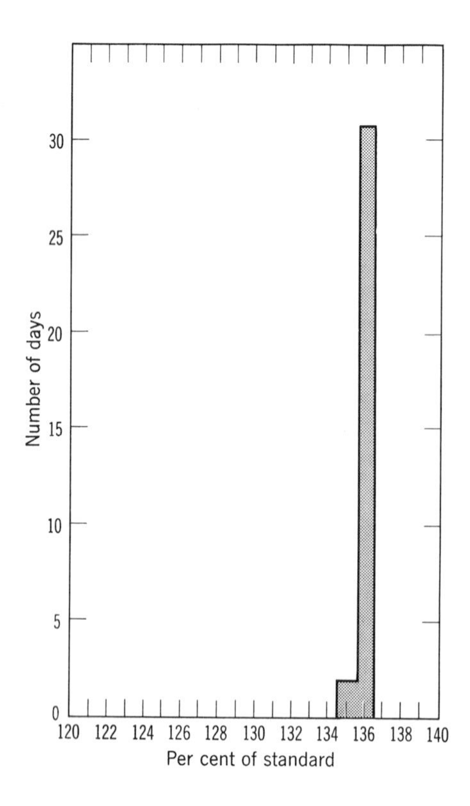

Figure C-4. An operator histogram, indicating the lack of variation in an operator's performance expected when output is being restricted.

assist in avoiding errors of interpretation, when the abscissa is rate of output, the high values should be located to the right. When the abscissa is performance time the high values should be located to the left.

If an operator histogram exhibits only a small amount of variation, as exemplified in Figure C-4, restriction of output is probably present. It is true that a histogram showing such uniformity could be obtained when actually there is little or no restriction, due to a sampling error or high effort on the part of the operator. But for an operator histogram the sample size is limited only by the number of days the present time standard has been in effect, so the chance that a person's performance will be erroneously typified is small. Furthermore, the presence of restriction does not necessarily mean there is a loose time standard. However, it is the author's opinion that causes of restricted output other than fear of discovery of a loose time standard will seldom cause the degree of uniformity referred to here.

The Performance Chart

Another technique that provides a graphic portrayal of the individual's output variability is a chronological plot of production rates as demonstrated in Figure C-5. The abscissa is chronological, successive days or lots, and the ordinate is production rate or some index of it. If the performance of more than one operator is to be plotted on this chart, it is suggested that the output of each be distinguished by color or symbol. Other useful information that may be included on the performance chart are the standard, the grand average performance over a prolonged period, and daily averages, as demonstrated in Figure C-5. It is suggested that the performance chart be interpreted in the manner recommended for an operator histogram.

Pooled Histogram

A frequency histogram is readily made from a performance chart; in fact it can be done quickly and conveniently on the same graph. First, construct a new ordinate scale, frequency of occurrence, as shown in Figure C-6. Next, determine from the points already plotted on the performance chart the frequency of occurrence of each production rate, and from this information plot a histogram on the new axes. If the range of production rates is not large and if there is enough data, each value on the scale may be treated as a cell. Otherwise they must be grouped into cells the size of which is determined

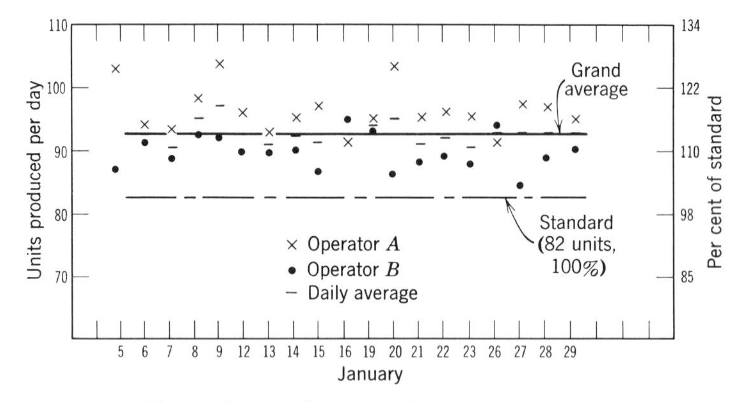

Figure C-5. Performance chart for an operation.

by the number and range of observations on the performance chart. When the output data for all individuals on the operation are combined into one histogram in this fashion, the result will be a composite picture of the variation between and within individuals on the job. This amounts to pooling the operator histograms, hence its title: pooled histogram.

When restriction of output exists, the pooled histogram will exhibit a small amount of variation and negative skewness. Therefore, when such characteristics are demonstrated by this histogram, restriction is probably being exerted, suggesting the possibility of a loose standard.

Detection of Loose Standards—A Summary

Regardless of the method of presenting the data, the pattern to look for is the same: a remarkably small amount of variation *within* and *between* individuals. Look for uniformity of performance among operators, or uniformity in the performance of each individual, and for a concentration of performance below some particular value in the output range (skewness). These characteristics should be expected when restriction of output is being exerted. Although such uniformity and skewness may be exhibited for reasons other than restriction, and although restriction may be due to something other than a loose standard, the presence of these characteristics appears to warrant investigation of the operation for causes of a loose time standard.

In attempting to detect a lack of variation in a person's performance, it is important that we give variation opportunity to appear if it is really there. Therefore the time interval used on the operator his-

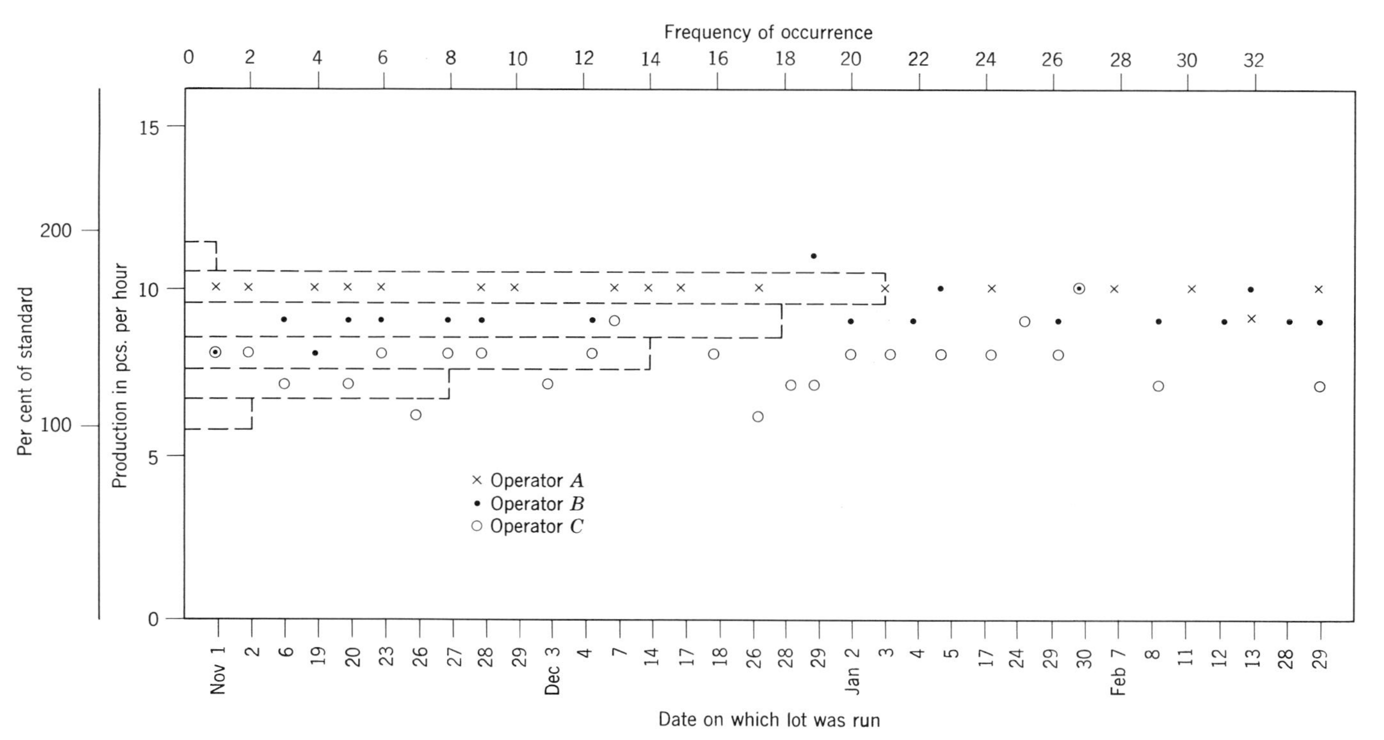

Figure C-6. A performance chart and pooled histogram for an operation performed by three employees.

togram or performance chart should not be longer than a day. If output for a longer period were to be used, we would not expect much variation in performance between successive periods even if the normal daily variations were present. Variations would have an opportunity to average out over these longer intervals. An output period of less than a day or lot, whichever is shorter, is not convenient nor is it desirable, for it seems that many people restrict production on a daily or lot basis.

When considering the economic feasibility of using these various charting devices for control and improvement of the time-standard structure, remember that some important benefits of a more consistent time-standard structure are in the area of human relations (reduced dissatisfaction, better morale, etc.). Such benefits are extremely difficult to evaluate and consequently they are obscure and remote in the ordinary manager's thinking. It is true also that some firms keep in tabular form the same data it has been suggested be kept in the form of a performance chart, and for the same purposes. If a management thinks this is worthwhile, it should be able to justify keeping the data in the graphic form suggested, for this greatly extends the interpretive value of that data by more effectively picturing variation patterns and trends.

Description of these evaluation procedures has required numerous statements of approximation, conjecture, and opinion. Perhaps even more apparent has been the use of qualitative statements where quantitative ones are desirable. Statements of this nature are partly the result of the subjectiveness involved in evaluating worker skill and effort, and partly the result of a dearth of information on the productivity variation to be expected within and between individuals. These ideas have been presented in spite of the deficiencies, because even in their present state these procedures provide the practitioner with means of improving the consistency of his time standard structure. Furthermore, this presentation points out information that is lacking and consequently the research that is needed to provide a more objective approach to standards evaluation and control.

appendix D

Table of t Values from Student's t-distribution, for $C = 0.90$

M	d.f.	t
5	4	2.13
6	5	2.02
7	6	1.94
8	7	1.90
9	8	1.86
10	9	1.83
15	14	1.76
20	19	1.73
25	24	1.71
30	29	1.70
Above 30	—	1.65

appendix E

Table of d_2 Factors for Estimation of Standard Deviation from the Sample Range*

$$s = \frac{R}{d_2}$$

M	d_2†
5	2.326
6	2.534
7	2.704
8	2.847
9	2.970
10	3.078
11	3.173
12	3.258
13	3.336
14	3.407
15	3.472
16	3.532
17	3.588
18	3.640
19	3.689
20	3.735

* See page 231 of text.

† These factors are based on the assumption that the samples are drawn from a normal population.

Index